T0181632

Lecture Notes of the Institute for Computer Sciences, Social Informatics and Telecommunications Engineering 536

The LNICST series publishes ICST's conferences, symposia and workshops.
LNICST reports state-of-the-art results in areas related to the scope of the Institute.
The type of material published includes

- Proceedings (published in time for the respective event)
- Other edited monographs (such as project reports or invited volumes)

LNICST topics span the following areas:

- General Computer Science
- E-Economy
- E-Medicine
- Knowledge Management
- Multimedia
- Operations, Management and Policy
- Social Informatics
- Systems

Prakash Pareek · Nishu Gupta · M. J. C. S. Reis
Editors

Cognitive Computing and Cyber Physical Systems

4th EAI International Conference, IC4S 2023
Bhimavaram, Andhra Pradesh, India, August 4–6, 2023
Proceedings, Part I

 Springer

Editors
Prakash Pareek ⓘ
Vishnu Institute of Technology
Bhimavaram, Andhra Pradesh, India

Nishu Gupta ⓘ
Norwegian University of Science
and Technology
Gjøvik, Norway

M. J. C. S. Reis ⓘ
University of Trás-os-Montes e Alto Douro
Vila Real, Portugal

ISSN 1867-8211   ISSN 1867-822X (electronic)
Lecture Notes of the Institute for Computer Sciences, Social Informatics
and Telecommunications Engineering
ISBN 978-3-031-48887-0   ISBN 978-3-031-48888-7 (eBook)
https://doi.org/10.1007/978-3-031-48888-7

This Springer imprint is published by the registered company Springer Nature Switzerland AG
The registered company address is: Gewerbestrasse 11, 6330 Cham, Switzerland

Paper in this product is recyclable.

Preface

We are delighted to introduce the proceedings of the fourth edition of the European Alliance for Innovation (EAI) International Conference on Cognitive Computing and Cyber Physical Systems (EAI IC4S 2023), hosted by Vishnu Institute of Technology, Bhimavaram, Andhra Pradesh, India during 4–6 August 2023 in hybrid mode. This conference has together researchers, developers and practitioners around the world who are leveraging and developing intelligent computing systems and cyber physical systems so that communication becomes smarter, quicker, less expensive and accessible in bundles. The theme of EAI IC4S 2023 was "Cognitive computing approaches with machine learning techniques and advanced communications".

The technical program of EAI IC4S 2023 consisted of 70 full papers, which were presented in online mode, i.e. on a web platform, and also in offline mode. The above papers were presented by the registered authors in fourteen technical sessions under five different tracks. The conference tracks were: Track 1 – Machine Learning and its Applications; Track 2 – Cyber Security and Signal Processing; Track 3 – Image Processing; Track 4 – Smart Power Systems; and Track 5 – Smart City Eco-system and Communications. Apart from the high-quality technical paper presentations, the technical program also featured two keynote speeches and one plenary talk. The two keynote speakers were Nishu Gupta from the Department of Electronic Systems, Faculty of Information Technology and Electrical Engineering, Norwegian University of Science and Technology (NTNU) in Gjøvik, Norway and Manuel J. Cabral S. Reis from UTAD University Engineering Department, Portugal. The plenary talk was presented by Anil Gupta, Associate Director, C-DAC, Pune, India on the role of cyber physical systems in assistive technology.

Coordination with the steering chair, Imrich Chlamtac, was essential for the success of the conference. We sincerely appreciate his constant support and guidance. Manuel J. Cabral S. Reis successfully served as general chair for this edition and helped the conference to proceed smoothly. It was also a great pleasure to work with such an excellent organizing committee team for their hard work in organizing and supporting the conference. In particular, the Organizing Committee chaired by Nishu Gupta, and the Technical Program Committee, chaired by Prakash Pareek coordinated and completed the peer-review process of technical papers and made a high-quality technical program. We are also grateful to Conference Managers Sara Csicsayova and Kristina Havlickova for their support and to all the authors who submitted their papers to the EAI IC4S 2023 conference.

We sincerely appreciate the management and administration of Vishnu Institute of Technology, Bhimavaram (VITB), Andhra Pradesh, India and especially Chairman of Sri Vishnu Educational Society (SVES), Shri K. V. Vishnu Raju, Vice Chairman of SVES, Sri Ravichandran Rajagopal, Secretary of SVES, Shri K. Aditya Vissam and D. Suryanarayana, Director and Principal of VITB, K. Srinivas, Vice Principal, VITB and

N. Padmavathy, Dean R&D, VITB for giving their support to us as the main host Institute of EAI IC4S 2023.

We strongly believe that EAI IC4S 2023 conference provided a good forum for all researchers, developers and practitioners to discuss scientific and technological aspects relevant to cognitive computing and cyber physical systems. We also expect that future EAI IC4S conferences will be as successful and stimulating as indicated by the contributions presented in this volume.

December 2023

Prakash Pareek
Nishu Gupta
M. J. C. S. Reis

Conference Organization

Steering Committee

Imrich Chlamtac   University of Trento, Italy

Organizing Committee

General Chair

Manuel J. Cabral S. Reis   UTAD University, Portugal

General Co-chairs

Mohammad Derawi   Norwegian University of Science and Technology, Norway
Ahmad Hoirul Basori   King Abdulaziz University, Saudi Arabia

Organizing Chair

Nishu Gupta   Norwegian University of Science and Technology, Norway

Local Chair

D. Suryanarayana   Vishnu Institute of Technology, India

Organizing Secretaries

Pravesh Kumar   Vaagdevi College of Engineering, India
Prakash Pareek   Vishnu Institute of Technology, India

Technical Program Committee Chair

Prakash Pareek   Vishnu Institute of Technology, India

Technical Program Committee Co-chair

Zhihan Lv   Uppsala University, Sweden

Web Chair

Ariel Soales Teres   Federal Institute of Maranhão, Brazil

Sponsorship and Exhibit Chair

Nishit Malviya   IIIT Ranchi, India

Publications Chair

Luy Nguyen   Vietnam National University Ho Chi Minh
                                 City-University of Technology, Vietnam

Publicity and Social Media Chair

Sumita Mishra   Amity University, India

Publicity and Social Media Co-chair

S. Mahaboob Hussain   Vishnu Institute of Technology, India

Technical Program Committee

Thanos Kakarountas   University of Thessaly, Greece
Anil Gupta   C-DAC, India
D. Suryanarayana   Vishnu Institute of Technology, India
D. J. Nagendra Kumar   Vishnu Institute of Technology, India
William Hurst   Wageningen University and Research,
                                 The Netherlands
Felix Härer   University of Fribourg, Switzerland
Ahmad Hoirul Basori   King Abdulaziz University,
                                 Kingdom of Saudi Arabia
Mukesh Sharma   NTU, Singapore
Anuj Abraham   A*STAR, Singapore
Sumathi Lakshmiranganatha   Los Alamos National Laboratory, USA
Jitendra Kumar Mishra   IIIT Ranchi, India

Lakhindar Murmu	IIIT Naya Raipur, India
Nishit Malviya	IIIT Ranchi, India
Daniele Riboni	University of Cagliari, Italy
Ashok Kumar	NIT Srinagar, India
Dipen Bepari	NIT Raipur, India
Shubhankar Majumdar	NIT Meghalaya, India
N. Padmavathy	Vishnu Institute of Technology, India
R. V. D. Ramarao	Vishnu Institute of Technology, India
Amit Kumar Dubey	Technology Innovation Institute, UAE
Dharmendra Kumar	Madan Mohan Malaviya University of Technology, India
R. Srinivasa Raju	Vishnu Institute of Technology, India
Venkata Naga Rani Bandaru	Vishnu Institute of Technology, India
G. K. Mohan Devarakonda	Vishnu Institute of Technology, India
Atul Kumar	IIITDM Jabalpur, India
Somen Bhattacharjee	IIIT Dharwad, India
Sumit Saha	NIT Rourkela, India
K. Srinivas	Vishnu Institute of Technology, India
Amit Bage	NIT Hamirpur, India
Sridevi Bonthu	Vishnu Institute of Technology, India
P. Sita Rama Murty	Vishnu Institute of Technology, India
Idamakanti Kasireddy	Vishnu Institute of Technology, India
Pragaspathy S.	Vishnu Institute of Technology, India
Abhinav Kumar	NIT Allahabad, India
Sadhana Kumari	BMS College of Engineering, India
Ashish Singh	Kalinga Institute of Industrial Technology, India
Vegesna S. M. Srinivasavarma	SSN Institutions, India
B. V. V. Satyanarayana	Vishnu Institute of Technology, India
G. Prasanna Kumar	Vishnu Institute of Technology, India
V. S. N. Narasimha Raju	Vishnu Institute of Technology, India
Sunil Saumya	IIIT Dharwad, India
Lokendra Singh	KL University, Vijaywada, India
A. Prabhakara Rao	KL University, Hyderabad, India
Ravi Ranjan	Government Engineering College, Vaishali, India
S. Sugumaran	Vishnu Institute of Technology, India
Akash Kumar Pradhan	MVGR College of Engineering, India
Abhishek Pahuja	KL University, Vijaywada, India
Ajay Kumar Kushwaha	Bharati Vidyapeeth (Deemed to be University), India
D. M. Dhane	Bharati Vidyapeeth (Deemed to be University), India
Naveen Kumar Maurya	Vishnu Institute of Technology, India

N. P. Nethravathi	Reva University, India
Sasidhar Babu S.	Reva University, India
Gopal Krishna	Presidency University, India
Mayur Shukla	Oriental College of Technology, India
Bhanu Pratap Singh	LNCT, India
Priyanka Bharti	Reva University, India
Laxmi B. Rannavare	Reva University, India
Vikram Palodiya	Sreenidhi Institute of Science and Technology, India
Madhumita Mishra	Reva University, India
D. R. Kumar Raja	Reva University, India
Ashwin Kumar U. M.	Reva University, India
Syed Jahangir Badashah	Sreenidhi Institute of Science and Technology, India
Vipul Agarwal	KL University, Vijaywada, India
Veeramni S.	Amrita University, India
Gireesh Gaurav Soni	SGSITS, India
Lalit Purohit	SGSITS, India
S. Mahaboob Hussain	Vishnu Institute of Technology, India
Megha Kuliha	SGSITS, India
R. C. Gurjar	SGSITS, India
Sumit Kumar Jindal	Vellore Institute of Technology, India
P. Ramani	SRM Institute of Science and Technology, India
T. J. Nagalakshmi	Saveetha University, India
Rajkishor Kumar	Vellore Institute of Technology, India
Avinash Chandra	Vellore Institute of Technology, India
Naveen Mishra	Vellore Institute of Technology, India
Abhishek Tripathi	Kalasalingam Academy of Research and Education, India
Tripta	Government Engineering College, Vaishali, India

Contents – Part I

Cyber Security and Signal Processing

Image Processing

Contents – Part II

Smart City Eco-System and Communications

Machine Learning and Its Applications

Schizophrenia Identification Through Deep Learning on Spectrogram Images

Amarana Prabhakara Rao[1] ID, G. Prasanna Kumar[2](✉) ID, Rakesh Ranjan[3] ID,
M. Venkata Subba Rao[4], M. Srinivasulu[2], and E. Sravya[2]

[1] Department of Electronics and Communication Engineering, Koneru Lakshmaiah Education Foundation, Bowrampet, Hyderabad 500043, Telangana, India
[2] Department of ECE, Vishnu Institute of Technology, Bhimavaram 534202, Andhra Pradesh, India
godiprasanna@gmail.com
[3] Department of ECE, NIT Patna, Patna, India
[4] Department of ECE, Shri Vishnu Engineering College for Women, Bhimavaram 534202, Andhra Pradesh, India

Abstract. Schizophrenia (SZ) is one of the mental disorder due to which many people are suffering around the world. People suffering with this disorder experience hallucinations, delusions, confusing speech and thinking patterns, etc. In a clinical environment, doctors judge Schizophrenia directly using electroencephalogram (EEG). Automatic detection of SZ is achieved in earlier works by using the time domain and frequency domain features extracted from the given EEG signals. These features are used to train various Machine Learning and Deep Learning approaches for the classification of SZ from the given EEG signal. The proposed work uses Short-Time Fourier Transform (STFT) for converting 1D EEG data into 2D spectrogram image data. This work proposes a simple Convolutional Neural Network (CNN) model for the efficient detection of SZ from the given spectrograms. Performance of the proposed CNN model is compared with various existing CNNs such as Alex net, VGG16, Resnet. Performance of these CNNs is evaluated in terms of accuracy, precision, recall and F1 Score. It is observed from the results that the proposed CNN performed better showing its potential for efficient detection of SZ.

Keywords: Schizophrenia · Electroencephalogram · Spectrogram · Short-time Fourier Transform · Convolutional Neural Networks · Deep Learning Approaches

1 Introduction

There are many mental disorders for human beings in which SZ is one of the important mental disorder due to which many people are suffering [1, 2]. It changes the way a person thinks and behaves. People suffering with this disorder experience hallucinations, delusions, their speech and thinking patterns are confusing, they want to disconnect

P. Pareek et al. (Eds.): IC4S 2023, LNICST 536, pp. 3–11, 2024.
https://doi.org/10.1007/978-3-031-48888-7_1

from people around them including their dear ones, not even care about their personal hygiene, etc. [3]. In a clinical environment, doctors judge Schizophrenia directly using electroencephalogram (EEG). EEG is the most commonly used signal to analyze the mental condition of human beings. EEG signals record various mental conditions of the brain such as mental stress and other disorders. EEGs are becoming popular in recent years in research and diagnosis of various neurological disorders such as Epilepsy, Schizophrenia, etc. These signals contain significant amount of information of higher dimensions and they show complex functioning of the brain and they are very difficult to analyze directly [1]. EEGs provide more detailed information over other existing methods as far as Schizophrenia is concerned [4]. Researchers have proposed both time features [5] and frequency features [6] extracted from EEG signals to detect the state changes in the brain and detect Schizophrenia.

In the past few years, computer aided diagnosis supported by machine learning (ML) algorithms have revolutionized the study of complex EEG signals using various time-domain and frequency domain features to identify the schizophrenia. Several researchers have used ML based framework in the diagnosis of various diseases such as epileptic Seizures, Schizophrenia, etc. [7]. Many DL approaches are invented today for classi-fication and segmentation in today's world, because DL gives the better performance compared to ML techniques. This work mainly focuses on DL approaches and proposes a simple CNN model for efficient detection of SZ.

First part of this paper is an Introduction section which gives an insight into the global burden of Schizophrenia and how it can be detected at an early stage by using EEG signals which can be analyzed efficiently with various computer aided diagnostic tools involving various ML and DL frameworks. Second part of the paper gives a brief review of the earlier works done in this domain. Next section – materials and methods present the details of data used to train the deep learning models and details of proposed deep learning model. In the next section, results are presented and a summary of these results are discussed in this section. In the last section conclusions are presented.

2   Literature Review

In this section, a summary of the research work done by the researches in detecting Schizophrenia using EEG signals analyzed by various traditional ML algorithms as well as trending DL models. In a recent study Miras et al. [8] have performed various linear and non-linear measures to extract seventeen features from the resting stage EEG signals. These selected features were used as inputs to five traditional ML based algorithms to analyze SZ. This study includes 31 patients out of which 20 were healthy controls and 11 were SZ patients. In another study, Ranjan et al. [9] applied Kruskal Wallis test to select 8 significant features out of the 24 features extracted from 16 channel EEG signals of 84 subjects. These features were applied to various classical ML algorithms and they have reported that ensemble bagging tree classifier performed better with an accuracy of 92.3%.

A hybrid framework combining brain-effective connectivity analysis [10] and deep learning is proposed for schizophrenia detection using EEG signals. Transfer Entropy measures causalities between EEG channels, forming connectivity images. These images are inputted into pre-trained CNN models (VGG-16, ResNet50V2, InceptionV3, EfficientNetB0, DenseNet121), and their deep features are processed by an LSTM model. The hybrid CNN-LSTM models achieve exceptional accuracy, with EfficientNetB0-LSTM reaching 99.90% average accuracy and 99.93% F1-score using 10-fold cross-validation. The method shows strong capability in detecting schizophrenia patients from healthy controls.

Researchers developed an interpretable machine learning method [11] for diagnosing schizophrenia based on DSM-5 criteria. The method integrates smoothly into existing clinical processes, providing clinicians with trust and understanding. By combining two attention mechanisms, attribute importance and interactivity are determined. The model demonstrates the robustness and real-world applicability through experiments with augmented test data. It achieves a high accuracy of 98% with 10-fold cross-validation, highlighting its effectiveness in assisting clinicians with schizophrenia diagnosis.

A study by Zulfikar Aslan [12] introduces a highly accurate method for automatically detecting schizophrenia from EEG records. It utilizes the Continuous Wavelet Transform to extract time-frequency features from the signals, achieving the highest accuracy in the literature. The VGG16 deep learning architecture is employed for feature extraction, resulting in classification accuracies of 98% for SZ patients and 99.5% for healthy individuals. Visualizations of the CNN network highlight frequency component differences between SZ patients and healthy individuals, facilitating easy interpretation. The performance of the model was evaluated in terms of accuracy, sensitivity, specificity, and F1 score. The results demonstrated that the deep learning approach achieved high accuracy and reliable classification performance, outperforming traditional machine learning methods and indicating the potential of CNN-based models in the accurate identification of schizophrenia using EEG signals.

ML based works trained with various algorithms of SVM, KNN and RF. These works achieved maximum accuracy of 92.5 [13]. The researchers extracted various features from the EEG data and applied a SVM classifier to differentiate between SZ patients and Normal people. The proposed framework achieved an accuracy of 90.3% in the classification task. In [14] several ML algorithms, including SVM, KNN, RF, and ANN, were trained and evaluated. The SVM algorithm achieved the accuracy, sensitivity and specificity of 86.5%, 85.7% and 87.2% respectively and an AUC-ROC of 0.907, indicating its effectiveness in distinguishing individuals with schizophrenia from healthy controls based on EEG signals.

The literature shows the work progress of EEG classification using ML and DL methods. The DL methods are superior than the ML methods. This work mainly focusses on DL method in detection of SZ using spectrogram-based EEGs.

This paper is organized as follows, Sect. 1 presents introduction and motivation of work, Sect. 2 describes the literature of the proposed work and some important findings in the existing work, Sect. 3 gives the detailed description on methodology, dataset used and the proposed CNN architecture. Results and comparisons are briefly described in Sect. 4 and Sect. 5 gives the conclusions and future scope of the work.

3 Proposed Methodology

The proposed methodology mainly has three important stages that is collecting EEG data set for Norm and SZ, converting the EEG 1D data into 2D image data using STFT which is called as spectrogram and customized CNN for classification. Details of proposed work is illustrated in a block diagram as shown in Fig. 1. The EEG data acquired from source has three conditions, condition-1: hitting a button causes an audible sound to be generated immediately, condition-2: Listening the same audible sound passively, and condition-3: hitting a button does not cause a sound to be generated.

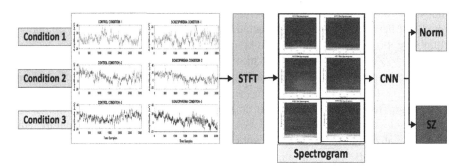

Fig. 1. Block Diagram of the proposed method

The next important phase of the work is converting 1D signal to 2D image. The EEG signals acquired from the patients and normal persons are converted into spectrogram using STFT. The STFT is a Fourier transform extension that offers information on the components of a signal in the frequency domain. Main idea behind the STFT is to divide a signal into short segments or windows and apply the Fourier transform to each window individually. By doing so, we can examine the frequency content of the signal with a small duration of time. Resulting STFT representation displays how the signal's frequency components evolve over different time intervals. It provides valuable information about the spectral content of a signal at different points in time, which is essential for understanding and manipulating signals with time-varying characteristics. The mathematical expression for STFT with an input x[n] is given below in Eq. (1).

$$STFT(x[n]) = X[m,f] = \sum_{n=-\infty}^{\infty} x[n]w[n-m]e^{-jfn} \tag{1}$$

where w[n] is the window function with a small duration and sliding along the time axis until the last point of the signal. This mechanism gives the time and frequency representation of a signal. The STFT of EEG signals under Norm and SZ is shown in Fig. 1.

Customized CNN

The proposed CNN is a customized CNN, which consists of input layer, hidden and fully connected layers. CNN mainly consists of convolution layers, which is the core component of CNN. Convolution layer is a fundamental building block in CNNs, which are a type of DL architecture commonly used for image and video processing tasks. The convolutional layer performs a convolution operation on the input data to extract local features and learn hierarchical representations.

There are filters or kernels in CNN, which are small-sized matrices. The output of each convolution layer is convolution sum of input with filter weights. This involves element-wise multiplication of the filter with a local receptive field of the input data, followed by summation of the multiplied values. Equation (2) shows the mathematical expression for convolution in discrete domain.

$$y[m, n] = \sum_{k=-\infty}^{\infty} \sum_{r=-\infty}^{\infty} x[k, r]h[m - k, n - r] \tag{2}$$

where h[m,n] is the filter weights of the convolution layer. During the convolution operation, the filter is slid over the input data in a sliding window manner, computing the element-wise products and summing them up to generate a single value in the corresponding location of the feature map. This process is repeated for every possible receptive field in the input, resulting in feature maps.

Max Pooling Layer
This layer is to extract the most salient features from each local neighborhood of the input feature map. It divides the feature map into non-overlapping rectangular parts known as pooling window or pooling regions, and then substitutes the highest number inside each region.

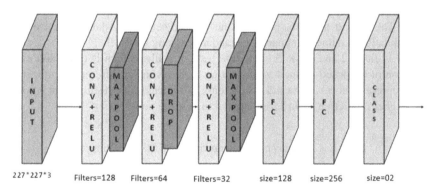

Fig. 2. Customized CNN for training and testing

The customized CNN is shown in Fig. 2, it consists of input layer of size $227 \times 227 \times 3$, followed by convolution layers and Relu activation function and for pooling max pooling is used. At the middle e network dropout layer in order to avoid the overfitting problem finally fully connected layers and classification layer of size 02 for binary classification.

The training and testing can be done by using ten-fold cross validation. The process of tenfold validation can be summarized as follows. Data Splitting: Initially the dataset portioned into ten subsets, or folds of roughly equal size. Each fold should ideally contain a representative sample of the data.

Model Training and Evaluation: The main module model training, here in 9th fold model is trained and reaming for validation. Ten times doing the same process validate the set once. In each iteration, the model is trained on a different combination of folds. Performance Measurement: After each training iteration, the model's performance is evaluated using a chosen performance metric on the validation set.

Average Performance: The performance results from the ten iterations are usually averaged to obtain a single performance metric that represents the model's performance across the entire dataset. This average performance metric is used to compare and assess the model's generalization ability. By utilizing ten-fold cross-validation, the potential bias in model evaluation caused by the specific partitioning of the data is reduced. It provides a more robust estimate, as it evaluates the model on multiple diverse subsets of the data.

4 Results

The proposed method uses a customized CNN for training and testing. The EEG data of 734 normal samples and 874 SZ effected samples are used for classification. The data split for training and validation is 0.9 and 0.1 respectively. The optimizer used in the proposed work is SGDM, RMS Prop and Adams optimizer. The number of Epochs is 6 with a BS of 50 is used as training parameters. The validations accuracy obtained after training process under various optimizers is tabulated in Table 4, Table 5 and Table 6. It shows that validation accuracy of adams optimizer achieves best results than other two, which is for Alex net 82.1, VGG16 83.1, Resnet 83.6 and for proposed customized net it is 86.3. This indicates that the proposed customized net under Adams optimizer gives higher validation accuracy compared to standard existing methods.

Table 1, Table 2 and Table 3 shows the testing results, which is confusion matrix comparison of proposed method with existing methods under different optimizers. In which 0 represents normal data and 1 represents SZ effected data. It shows that proposed net has higher prediction accuracy compared to existing nets.

Table 1. Confusion matrix comparison with SGDM optimizer

	Alex net		VGG16		Resnet		Proposed CNN	
	0	1	0	1	0	1	0	1
0	128	21	126	23	125	24	130	19
1	68	91	67	92	64	95	62	97

Table 2. Confusion matrix comparison with RMSprop optimizer

		Alex net		VGG16		Resnet		Proposed CNN	
		0	1	0	1	0	1	0	1
0		142	7	143	6	146	3	149	0
1		78	81	77	82	2	87	68	91

Table 3. Confusion matrix comparison with Adams optimizer

		Alex net		VGG16		Resnet		Proposed CNN	
		0	1	0	1	0	1	0	1
0		143	6	144	5	145	4	148	1
1		2	157	1	158	2	157	1	158

The performance metrics for comparing proposed CNN with existing CNN is by the following metrics. These parameters are calculated from confusion matrix. These are formulated in Eqs. (3–6) as follows:

$$\text{Accuracy} = \frac{TP + TN}{TP + TN + FP + FN} \tag{3}$$

$$\text{Precision} = \frac{TP}{TP + FP} \tag{4}$$

$$\text{Recall} = \frac{TP}{TP + FN} \tag{5}$$

$$\text{F1 Score} = \frac{2(\text{Precision} * \text{Recall})}{\text{Precision} + \text{Recall}} \tag{6}$$

The performance metric comparison is tabulated in Table 4, Table 5 and Table 6. The following are observations made from results. The proposed customized CNN has higher validation accuracy, test accuracy, precision, Recall and F1 score compared to existing CNNs.

Table 4. Performance metric comparison with SGDM optimizer

	Validation Accuracy	Test Accuracy	Precision	Recall	F1 score
Alexnet	65.75	71.11	85.91	65.30	74.20
VGG16	67.32	70.77	84.56	65.28	73.67
ResNet50	66.28	71.42	83.86	66.13	73.94
CNN	**72.50**	**73.71**	**87.24**	**67.71**	**76.24**

Table 5. Performance metric comparison with RMS prop optimizer

	Validation Accuracy	Test Accuracy	Precision	Recall	F1 score
Alexnet	66.84	72.40	95.31	64.54	76.96
VGG16	68.43	73.03	95.97	65.28	77.70
ResNet50	72.21	75.64	97.78	66.97	79.49
CNN	**80.52**	**77.92**	**100**	**67.71**	**80.74**

Table 6. Performance metric comparison with Adams optimizer

	Validation Accuracy	Test Accuracy	Precision	Recall	F1 score
Alexnet	82.1	97.41	95.73	98.62	96.54
VGG16	83.3	98.05	96.64	99.31	97.95
ResNet50	83.6	98.13	97.2	98.82	98.34
Proposed CNN	**86.3**	**99.35**	**99.32**	**99.31**	**99.31**

The following are the advantages of the proposed method compared to existing networks

1. Achieves higher validation accuracy of 86.3.
2. Achieves higher testing accuracy of 99.35 with a three-convolution layer network
3. The proposed model is very simple and training time is also low
4. The proposed method has a potential to replace traditional machine learning frame work.

5 Conclusions

This work mainly focuses on identification of SZ at early stages using deep learning models. The EEG signals of Normal and SZ samples are collected from an open-source data set and these EEG signals are converted into image spectrogram using STFT. The spectrogram of the Normal and SZ samples is trained by using standard CNNs such as Alexnet, VGG16 and Resnet. This paper proposes a simple customized CNN with three convolution layers for training, which is less complex and high speed. The proposed CNN gives better validation accuracy in training and testing as compared to the existing CNN models. The results exhibited the potential of the proposed CNN model in detecting SZ efficiently. In future, this work can be extended in two ways: (i) By taking scalogram of EEG data for training the proposed CNN and (ii) By classifying multiple SZ conditions using Spectrogram and scalogram.

References

1. Sun, J., et al.: A hybrid deep neural network for classification of schizophrenia using EEG data. Sci. Rep. **11**, 1–16 (2021)
2. Prabhakara Rao, A., Prasanna Kumar, G., Ranjan, R.: Performance comparison of classification models for identification of breast lesions in ultrasound images BT - pattern recognition and data analysis with applications. In: Gupta, D., Goswami, R.S., Banerjee, S., Tanveer, M., Pachori, R.B. (eds.) Pattern Recognition and Data Analysis with Applications. LNEE, vol. 888, pp. 689–699. Springer, Heidelberg (2022). https://doi.org/10.1007/978-981-19-1520-8_56
3. Matalam, C.L., Hembra, M.S.: Community-based mental health project in Davao region. SPMC J. Health Care Serv. **8**(2), 5 (2022)
4. Miyauchi, T., et al.: Computerized EEG in schizophrenic patients. Biol. Psychiatry **28**, 488–494 (1990)
5. Bonita, J.D., et al.: Time domain measures of inter-channel EEG correlations: a comparison of linear, nonparametric and nonlinear measures. Cogn. Neurodyn. **8**, 1–15 (2014)
6. Singh, K., Singh, S., Malhotra, J.: Spectral features based convolutional neural network for accurate and prompt identification of schizophrenic patients. Proc. Inst. Mech. Eng. Part H J. Eng. Med. **235**, 167–184 (2021)
7. Rao, A.P., Bhaskar, J., Kumar, G.P.: Machine learning framework for identification of abnormal EEG signal. In: Gupta, N., Pareek, P., Reis, M. (eds.) IC4S 2022. LNICST, vol. 472, pp. 42–54. Springer, Cham (2023). https://doi.org/10.1007/978-3-031-28975-0_4
8. de Miras, J.R., Ibáñez-Molina, A.J., Soriano, M.F., Iglesias-Parro, S.: Schizophrenia classification using machine learning on resting state EEG signal. Biomed. Signal Process. Control **79**, 104233 (2023)
9. Ranjan, R., Sahana, B.C.: A machine learning framework for automatic diagnosis of schizophrenia using EEG signals. In: 2022 IEEE 19th India Council International Conference (INDICON), pp. 1–6. IEEE (2022)
10. Bagherzadeh, S., Shahabi, M.S., Shalbaf, A.: Detection of schizophrenia using hybrid of deep learning and brain effective connectivity image from electroencephalogram signal. Comput. Biol. Med. **146**, 105570 (2022)
11. Organisciak, D., Shum, H.P.H., Nwoye, E., Woo, W.L.: RobIn: a robust interpretable deep network for schizophrenia diagnosis. Expert Syst. Appl. **201**, 117158 (2022)
12. Aslan, Z., Akin, M.: A deep learning approach in automated detection of schizophrenia using scalogram images of EEG signals. Phys. Eng. Sci. Med. **45**, 83–96 (2022)
13. Buettner, R., et al.: High-performance exclusion of schizophrenia using a novel machine learning method on EEG data. In: 2019 IEEE International Conference on E-Health Networking, Application & Services (HealthCom), pp. 1–6. IEEE (2019)
14. Siuly, S., Khare, S.K., Bajaj, V., Wang, H., Zhang, Y.: A computerized method for automatic detection of schizophrenia using EEG signals IEEE Trans. Neural Syst. Rehabil. Eng. **28**(2390), 2400 (2020)

An Optimized Ensemble Machine Learning Framework for Multi-class Classification of Date Fruits by Integrating Feature Selection Techniques

V. V. R. Maheswara Rao[1]([⊠]) [iD], N. Silpa[1] [iD], Shiva Shankar Reddy[2] [iD], S. Mahaboob Hussain[3] [iD], Sridevi Bonthu[3] [iD], and Padma Jyothi Uppalapati[3] [iD]

[1] Shri Vishnu Engineering College for Women (A), Bhimavaram, India
mahesh_vvr@yahoo.com, nrusimhadri.silpa@gmail.com
[2] S.R.K.R. Engineering College (A), Bhimavaram, India
[3] Vishnu Institute of Technology, Bhimavaram, India
mahaboobhussain.smh@gmail.com

Abstract. Date fruits are widely consumed and valued for their nutritional properties and economic importance. Accurate classification of date fruits is crucial for quality control, sorting, and grading processes in the food industry. However, the classification of date fruits poses several challenges due to variations in their external visible physical attributes and the presence of multiple cultivars. To address the issue of high-dimensional feature space, the authors, in this research study propose an optimized ensemble machine learning framework for multi-class classification of date fruits by integrating feature selection techniques. The objective is to develop a reliable and efficient system that can accurately classify date fruits into different classes based on their physical attributes. Initially, the framework applies various feature selection techniques to identify the most relevant and discriminative features for classification. Next, it employs variants of ensemble machine learning techniques to perform multi-class classification. We utilize popular ensemble methods such as Boosted Trees, Bagged Trees, RUSBoosted Trees and Optimized Ensemble to improve the accuracy and optimization of the classification model. By integrating feature selection techniques, the proposed optimized ensemble classifier effectively handles the challenges of high-dimensional feature space and variations in date fruit characteristics. The results confirm the superiority of the proposed framework, highlighting its potential for practical applications in the food industry.

Keywords: Date Fruit Classification · Machine Learning · Feature Selection · Ensemble Classifier · Neural Networks

1 Introduction and Related Work

Date fruits are widely consumed and recognized for their nutritional value and health benefits. The accurate classification of date fruits [1–4] into different categories is crucial for quality control, marketability, and efficient supply chain management. However,

P. Pareek et al. (Eds.): IC4S 2023, LNICST 536, pp. 12–27, 2024.
https://doi.org/10.1007/978-3-031-48888-7_2

the process of manually classifying date fruits can be labor-intensive, time-consuming, and prone to human errors. Therefore, there is a need for automated systems that can effectively classify date fruits into multiple classes.

In recent years, machine learning techniques [5–12] have shown great potential in solving complex classification problems [13–22]. These techniques enable computers to learn from data and make accurate predictions or decisions. In the context of date fruit classification, machine learning algorithms can analyze various features of date fruits, such as size, color, texture, and shape, to differentiate between different classes. This research paper presents a robust ensemble machine learning framework specifically designed for the multi-class classification of date fruits. The framework integrates feature selection techniques to enhance the classification performance by reducing the dimensionality of the data and focusing on the most informative features.

Feature selection [23–27] is a critical step in the classification process as it helps identify the most relevant features that contribute to accurate classification. By eliminating irrelevant or redundant features, the framework can improve computational efficiency and mitigate the curse of dimensionality of the classification models. Various feature selection techniques, including filter methods, wrapper methods, and embedded methods, are explored and integrated into the framework.

The ensemble learning approach [28, 29] is adopted to further enhance the classification accuracy and robustness of the framework. Ensemble learning combines multiple base classifiers to make collective predictions, leveraging the diversity of classifiers to achieve better overall performance. Different ensemble techniques are investigated and integrated into the framework.

To evaluate the effectiveness of the proposed framework, a comprehensive dataset of date fruit images is collected, encompassing different varieties, ripeness stages, lighting conditions, and imaging angles. The dataset is carefully labelled and pre-processed to ensure high-quality data for training and testing the classification models. The experimental results demonstrate the superiority of the proposed framework in accurately classifying date fruits into multiple classes. The integration of feature selection techniques significantly improves the efficiency and effectiveness of the classification process by selecting the most informative features. The ensemble learning approach further enhances the performance by leveraging the collective intelligence of multiple classifiers. The framework's robustness is also evaluated by conducting experiments with different base classifiers and ensemble techniques.

The application of this research extends beyond date fruit classification. The proposed framework can be adapted and applied to other multi-class classification problems in various domains. By providing an automated and accurate solution for date fruit classification, the framework contributes to improving quality control and management in the date fruit industry.

The remaining portion is organized as follows. Section 2 describes the proposed optimized date fruit multi-class classification framework. Section 3 discusses the experimental results and evaluation of our suggested framework. Finally, Sect. 4 finishes the work with concluding observations.

2   Proposed ML Based Optimized Ensemble Date Fruit Classifier (ML-OEDFS)

The primary focus of this research paper is the introduction of a novel Machine Learning based Optimized Ensemble Date Fruit Classifier (ML-OEDFC) for accurately classifying date fruits into different categories.

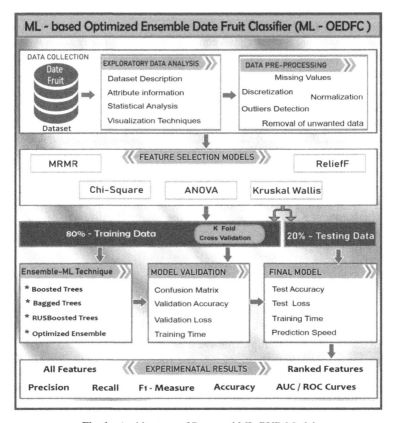

Fig. 1. Architecture of Proposed ML-RVR Model

The proposed system presented in Fig. 1, integrates feature selection techniques to enhance the classification performance by reducing data dimensionality and emphasizing the most informative features. The overarching goal of this work is to develop an automated system that effectively categorizes date fruits, leading to improved quality control, marketability within the date fruit industry.

2.1   ML-OEDFC: Data Collection and Exploration of Dataset

The proposed ML-RDFC leverages the Muratkoklu Date Fruit dataset repository [1] for research purposes to gain insights into the characteristics and relationships between different dry fruits. The dataset consists of 34 attributes that can be classified into seven distinct classes. In the research investigations, data collection is conducted to analyze the attributes associated with dry fruit category identification systems.

The model created visual representations, including Pie chart and correlation matrix to explore the data further. It is observed a normal distribution and most date fruits had a similar size range as presented in Fig. 2.

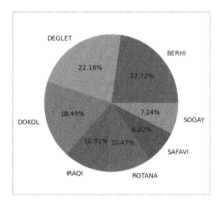

Fig. 2. Distribution of Data across all varieties of Date Fruits

Figure 3 displays the correlation matrix of the morphological attributes employed in this study. The matrix effectively visualizes the interconnectedness and interdependencies among the different morphological attributes utilized for identifying Date fruit

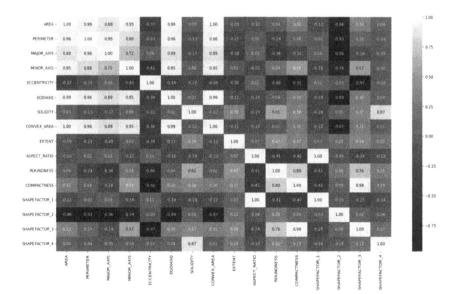

Fig. 3. Correlation Matrix of Morphological Attributes

varieties. Upon careful analysis of this matrix, it becomes evident that there exists a substantial and consistent association between the various qualities.

2.2 ML-OEDFC: Data Pre-processing

The collected dataset has already undergone stages of date fruits images collection, rotation, flipping and scaling to enhance the quality of dataset. In addition, the researchers performed resizing, normalization, noisy reduction, and edge detection to extract relevant features. For the proposed study, the processed dataset is taken as input and split the dataset into training, validation, and testing sets to evaluate the classifier's performance.

2.3 ML-OEDFC: Feature Selection Techniques

Feature selection plays a crucial role in optimizing the performance of a Date fruit classifier. In this study, various feature selection techniques were explored to identify the most relevant and informative attributes for accurate classification.

One commonly used technique is the Filter approach, which relies on statistical measures to assess the relevance of each feature independently of the classifier. These techniques evaluate the relationship between the attributes and the target class, allowing the selection of features that contribute the most to classification accuracy. Another approach is the Wrapper method, which utilizes the classifier's performance as the criterion for feature selection. This method involves creating subsets of features and evaluating their impact on the classifier's accuracy. Embedded methods integrate feature selection within the learning algorithm itself. They select features during the model training process based on their importance in improving the model's performance. Additionally, hybrid approaches that combine multiple feature selection techniques have been proposed. These approaches leverage the strengths of different methods to overcome their individual limitations, resulting in improved feature subsets and classification performance.

By applying these feature selection techniques to a Date fruit classifier, the proposed study aim to identify the most discriminative attributes that contribute significantly for accurate classification. An algorithm for performing feature selection using ML-OEDFC framework and ranking the features based on different feature selection methods. The algorithm also identifies the best feature selection method for a given classifier.

Algorithm: ML-OEDFC Feature Selection and Classifier Evaluation.

Input: Selected Date Fruits Dataset.
Classifiers - Boosted Trees(BT), Bagged Trees(BGT), RUSBoosted Trees (RUSBT), Optimized Ensemble (OE)
Output: Ranked Features by Feature Selection Methods and Best suitable Feature Selection Method for each classifier.

Step 1. Load the Selected Date Fruits Dataset into memory.
Step 2. Preprocess the dataset (e.g., handle missing values, discretization, normalization, encode categorical variables, outliers detection etc.).
Step 3. Split the dataset into input features (X) and target class(C).

Step 4. Initialize an empty dictionary to store the feature scores for each feature selection method.

Step 5. For each feature selection method in the list of feature selection methods (MRMR, Chi2, ReliefF, ANOVA, Kruskal Wallis), do the following steps:
   a. Apply the specific feature selection method to the input Date Fruits features X and target class C.
   b. Retrieve the feature scores or rankings obtained from the fea-ture selection method.
   c. Store the feature scores or rankings in the dictionary with the feature selection method as the key.

Step 6. Rank the features based on their scores obtained from each feature selection method.

Step 7. For each feature, calculate the average rank across all feature selec-tion methods.

Step 8. Sort the features based on their average rank in ascending order.

Step 9. Initialize an empty list to store the best feature selection methods for each classifier.

Step 10. For each classifier in the list of classifiers (BT, BGT, RUSBT, OE), do the following steps:
   a. Train the classifier using the Date Fruits input features selected based on the top-ranked features.
   b. Evaluate the classifier's performance using appropriate evalua-tion met-rics (e.g., validation accuracy test accuracy, precision, recall, F1-score, AUC/ROC curve etc.).
   c. Store the classifier's performance metrics.

Step 11. Identify the best suitable feature selection method for each classifier based on their performance metrics.

Step 12. Return the ranked features and the best suitable feature selection method for each classifier as the output.

The algorithm begins by importing the dataset and, if necessary, preprocessing it. The system then applies each feature selection method to the dataset and stores the feature scores or rankings that result. The features are then ranked according to their average position across all selection methodologies. For each classifier, the algorithm trains the model with the most highly ranked features and evaluates its performance using the most pertinent metrics. Based on each classifier's respective performance metrics, the algorithm then determines the optimal feature selection method for each classifier.

2.4 ML-OEDFC: Training Ensemble Machine Learning Models

In ML-OEDFC, the training process involves utilizing variants of ensemble machine learning models. Ensemble models are renowned for their ability to enhance predictive performance by combining the predictions of multiple base models. ML-OEDFC takes advantage of ensemble techniques to create a robust and accurate ensemble.

ML-OEDFC employs multiple ensemble machine learning model variants to improve its performance. Bagging independently trains multiple models on distinct subsets of the training data, and then combines their predictions to produce the final ensemble out-put. Boosting trains models sequentially, with an emphasis on correcting prior errors. Combining random under-sampling with boosting, RUSBoosted Tree addresses class imbalance. It enhances classification performance and balances the dataset. Using techniques such as genetic algorithms, Optimized Ensemble optimizes the combination of base classifiers to improve overall accuracy and generalization. These variants enhance the capabilities of ML-OEDFC by minimizing overfitting, compensating for class imbalance, and optimizing ensemble performance.

The inclusion of all variants of ensemble learning models in ML-OEDFC expands the ensemble model repertoire, enabling the classifier to tackle specific challenges related to class imbalance and ensemble optimization. These variants provide additional flexibility and adaptability, allowing ML-OEDFC to deliver superior performance in diverse datasets and classification scenarios.

2.5   ML-OEDFC: K-Fold Cross Validation of Proposed Model

After training the ML-OEDFC model for classifying the date fruits with the chosen ensemble variants and feature selection techniques, it is necessary to assess its performance and generalization capabilities. One common strategy used for this purpose is k-fold cross-validation, where the dataset is divided into k subsets. In the case of ML-OEDFC, k is set to 5 and the percentage of training data is 80.

The k-fold cross-validation procedure begins with the dataset being divided into k folds of equal size. The model is then trained and evaluated k times, with a different fold serving as the validation set and the remaining folds serving as the training set each time. This enables a comprehensive evaluation of the performance of the model across various subsets of the data.

For each iteration of k-fold cross-validation, the ML-OEDFC model is trained on the training set and the corresponding validation set is used to calculate the performance metrics. Depending on the classification assignment, the metrics typically include accuracy, precision, recall, F1-score, or any other pertinent evaluation measures. At the conclusion of k-fold cross-validation, the performance metrics from each iteration are averaged to produce a more accurate estimation of the model's performance. This mitigates any bias imposed by a single validation set. It provides insight into the model's ability to generalize to new data and aids in choosing the opti-mal ensemble variant and feature selection method combination.

2.6   ML-OEDFC: Testing with Final Model

After completing the ML-OEDFC validation procedure using k-fold cross-validation, the model is evaluated using an independent test dataset. This dataset is distinct from the training and validation sets and is used to evaluate the efficacy of the model. The remaining 20% of the dataset is utilized to assess the ML-OEDFC model, and its predictions are compared against the actual or reference labels. Various evaluation metrics, including accuracy, precision, recall, and F1-score, are calculated to quantify the performance of the model on the test set. These metrics provide an objective evaluation of the model's classification accuracy for date fruits.

The ML-OEDFC model has effectively learned the underlying patterns and characteristics of the data if it achieves satisfactory results on the test dataset. The model benefits from the combined predictive potential of multiple base models and the ability to zero in on the most informative features by utilizing ensemble learners and feature selection techniques. This ultimately leads to a final model that can classify date fruits with high accuracy.

3   Experimental Results and Analysis

In order to evaluate the performance of the proposed ML-OEDFC in accurately classifying date fruits, a comprehensive series of experiments was conducted, and the significant results are presented in this section. The dataset utilized in this study consists of 34 distinct features derived from three primary categories: morphological, shape, and color characteristics. These characteristics were meticulously chosen to capture useful details about date fruits and to provide an exhaustive representation of their characteristics.

Four variants of ensemble machine learning models were devised in order to facilitate the classification process. These variants were created with the intent of enhancing the functionality of the ML-OEDFC algorithm. The objective of creating multiple variants was to compare their efficacy and identify the most effective model for accurately classifying date fruits.

K-fold cross-validation was utilized to ensure robustness and evaluate the generalization capabilities of the ML-OEDFC models. This method divides the dataset into K subsets or folds, where K is the number of subsets. This division is responsible for training and validating ML models using various combinations of training and testing data. By performing cross-validation, we hope to evaluate the consistency and reliability of the proposed ML-OEDFC models across diverse data subsets.

To assess the classification performance of the proposed ML-OEDFC variants on date fruits, a confusion matrix was developed as a fundamental tool for evaluating the classification models' accuracy. The confusion matrix shown in Table 1 provides an exhaustive examination of the predictions made by the models, allowing for a thorough analysis of the classification outcomes. Overall, the data show that different algorithms perform differently for each rice varieties, and the method used has a considerable impact on classification accuracy. The "BT" method produces good results for most date fruit varieties, and the "OE" algorithm performs very well across numerous date fruit types.

Table 1. Confusion Matrix for Classifying Date Fruit Types using ML-OEDFC Classifiers.

Date Fruit	Algorithm	BERHI	DEGLET	DOKOL	IRAQI	ROTANA	SAFARI	SOGAY
BERHI	BT	41	0	0	10	2	0	2
	BGT	41	0	0	8	1	0	1
	RUSBT	37	0	0	7	2	0	2
	OE	38	0	0	9	0	0	0
DEGLET	BT	0	51	9	0	2	0	8
	BGT	0	55	10	0	1	1	13
	RUSBT	0	55	13	0	1	0	10
	OE	0	54	8	0	1	0	9
DOKOL	BT	0	11	153	0	0	0	1
	BGT	0	9	153	0	0	0	0
	RUSBT	0	8	148	0	0	0	0
	OE	0	8	155	0	0	0	0
IRAQI	BT	9	1	0	44	1	0	2
	BGT	9	0	0	48	1	0	1
	RUSBT	12	0	0	49	1	0	2
	OE	11	0	0	48	0	0	0
ROTANA	BT	1	1	0	3	124	0	3
	BGT	1	1	0	1	125	0	3
	RUSBT	1	1	0	2	123	0	2
	OE	0	1	0	1	128	0	2
SAFARI	BT	0	0	2	1	0	156	1
	BGT	0	0	0	1	0	157	1
	RUSBT	0	0	1	0	0	156	1
	OE	0	0	0	0	0	158	1
SOGAY	BT	1	14	0	0	4	3	58
	BGT	1	13	1	0	5	1	56
	RUSBT	2	14	2	0	6	3	58
	OE	3	15	1	0	4	1	63

A number of tests were carried out in order to assess the efficacy of different ensemble machine learning approaches in categorizing date fruits. The findings, which are summarized in Table 2, reveal that the "OE" approach outperforms alternative ensemble ML techniques. It exhibits greater validation and test accuracy shown in Fig. 4, indicating that it is useful in properly classifying date fruits.

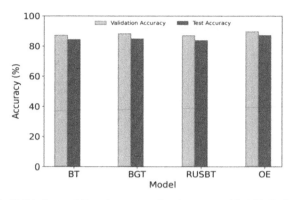

Fig. 4. Validation and Test Accuracy of various ensemble ML Techniques

Table 2. Performance of Ensemble ML Techniques

Ensemble ML technique	Training time (Sec)	Validation accuracy (Sec)	Test accuracy (Sec)
BT	6.72 s	87.2	84.4
BGT	3.64 s	88.3	84.9
RUSBT	2.76 s	87.1	83.8
OE	11.82 s	89.6	87.2

The ML-OEDFC algorithm performed an experiment using test data and generated ROC curve shown in Fig. 5. By plotting the true positive rate against the false positive rate, the ML-OEDFC algorithm provides a visual representation of its classification performance in distinguishing between different classes in the test data.

To leverage the advantages of feature selection methods for outperformed OE classification model, the proposed approach incorporates MRMR, Chi2, ReliefF, ANOVA, and Kruskal methods. Through experimentation, the rank of each feature is determined using these algorithms, and the results are tabulated in Table 3. This analysis enables the assessment of the significance and relevance of each feature, facilitating the optimization of classification performance.

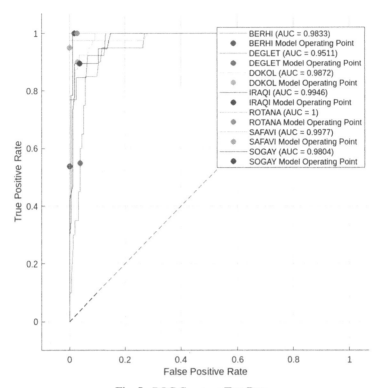

Fig. 5. ROC Curve on Test Data

With the employed feature selection methods, no. of experiments is conducted to find the impact on performance of classifiers mainly in terms of accuracy. The results are tabled in Table 4 and graph representation given in Fig. 6.

According to the observed results in the table, the MRMR feature selection method consistently performs well regardless of the size of the feature subset. It obtains a high degree of accuracy, ranging from 84.4% to 87.2%, demonstrating its effectiveness in selecting pertinent features. The Chi2 method, on the other hand, demonstrates variable efficacy, with 87.2% accuracy for 30 features and 77.7% accuracy for 10 features. With accuracies ranging from 82.1% to 86.6%, the ReliefF method exhibits a relatively stable performance. ANOVA also yields promising outcomes, with 89.0% accuracy for 30 features. The efficacy of the Kruskal method is comparable, ranging from 78.2 to 86.6%. Across a variety of feature subset sizes, the MRMR and ANOVA methods stand out as effective feature selection strategies.

Table 3. Ranking of Attributes by Feature Selection Algorithms

S. no	Feature	MRMR	Chi2	ReliefF	ANOVA	Kruskal
1	AREA	R28	R04	R02	R04	R04
2	PERIMETER	R17	R06	R05	R06	R07
3	MAJOR_AXIS	R11	R10	R08	R07	R08
4	MINOR_AXIS	R25	R01	R03	R02	R01
5	ECCENTRICITY	R33	R19	R24	R23	R22
6	EQDIASQ	R20	R05	R04	R01	R03
7	SOLIDITY	R15	R27	R33	R33	R30
8	CONVEX_AREA	R07	R03	R01	R03	R05
9	EXTENT	R08	R34	R32	R32	R34
10	ASPECT_RATIO	R23	R18	R21	R25	R21
11	ROUNDNESS	R05	R17	R26	R26	R26
12	COMPACTNESS	R12	R21	R20	R22	R24
13	SHAPEFACTOR_1	R01	R02	R09	R05	R02
14	SHAPEFACTOR_2	R34	R11	R16	R12	R09
15	SHAPEFACTOR_3	R30	R20	R19	R21	R23
16	SHAPEFACTOR_4	R03	R33	R34	R34	R33
17	MeanRR	R29	R08	R06	R09	R15
18	MeanRG	R32	R13	R10	R14	R18
19	MeanRB	R31	R23	R15	R16	R13
20	StdDevRR	R26	R30	R27	R29	R29
21	StdDevRG	R02	R32	R30	R31	R32
22	StdDevRB	R21	R31	R31	R30	R31
23	SkewRR	R19	R25	R23	R20	R20
24	SkewRG	R27	R15	R17	R10	R11
25	SkewRB	R06	R26	R25	R24	R25
26	KurtosisRR	R24	R29	R28	R28	R28
27	KurtosisRG	R10	R22	R18	R17	R19
28	KurtosisRB	R16	R28	R29	R27	R27
29	EntropyRR	R18	R07	R12	R11	R06
30	EntropyRG	R09	R12	R13	R18	R10
31	EntropyRB	R14	R16	R22	R19	R16
32	ALLdaub4RR	R13	R09	R07	R08	R14

(continued)

Table 3. (*continued*)

S. no	Feature	MRMR	Chi2	ReliefF	ANOVA	Kruskal
33	ALLdaub4RG	R22	R14	R11	R13	R17
34	ALLdaub4RB	R04	R24	R14	R15	R12

Table 4. Results of Classifiers on all Attributes

Feature selection method	All features	30 Features	25 Features	20 Features	15 Features	10 Features
MRMR	86.6	86.0	87.2	86.6	84.4	87.2
Chi2	86.0	87.2	84.4	86.0	84.4	77.7
ReliefF	86.0	86.6	85.5	82.7	83.8	82.1
ANOVA	88.3	89.0	86.0	82.1	81.6	81.0
Kruskal	86.0	85.5	86.6	83.8	83.8	78.2

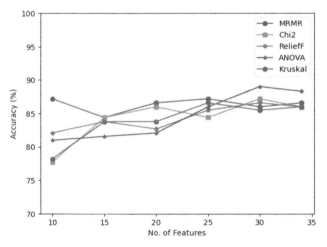

Fig. 6. Accuracy of Classifiers

The results indicate that a combination of the OE with MRMR and OE with ANOVA methods, can accurately classify date fruits. The MRMR method consistently achieves high performance across a variety of subset sizes, demonstrating its capacity to select pertinent features. In the meantime, ANOVA yields optimistic results, with an accuracy of 89.0% for 30 features. Moreover, the incorporation of feature selection methods can improve classification accuracy and also potentially reduce processing time by focusing on the most informative features, thereby enhancing the efficacy of the date fruit classification process.

4   Conclusion

In this study, a machine learning-based framework was proposed with the aim of automatically classifying date fruits with high accuracy. Specifically, variants of ensemble machine learning techniques were used, integrating feature selection techniques. The results indicated that the optimized ensemble variant performed better when incorporating the MRMR and ANOVA feature selection methods. These findings highlight the effectiveness of integrating these feature selection techniques into the ensemble framework. The proposed framework offers a promising solution for accurate and efficient classification of date fruits, with potential applications in the food industry and beyond. This resulted in the development of a software application that allowed users to categorize and obtain information about date fruits. The ML based framework could improve the success rates of classifying a variety of objects, such as date fruits, vegetables, fruits, and grains. This research endeavor not only contributes to the advancement of date fruit classification but also holds potential for broader applications in other domains requiring multi-class classification systems.

References

1. Koklu, M., Kursun, R., Taspinar, Y.S., Cinar, I.: Classification of date fruits into genetic varieties using image analysis. Math. Probl. Eng. **2021**, 1–13 (2021)
2. Khalid, S., Khalid, N., Khan, R.S., Ahmed, H., Ahmad, A.: A review on chemistry and pharmacology of AJWA date fruit and pit. Trends Food Sci. Technol. **63**, 60–69 (2017)
3. Aiadi, O., Kherfi, M.L.: A new method for automatic date fruit classification. Int. J. Comput. Vis. Robot. **7**(6), 692–711 (2017)
4. Hossain, M.S., Muhammad, G., Amin, S.U.: Improving consumer satisfaction in smart cities using edge computing and caching: a case study of date fruits classification. Futur. Gener. Comput. Syst. **88**, 333–341 (2018)
5. Liakos, K., Busato, P., Moshou, D., Pearson, S., Bochtis, D.: Machine learning in agriculture: a review. Sensors **18**(8), 2674 (2018)
6. Silpa, N., Maheswara Rao, V.V.R.: Machine learning-based optimal segmentation system for web data using a Genetic approach. J. Theor. Appl. Inf. Technol. **100**(11), 3552–3561 (2022)
7. Kovvuri, A.R., Uppalapati, P.J., Bonthu, S., Kandula, N.R.: Water level forecasting in reservoirs using time series analysis – auto ARIMA model. In: Gupta, N., Pareek, P., Reis, M. (eds.) Cognitive Computing and Cyber Physical Systems. IC4S 2022. Lecture Notes of the Institute for Computer Sciences, Social Informatics and Telecommunications Engineering, vol. 472. Springer, Cham (2023). https://doi.org/10.1007/978-3-031-28975-0_16
8. Maheswara Rao, V.V.R., Silpa, N., Gadiraju, M., Shankar, R.S., Vijaya Kumar, D., Brahma Rao, K.B.V.: An optimal machine learning model based on selective reinforced Markov decision to predict web browsing patterns. J. Theor. Appl. Inf. Technol. **101**(2), 859–873 (2023)

9. Reddy, S.S., Gadiraju, M., Maheswara Rao, V.V.R.: Analyzing student reviews on teacher performance using long short-term memory. In: Raj, J.S., Kamel, K., Lafata, P. (eds.) Innovative Data Communication Technologies and Application. Lecture Notes on Data Engineering and Communications Technologies, vol 96. Springer, Singapore (2022). https://doi.org/10.1007/978-981-16-7167-8_39

10. Ahmed, S.S., Uppalapati, P.J., Ayesha, S., Hussain, S.M., Narasimharao, K., Silpa, N.: Assessing public sentiment towards digital India through Twitter sentiment analysis: a comparative study. In: 7th International Conference on Intelligent Computing and Control Systems (ICICCS), pp. 955–959. IEEE (2023)

11. Silpa, N., Rao, V.M.: Enriched big data pre-processing model with machine learning approach to investigate web user usage behaviour. Indian J. Comput. Sci. Eng. [Internet] 12(5), 1248–1256 (2021)

12. Reddy, S.S., Gadiraju, M., Preethi, N.M., Rao, V.M.: A novel approach for prediction of gestational diabetes based on clinical signs and risk factors. EAI Endorsed Trans. Scalable Inf. Syst. 10(3), e8 (2023)

13. Cinar, I., Koklu, M.: Identification of rice varieties using machine learning algorithms. J. Agricult. Sci. 28(2), 307–325 (2022)

14. Cinar, I., Koklu, M.: Classification of rice varieties using artificial intelligence methods. Int. J. Intell. Syst. Appl. Eng. 7(3), 188–194 (2019)

15. Behera, S.K., Rath, A., Mahapatra, A., Sethy, P.K.: Identification, classification & grading of fruits using machine learning & computer intelligence: a review. J. Ambient Intell. Human. Comput. 1–11 (2020)

16. Tharwat, A.: Classification assessment methods. Appl. Comput. Inform. 17(1), 168–192 (2020)

17. Oliveira, M.M., Cerqueira, B.V., Barbon, S., Barbin, D.F.: Classification of fermented cocoa beans (cut test) using computer vision. J. Food Compos. Anal. 97, Article ID 103771 (2021)

18. Tahir, M.A.U.H., Asghar, S., Manzoor, A., Noor, M.A.: A classification model for class imbalance dataset using genetic programming. IEEE Access 7, 71013–71037 (2019)

19. Koklu, M., Ozkan, I.A.: Multiclass classification of dry beans using computer vision and machine learning techniques. Comput. Electron. Agricult. 174, Article ID 105507 (2020)

20. Maheswara Rao, V.V.R., Silpa, N., Mahesh, G., Reddy, S.S.: An enhanced machine learning classification system to investigate the status of micronutrients in rural women. In: Mahapatra, R.P., Peddoju, S.K., Roy, S., Parwekar, P., Goel, L. (eds.) Proceedings of International Conference on Recent Trends in Computing . Lecture Notes in Networks and Systems, vol. 341. Springer, Singapore (2022). https://doi.org/10.1007/978-981-16-7118-0_4

21. Jyothi, U.P., Dabbiru, M., Bonthu, S., Dayal, A., Kandula, N.R.: Comparative analysis of classification methods to predict diabetes mellitus on noisy data. In: Doriya, R., Soni, B., Shukla, A., Gao, X.Z. (eds.) Machine Learning, Image Processing, Network Security and Data Sciences. Lecture Notes in Electrical Engineering, vol. 946. Springer, Singapore (2023). https://doi.org/10.1007/978-981-19-5868-7_23

22. Maheswara Rao, V.V.R., Silpa, N., Gadiraju, M., Reddy, S.S., Bonthu, S., Kurada, R.R.: A plausible RNN-LSTM based profession recommendation system by predicting human personality types on social media forums. In: Proceedings of the 7th International Conference on Computing Methodologies and Communication (ICCMC), Erode, India, pp. 850–855 (2023)

23. Vyas, A.M., Talati, B., Naik, S.: Colour feature extraction techniques of fruits: a survey. Int. J. Comput. Appl. 83(15) (2013)

24. Korohou, T., Okinda, C., Li, H., et al.: Wheat grain yield estimation based on image morphological properties and wheat biomass. J. Sens. 2020, Article ID 1571936, 1–11 (2020)

25. Subbarao, M.V., Sindhu, J.T.S., Reddy, Y.C.A.P., Ravuri, V., Vasavi, K.P., Ram, G.C. : Performance analysis of feature selection algorithms in the classification of dry beans using KNN and neural networks. In: Proceedings of the International Conference on Sustainable Computing and Data Communication Systems (ICSCDS), Erode, India, pp. 539–545 (2023)
26. Wang, G., Lauri, F., Hassani, A.H.E.: Feature selection by mRMR method for heart disease diagnosis. IEEE Access **10**, 100786–100796 (2022)
27. Wang, Z., et al.: Segmentalized mRMR features and cost-sensitive ELM With fixed inputs for fault diagnosis of high-speed railway turnouts. IEEE Trans. Intell. Transp. Syst. **24**(5), 4975–4987 (2023)
28. Lopes, J.F., Ludwig, L., Barbin, D.F., Grossmann, M.V.E., Barbon, S.: Computer vision classification of barley flour based on spatial pyramid partition ensemble. Sensors **19**(13), 2953 (2019)
29. Subbarao, M.V., Ram, G.C., Varma, D.R. : Performance analysis of pistachio species classification using support vector machine and ensemble classifiers. In: Proceedings of the International Conference on Recent Trends in Electronics and Communication (ICRTEC), Mysore, India, 2023, pp. 1–6 (2023)

Harmonizing Insights: Python-Based Data Analysis of Spotify's Musical Tapestry

Deepesh Trivedi[1], Manas Saxena[1], S. S. P. M. Sharma B[2] (iD),
and Indrajeet Kumar[3]([✉]) (iD)

[1] School of CSIT, Symbiosis University of Applied Sciences, 453112 Indore, India
[2] School of MT, Symbiosis University of Applied Sciences, Indore, India
[3] School of CSIT, Symbiosis University of Applied Sciences, Indore, India
indrajeet.kumar@suas.ac.in

Abstract. This research paper analysis Spotify data using Python to investigate the characteristics contributing to song popularity. The objectives are to assess the popularity index, identify key attributes of popular songs, and develop a model for predicting song popularity based on current characteristics. The analysis involves data cleaning, exploratory data analysis, and visualization using Python libraries. With over 381 million monthly active users, Spotify provides a rich dataset for understanding music listening habits. Previous studies have explored Spotify's technologies and popularity, enhancing understanding of its protocols and user behavior. This research paper aims to uncover patterns and relationships within the data by applying statistical and machine-learning techniques. The findings will inform actionable recommendations and contribute to a better understanding of music consumption patterns and preferences.

Keywords: Analysis · exploratory data analysis · machine-learning techniques · spotify

1 Introduction

Data analysis has become essential for gaining valuable insights and understanding user behavior in various industries. In the realm of music streaming, Spotify stands out as one of the leading platforms with a vast amount of user data [1]. By leveraging data analysis techniques, we can delve into Spotify's rich dataset to uncover patterns, identify popular song attributes, and gain a deeper understanding of listener preferences [2– 5]. Python, a versatile programming language, provides a powerful ecosystem of libraries and tools for data analysis. With its robust capabilities, Python has become a popular choice for conducting data analysis on platforms like Spotify [6–8]. By utilizing Python's data manipulation, visualization, and machine learning libraries, we can efficiently clean, explore, and analyze Spotify data to extract meaningful insights [9].

The objective of data analysis on Spotify using Python is to explore the relationships between various song characteristics and their popularity, predict song popularity based

P. Pareek et al. (Eds.): IC4S 2023, LNICST 536, pp. 28–44, 2024.
https://doi.org/10.1007/978-3-031-48888-7_3

on specific attributes, and provide actionable recommendations for music professionals and enthusiasts [10–13]. By applying statistical techniques, machine learning algorithms, and data visualization, we can uncover trends, patterns, and correlations that contribute to a better understanding of music consumption habits [14]. Overall, data analysis on Spotify using Python opens up a world of possibilities to uncover hidden trends and patterns within the vast music library. It empowers us to make informed decisions, optimize user experiences, and enhance the overall music streaming ecosystem [15]. Data analysis on Spotify using Python has gained significant attention in the field of music analysis, enabling researchers and analysts to explore the vast amount of data generated by the platform [16].

In recent times, music streaming platforms like Wynk Music, Apple Music, and Spotify have witnessed an overwhelming influx of users and artists. These platforms serve as a hub for artists to upload their audio tracks [17], while also allowing users to discover and listen to their favorite songs [18]. In light of this scenario, the development of a predictive system capable of gauging the popularity of artists becomes crucial [19–22]. The application of Multiple-Input Multiple-Output (MIMO) technology in the context of music streaming services, particularly for Spotify, has been an area of interest in recent research. MIMO, a well-established technique in wireless communications, involves using multiple antennas at both the transmitter and receiver to improve data throughput and link reliability [23–26].

Fig. 1. Illustrating the Efficacy of Popularity Estimation: Demonstrating the Value in Assessing Artist Popularity

In the context of Spotify's music streaming platform, researchers and experts have explored how MIMO can enhance the user experience, optimize network performance, and address the challenges related to audio streaming [27, 28]. One of the primary focuses of MIMO for Spotify is to improve audio quality during music playback. By leveraging multiple antennas, MIMO can mitigate channel fading and reduce the impact of signal degradation, resulting in more stable and consistent audio streaming [29]. This enhancement is crucial for providing listeners with a seamless and immersive music experience without disruptions or audio artifacts [30]. Such a system can greatly benefit artists in planning strategies to enhance their visibility and reach. By leveraging the predictive capabilities of this system (as depicted in Fig. 1), artists can gain insights into the potential impact of uploading their audio tracks or tweaking elements like the content of their biographies [31]. They can simulate the potential popularity achieved through the predicted system, enabling them to make informed decisions regarding their promotional efforts. This growing demand for artist popularity prediction in music streaming services emphasizes the need for effective predictive models to cater to these requirements [32–36].

1.1 Online Music Services

According to Hall (2018), one of the most popular on-demand music services with a large user base is Spotify, which allows listeners to stream full-length content over the Internet without the need for purchasing or downloading. As of July 2017, Spotify had 60 million subscribers, and by January 2018, the number increased to 70 million (Hall 2018) [5]. The extensive repertoire of Spotify includes over 30 million songs, contributing to its widespread adoption and popularity (Hall, 2018). Previous studies conducted by Kreitz (2010), Loiacono (2014), and Verkoelen have examined various aspects of Spotify's technologies and its user base.

In a separate study, researchers focused on investigating the protocols and peer-to-peer architecture of Spotify to gain insights into its functioning and user interactions. They also explored the impact of the peer-to-peer network on user access patterns, properly referencing the specific report or publication where this study is mentioned will ensure accurate attribution.

1.2 Melody in the Machine: Harnessing Machine Learning to Forecast Chart-Topping Hits

The availability of a vast amount of digital music online and advancements in technology have significantly influenced music consumption habits. Kaminskas and Ricci (2012) suggest that users now search for specific music collections and rely on automatic playlist recommendations. In the field of Music Information Retrieval, researchers have been studying these concepts (Kaminskas & Ricci 2012).

This research aims to investigate the potential of utilizing 13 audio factors to predict the success of songs. The study employs four distinct machine learning techniques, namely logistic regression, K-nearest neighbors, Gaussian Naïve Bayes, and Support Vector Machine. The objective is to analyze how these techniques can effectively predict the success or popularity of a given song based on its audio characteristics. The researcher compared the results obtained from these models using the available data (Table 1).

Table 1. Model comparisons result.

S. no	Model	Accuracy
1	K-nearest Neighbours	52.00%
2	Logistic Regression	58.27%
3	Support Vector Machine	51.98%
4	Gaussian Naïve Bayes	60.50%

In this research paper, we aim to collect and clean Spotify data, perform exploratory data analysis, develop models to predict song popularity and create visualizations to effectively communicate our findings. Through this analysis, we can gain valuable insights into listener preferences, identify key factors driving song popularity, and make data-driven decisions for music curation, recommendation systems, and marketing strategies.

2 System Model

To build a system model for data analysis on Spotify using Python, you can follow these general steps:

2.1 Data Collection

The datasets used in this research are obtained from Kaggle. Kaggle is a popular online platform that hosts a wide range of datasets contributed by the data science community. It serves as a repository for diverse datasets across various domains, including music, finance, healthcare, and more. Researchers often rely on Kaggle to access high-quality datasets that are readily available for analysis and experimentation. In this particular study, the researchers downloaded the necessary datasets from Kaggle to conduct their analysis on predicting song success using machine learning techniques.

The Fig. 2 represents a visual depiction of the valuable insights obtained from analyzing the datasets provided by Spotify. These datasets contain a wealth of information related to the music available on the Spotify platform, including details about tracks, artists, genres, popularity, audio features, and more. The figure showcases the process of extracting meaningful insights from the Spotify datasets through data analysis and exploration techniques. It signifies the exploration of patterns, trends, and relationships within the music data, leading to a deeper understanding of the soundscape offered by Spotify. By studying the Spotify datasets, researchers, analysts, and music enthusiasts can gain valuable insights into various aspects of music consumption, artist popularity, genre preferences, and user behavior. These insights can be used for diverse purposes, such as improving recommendation algorithms, understanding audience preferences, identifying emerging trends, and supporting decision-making in the music industry.

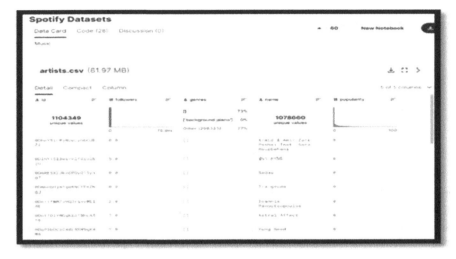

Fig. 2. Sounds of Spotify: Insights from the Spotify Datasets

Fig. 3. Harmonizing the Spotify Soundscape: Unveiling Insights from the Spotify Tracks Database

The Fig. 3 represents a database specifically dedicated to storing and organizing information about tracks from the music streaming platform, Spotify. This database contains a comprehensive collection of data related to various songs available on Spotify, including details such as track titles, artist names, album information, release dates, genres, audio features, and other relevant attributes. The Spotify Tracks DB serves as a valuable resource for researchers, analysts, and music enthusiasts who are interested in exploring and studying the vast musical landscape found on Spotify. It provides a structured and organized repository of track-related information, enabling users to query and analyze the data for various purposes. Researchers can leverage this database to investigate trends, patterns, and relationships within the music catalog, while analysts can derive insights to support decision-making in areas such as playlist curation, artist promotion, and user recommendation systems. Overall, the "Spotify Tracks DB" figure symbolizes the wealth of data available within the database, serving as a foundation for in-depth exploration and analysis of Spotify's vast music collection.

2.2 Data Pre-Processing

(A) To identify null values in the dataset

We have used the 'isnull()' function provided by the Pandas library. This function allows you to check for the existence of missing values within the dataset. In Fig. 4, the data frame is passed as an argument to the 'isnull()' function, which identifies the null values. By using the 'sum ()' function, we can calculate the total number of columns in the dataset that contain null values.

```
id                      0
name                   71
popularity              0
duration_ms             0
explicit                0
artists                 0
id_artists              0
release_date            0
danceability            0
energy                  0
key                     0
loudness                0
mode                    0
speechiness             0
acousticness            0
instrumentalness        0
liveness                0
valence                 0
tempo                   0
time_signature          0
dtype: int64
```

Fig. 4. Null value in data frame

By examining all the columns in the dataset, it was observed that the "song name" column contains a total of 71 null values.

(B) to determine the total number of rows and columns in the dataset, as well as inspect the data types and memory usage, the "Info ()" method can be employed

This method provides a concise summary of the dataset, displaying the column names, number of non-null values, data types, and approximate memory usage. By using the "info()" method, you can obtain this information efficiently (Fig. 5).

```
<class 'pandas.core.frame.DataFrame'>
RangeIndex: 586672 entries, 0 to 586671
Data columns (total 20 columns):
 #   Column            Non-Null Count   Dtype
---  ------            --------------   -----
 0   id                586672 non-null  object
 1   name              586601 non-null  object
 2   popularity        586672 non-null  int64
 3   duration_ms       586672 non-null  int64
 4   explicit          586672 non-null  int64
 5   artists           586672 non-null  object
 6   id_artists        586672 non-null  object
 7   release_date      586672 non-null  object
 8   danceability      586672 non-null  float64
 9   energy            586672 non-null  float64
 10  key               586672 non-null  int64
 11  loudness          586672 non-null  float64
 12  mode              586672 non-null  int64
 13  speechiness       586672 non-null  float64
 14  acousticness      586672 non-null  float64
 15  instrumentalness  586672 non-null  float64
 16  liveness          586672 non-null  float64
 17  valence           586672 non-null  float64
 18  tempo             586672 non-null  float64
 19  time_signature    586672 non-null  int64
dtypes: float64(9), int64(6), object(5)
memory usage: 89.5+ MB
```

Fig. 5. Total number of rows and columns in dataset, as well as inspect the data types and memory usage.

(C) To retrieve a list of the ten least popular songs from the Spotify dataset

We have employed the "sort_values ()" function to arrange the data in ascending order based on the popularity column. This will allow you to identify the songs with the lowest popularity scores (Fig. 6).

	id	name	popularity	duration_ms	explicit	artists	id_artists	release_date	danceability	energy
546130	[illegible]	Newspaper Reports On Abner, 26 February 1936	0	898578	0	['Norris Goff', 'Chester Lauck', 'Carlton Bric...	[illegible]	1935-02-20	0.595	0.262
546222	[illegible]	[illegible]	0	188440	0	['Hibari Misora']	[illegible]	1949	0.418	0.388
546221	[illegible]	[illegible]	0	173467	0	['Hibari Misora']	[illegible]	1949	0.842	0.178
546220	[illegible]	[illegible] CHOCOLO	0	205280	0	['Hibari Misora']	[illegible]	1949	0.606	0.467
546219	[illegible]	[illegible]	0	185733	0	['Hibari Misora']	[illegible]	1949	0.380	0.388
546218	[illegible]	[illegible]	0	183427	0	['Hibari Misora']	[illegible]	1949	0.631	0.249
546217	[illegible]	Screen Director's Playhouse, Music For Million...	0	178707	0	['Wilfrid Herbert', 'June Allyson', 'Joseph Kear...	[illegible]	1949-04-10	0.533	0.317
546216	[illegible]	[illegible]	0	162147	0	['Hibari Misora']	[illegible]	1949	0.529	0.546
546215	[illegible]	Screen Director's Playhouse, Trade Winds direc...	0	[illegible]	0	['Wally Maher', 'Tay Garnett', 'Lurene Tuttle'...	[illegible]	1949-05-19	0.599	0.321

Fig. 6. Total number of rows and columns in dataset, as well as inspect the data types and memory usage.

2.3 Data Exploration

To obtain descriptive statistics for numerical variables within the dataset, you can use the "describe ()" function. Additionally, applying the "transpose ()" function will provide a more convenient format for the summary statistics. By using these functions, you can gain insights into the central tendency, dispersion, and distribution of the numerical variables in the dataset (Figs. 7 and 8).

	count	mean	std	min	25%	50%	75%	max
popularity	586672.0	27.570053	18.370642	0.0	13.0000	27.000000	41.00000	100.000
duration_ms	586672.0	230051.167286	126526.087418	3344.0	175093.0000	214893.000000	263867.00000	5621218.000
explicit	586672.0	0.044086	0.205286	0.0	0.0000	0.000000	0.00000	1.000
danceability	586672.0	0.563594	0.166103	0.0	0.4530	0.577000	0.68600	0.991
energy	586672.0	0.542036	0.251923	0.0	0.3430	0.549000	0.74800	1.000
key	586672.0	5.221603	3.519423	0.0	2.0000	5.000000	8.00000	11.000
loudness	586672.0	-10.206067	5.089328	-60.0	-12.8910	-9.243000	-6.48200	5.376
mode	586672.0	0.658797	0.474114	0.0	0.0000	1.000000	1.00000	1.000
speechiness	586672.0	0.104864	0.179893	0.0	0.0340	0.044300	0.07630	0.971
acousticness	586672.0	0.449863	0.348837	0.0	0.0969	0.422000	0.78500	0.996
instrumentalness	586672.0	0.113451	0.266868	0.0	0.0000	0.000024	0.00955	1.000
liveness	586672.0	0.213935	0.184326	0.0	0.0983	0.139000	0.27800	1.000
valence	586672.0	0.552292	0.257671	0.0	0.3460	0.564000	0.76900	1.000
tempo	586672.0	118.464857	29.764108	0.0	95.6000	117.384000	136.32100	246.381
time_signature	586672.0	3.873382	0.473162	0.0	4.0000	4.000000	4.00000	5.000

Fig. 7. Descriptive Statistics on the dataset.

(A) To find the top ten popular songs with a popularity score greater than 90.

	id	name	popularity	duration_ms	explicit	artists	id_artists	release_date	danceability
93802	[illegible]	Peaches (feat. Daniel Caesar & Giveon)	100	198082	1	['Justin Bieber', 'Daniel Caesar', 'Giveon']	[illegible]	2021-03-19	0.677
93803	[illegible]	drivers license	99	242014	1	['Olivia Rodrigo']	[illegible]	2021-01-08	0.585
93804	[illegible]	Astronaut In The Ocean	98	132780	0	['Masked Wolf']	[illegible]	2021-01-06	0.778
92810	[illegible]	Save Your Tears	97	215627	1	['The Weeknd']	[illegible]	2020-03-20	0.680
92811	[illegible]	telepatía	97	160191	0	['Kali Uchis']	[illegible]	2020-12-04	0.653
92813	[illegible]	Blinding Lights	96	200040	0	['The Weeknd']	[illegible]	2020-03-20	0.514
93805	[illegible]	Leave The Door Open	96	242096	0	['Bruno Mars', 'Anderson .Paak', 'Silk Sonic']	[illegible]	2021-03-05	0.586
92814	[illegible]	The Business	95	164000	0	['Tiësto']	[illegible]	2020-09-16	0.798
91866	[illegible]	Streets	94	226987	1	['Doja Cat']	[illegible]	2019-11-07	0.749
92816	[illegible]	Heartbreak Anniversary	94	198371	0	['Giveon']	[illegible]	2020-03-27	0.449
92817	[illegible]	WITHOUT YOU	94	161385	1	['The Kid LAROI']	[illegible]	2020-11-06	0.662
92819	[illegible]	Bandido	94	232853	0	['Myke Towers', 'Juhn']	[illegible]	2020-12-10	0.713
93807	[illegible]	Friday (feat. Mufasa & Hypeman) - Dopamine Re...	94	169153	0	['Riton', 'Nightcrawlers', 'Mufasa & Hypeman', ...	[illegible]	2021-01-15	0.824

Fig. 8. Shows the top ten popular songs with a popularity score greater than 90

(B) To set the release date column as the index column in the dataset.

We have used the "set_index()" function. By applying this function, you can designate the release date column as the new index for the dataset in Fig. 9.

release_date	id	name	popularity	duration_ms	explicit	artists	id_artists	danceability	energy	key
1922-02-22	35iwgR4jXet0318WEWsa1Q	Carve	6	126903	0	['Uli']	['45tt06XoiDlio4L9EVpls']	0.645	0.4450	0
1922-06-01	021t14adgPcrDgSk7JTbKY	Capitulo 2.16 - Banquero Anarquista	0	98200	0	['Fernando Pessoa']	['14jtPCOoNZwquk5wd9DxrY']	0.695	0.2630	0
1922-03-21	07ASyahtSncedVUAZkNnc	Vivo para Quererte - Remasterizado	0	181640	0	['Ignacio Corsini']	['5LiOoJbxVSAMkBS2tt8m3X2']	0.434	0.1770	1
1922-03-21	08FmqUhxtyLTn6pAh6bk45	El Prisionero - Remasterizado	0	176907	0	['Ignacio Corsini']	['5LiOoJbxVSAMkBS2tt8m3X2']	0.321	0.0946	7
1922	08y9GloqCWfOGsKdwojr5e	Lady of the Evening	0	163080	0	['Dick Haymes']	['3BiJGZsyX9sJchTqcSA7Su']	0.402	0.1580	3

Fig. 9. Shows the release date as the new index.

(C) To obtain the name of the artist present in the 18th row of the dataset.

We can utilize the "iloc[]" method. This method allows you to filter and retrieve specific information from the dataset based on its index location. By specifying the index location as 18, you can extract the artist's name from the corresponding row.

```
Out[9]: artists    ['Victor Boucher']
        Name: 1922-01-01 00:00:00, dtype: object
```

Fig. 10. Shows the Victor Boucher information.

By using the "iloc[]" method and referencing the 18th row in the dataset, we identified that the artist's name associated with that particular row is Victor Boucher shown in Fig. 10.

(D) To convert the duration of songs from milliseconds to seconds.

We can perform the necessary calculation and update the duration column in the dataset. Afterward, you can print the column headers to confirm that the duration has been successfully converted to seconds shown in Fig. 11.

```
Out[11]:  release_date
          1922-02-22    127
          1922-01-06     98
          1922-03-21    182
          1922-03-21    177
          1922-01-01    163
          Name: duration, dtype: int64
```

Fig. 11. Shows that songs are present in seconds

(E) Correlation Map.
Let's create a correlation map as our first visualization. To begin, we will remove three unnecessary columns, namely "mode," "explicit," and an unnamed column. We will calculate the Pearson correlation coefficient for the remaining variables. For the correlation map, we will set the figure size to (14,6) and utilize the "heatmap()" function from the seaborn (sns) library. Additionally, we will enable annotations by setting "annotation = True". To format the data values in each cell, we will use "fmt = ".1g"". Lastly, we can choose a color map (cmap) from the seaborn documentation to customize the appearance of the correlation map.

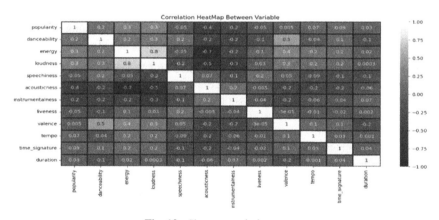

Fig. 12. Shows correlation map.

Upon executing the provided code, the correlation map was generated shown in Fig. 12. The color-coded scale on the right side represents the range from −1 to +1. Negative values near −1 indicate variables with minimal or negative correlation, while positive values greater than 0.0 indicate variables with a positive correlation.

2.4 Model Building

Model building refers to the process of constructing and developing a representation of a system or phenomenon using various techniques and methodologies. It involves creating a simplified version or a conceptual framework that captures the essential characteristics and relationships of the subject under study.

In the context of machine learning and data analysis, model building specifically refers to the construction of mathematical or statistical models that can make predictions, classify data, or uncover patterns and insights from available data. These models are trained on existing data, and their purpose is to generalize and make accurate predictions on new, unseen data.

The process of model building typically involves several steps. First, the problem at hand is defined, and the data relevant to the problem is collected and prepared for analysis. Then, an appropriate modeling technique is selected, considering the nature of the problem and the available data.

Next, the model structure and parameters are defined, and the training data is used to estimate these parameters. This involves using optimization algorithms to find the best fit between the model and the training data, minimizing the error or maximizing the likelihood of the observed data.

Once the model is trained, it is evaluated using validation data to assess its performance and generalization capabilities. This step helps in detecting and addressing potential issues such as overfitting or underfitting, which can occur when the model either memorizes the training data too closely or fails to capture its underlying patterns.

After evaluation, if the model performs well, it can be deployed to make predictions or generate insights on new, unseen data. If the model's performance is not satisfactory, further iterations and refinements may be required, such as adjusting the model's structure, exploring different algorithms, or collecting additional data.

(A) To create a Regression Plot Between Loudness and Energy.

We have utilize the "regplot()" function from the seaborn library. This function enables us to generate a scatter plot with a regression line representing the relationship between the two variables.

The regression plot has been generated shown in Fig. 13, illustrating a significant positive correlation between "Loudness" and "Energy." It is evident from the plot that all the data points or songs are oriented in the same direction. When the energy of a song increases, its loudness also tends to increase. Conversely, if the loudness decreases, the energy of the track also decreases.

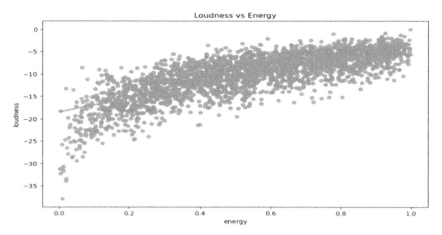

Fig. 13. Shows the regression plot between loudness and energy.

(B) To create the relationship between "Popularity" and "Acousticness".

We can generate a regression plot that displays a regression line. This plot will provide insights into the correlation between the two variables.

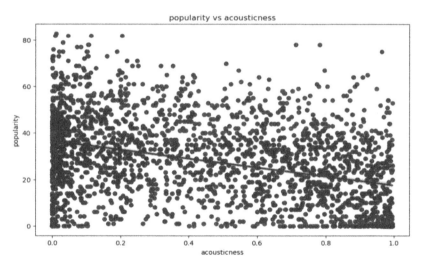

Fig. 14. Shows the relationship between Popularity and Acousticness.

In the regression plot, the downward-sloping blue regression line indicates an inverse relationship between "Acousticness" and "Popularity" shown in Fig. 14. This means that as the acousticness of a song increases, its popularity tends to decrease. Conversely, if the popularity of a song increases, the acousticness tends to decrease.

(C) To visualize the duration of songs for each year.
We can employ the seaborn library and utilize the "lineplot()" function. This line graph will provide a visual representation of how the song durations have varied over different years.

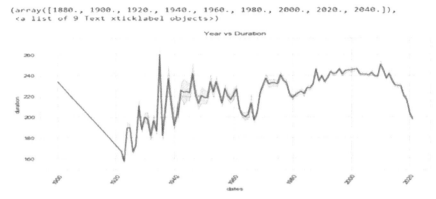

Fig. 15. Shows the duration of songs for each year.

After generating the line plot shown in Fig. 15, we can observe the duration of songs over time. The X-axis represents the years, while the Y-axis represents the duration of songs. Notably, songs from the 1920s to the 1960s were generally shorter in duration. Subsequently, there was a steady increase in song duration until around 2010. However, from 2010 onwards, there was a decline in song duration once again.

3   Simulation Results

These simulations involve creating mathematical or statistical models that capture the key dynamics and variables at play within the Spotify ecosystem. These variables can include user preferences, listening habits, music attributes, social interactions, and more. By incorporating these factors into the model, researchers and data scientists can simulate and study different scenarios to gain insights into how the platform operates and how users interact with it.

(A) To visualize the duration of songs with respect to different genres.
We have utilized the seaborn library and employ the "barplot()" function. This function allows you to create a horizontal bar plot, where each genre is represented on the y-axis, and the duration of the songs is depicted on the x-axis.

The horizontal bar plot displays the genres on the Y-axis and the song durations in milliseconds on the X-axis. Upon analyzing the data shown in Fig. 15, we can observe that the classical and world genres tend to have longer song durations, while children's music exhibits shorter song durations. Find top five genres by Popularity and pot a bar plot for the same.

`Text(0, 0.5, 'Genres')`

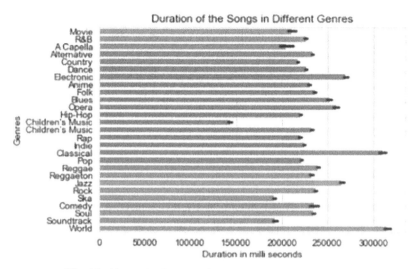

Fig. 16. Shows the duration of the songs in different genres.

(B) Units To find top five genres by Popularity and pot a barplot for the same.

`[Text(0.5, 1.0, 'Top 5 Genres by Popularity')]`

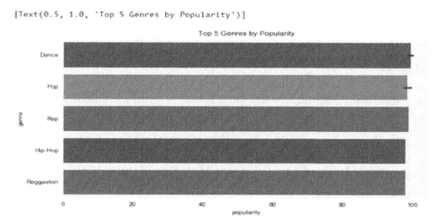

Fig. 17. Shows the top five genres.

According to the level of popularity, we have determined the most popular music genres to be Dance, Pop, Rap, Hip-Hop, and Reggaeton, as depicted in Fig. 16. This indicates that these genres have garnered significant attention and a large fan base among listeners. The term "popularity" here refers to the measure of how widely these genres are appreciated and enjoyed by the audience. It takes into account factors such as radio airplay, streaming numbers, sales, concert attendance, and overall cultural impact. Based on these indicators, the mentioned genres have emerged as the most prominent and influential in the current music landscape. Figure 17 visually represents this information, possibly through a graph or chart, highlighting the relative positions or proportions of each genre in terms of their popularity. It provides a clear visual representation of the data, allowing viewers to quickly grasp the dominance and significance of these specific genres. By identifying these top genres, it becomes easier for industry professionals, music enthusiasts, and researchers to understand the trends and preferences of music listeners. This information can be valuable for various purposes, such as marketing and promotion strategies, radio programming, playlist curation, and even predicting future music trends.

4 Conclusion

Data analysis on Spotify using Python offers valuable insights into user preferences, music trends, and song popularity. Python's libraries enable effective data collection, pre-processing, analysis, and visualization. Data visualization provides intuitive representations, aiding the communication of insights to stakeholders. Integration of Spotify's API or public datasets ensures access to reliable and diverse data sources. Data analysis on Spotify using Python empowers analysts to understand user preferences, identify popular song attributes, and make informed decisions. It enables personalized recommendations, targeted marketing strategies, and improved user experiences. By leveraging data analysis on Spotify, organizations can gain a competitive edge in the music industry. They can tailor their offerings to match user preferences, optimize marketing campaigns, and enhance the overall user experience.

References

1. Lawrence, D.L.: Addressing the value gap in the age of digital music streaming. In: Vand, J., Transnat'l, L., 52, 511. Clerk Maxwell, J. (eds.) A Treatise on Electricity and Magnetism, 3rd edn., vol. 2, pp. 68–73. Oxford, Clarendon, 1892 (2019)
2. Sciandra, M., Spera, I.C.: A model-based approach to Spotify data analysis: a Beta GLMM. J. Appl. Stat. **49**(1), 214–229 (2022)
3. Pérez-Verdejo, J.M., Piña-García, C.A., Ojeda, M.M., Rivera-Lara, A., Méndez-Morales, L.: The rhythm of Mexico: an exploratory data analysis of Spotify's top 50. J. Comput. Soc. Sci. **4**, 147–161 (2021)
4. Budzinski, O., Gaenssle, S., Lindstädt-Dreusicke, N.: Data (r) evolution: the economics of algorithmic search and recommender services. In: Handbook on Digital Business Ecosystems, pp. 349–366. Edward Elgar Publishing (2022)
5. Skog, D., Wimelius, H., Sandberg, J.: Digital service platform evolution: how Spotify leveraged boundary resources to become a global leader in music streaming (2018)
6. Hujran, O., Alikaj, A., Durrani, U.K., Al-Dmour, N.: Big data and its effect on the music industry. In: Proceedings of the 3rd International Conference on Software Engineering and Information Management, pp. 5–9, January 2020
7. Schulz, W.L., Durant, T.J., Siddon, A.J., Torres, R.: Use of application containers and workflows for genomic data analysis. J. Pathol. Inform. **7**(1), 53 (2016)

8. Salameh, A., Bass, J.: Influential factors of aligning Spotify squads in mission-critical and offshore research papers–a longitudinal embedded case study. In Product-Focused Software Process Improvement: 19th International Conference, PROFES 2018, Wolfsburg, Germany, 28–30 November 2018, Proceedings 19, pp. 199–215. Springer International Publishing (2018)

9. Lin, Y.C., Tsai, H.N., Lee, Y.C.: The effects of product categories, brand alliance fitness and personality traits on customer's brand attitude and purchase intentions: a case of Spotify. J. Stat. Manag. Syst. **23**(3), 677–693 (2020)

10. Mobasher, B., Dettori, L., Raicu, D., Settimi, R., Sonboli, N., Stettler, M.: Data science summer academy for chicago public school students. ACM SIGKDD Explor. Newsl **21**(1), 49–52 (2019)

11. Caveness, E., GC, P.S., Peng, Z., Polyzotis, N., Roy, S., Zinkevich, M.: Tensorflow data validation: data analysis and validation in continuous ml pipelines. In: Proceedings of the 2020 ACM SIGMOD International Conference on Management of Data, pp. 2793–2796, June 2020

12. Chen, C.C., Leon, S., Nakayama, M.: Are you hooked on paid music streaming?: an investigation into the millennial generation. Int. J. E-Business Res. (IJEBR) **14**(1), 1–20 (2018)

13. Fajana, O., Owenson, G., Cocea, M.: Torbot stalker: detecting tor botnets through intelligent circuit data analysis. In: 2018 IEEE 17th International Symposium on Network Computing and Applications (NCA), pp. 1–8. IEEE, November 2018

14. Pedrero-Esteban, L.M., Barrios-Rubio, A., Medina-Ávila, V.: Teenagers, smartphones and digital audio consumption in the age of Spotify. Comunicar. Media Edu. Res. J. **27**(2) (2019)

15. Isson, J.P.: Unstructured Data Analysis: How to Improve Customer Acquisition, Customer Retention, and Fraud Detection and Prevention. John Wiley & Sons (2018)

16. Peters, K., et al.: PhenoMeNal: processing and analysis of metabolomics data in the cloud. Gigascience **8**(2), giy149 (2019)

17. Belcastro, L., Marozzo, F., Talia, D.: Programming models and systems for big data analysis. Int. J. Parallel Emergent Distrib. Syst. **34**(6), 632–652 (2019)

18. Trabucchi, D., Buganza, T., Dell'Era, C., Pellizzoni, E.: Exploring the inbound and outbound strategies enabled by user generated big data: evidence from leading smartphone applications. Creativity Innov. Manag. **27**(1), 42–55 (2018)

19. Poldrack, R.A., Gorgolewski, K.J., Varoquaux, G.: Computational and informatic advances for reproducible data analysis in neuroimaging. Ann. Rev. Biomed. Data Sci. **2**, 119–138 (2019)

20. Ochi, V., Estrada, R., Gaji, T., Gadea, W., Duong, E.: Spotify danceability and popularity analysis using sap. arXiv preprint arXiv:2108.02370 (2021)

21. Smite, D., Moe, N.B., Floryan, M., Levinta, G., Chatzipetrou, P: Spotify guilds. Commun. ACM **63**(3), 56–61 (2020)

22. Ramos, E.F., Blind, K.: Data portability effects on data-driven innovation of online platforms: analyzing Spotify. Telecommun. Policy **44**(9), 102026 (2020)

23. Kumar, I., Mishra, M.K., Mishra, R.K.: Performance analysis of NOMA downlink for next - generation 5G network with statistical channel state information. Ingénierie des Systèmes d'Information, **26**(4), 417–423 (2021). https://doi.org/10.18280/isi.260410

24. Shankar, R., Kumar, I., Mishra, R.K.: Pairwise error probability analysis of dual hop relaying network over time selective Nakagami-m fading channel with imperfect CSI and node mobility. Traitement du Signal **36**(3), 281–295 (2019). . https://doi.org/10.18280/ts.360312

25. Kumar, I., Kumar, A., Kumar Mishra, R.: Performance analysis of cooperative NOMA system for defense application with relay selection in a hostile environment. J. Def. Model. Simul. (2022). https://doi.org/10.1177/15485129221079721

26. Ashish, I.K., Mishra, R.K.: Performance analysis for wireless non-orthogonal multiple access downlink systems. In: 2020 International Conference on Emerging Frontiers in Electrical and Electronic Technologies (ICEFEET), Patna, India, pp. 1–6 (2020). https://doi.org/10.1109/ICEFEET49149.2020.9186987

27. Kumar, I., Mishra, R.K.: An investigation of spectral efficiency in linear MRC and MMSE detectors with perfect and imperfect CSI for massive MIMO systems. Traitement du Signal **38**(2), 495–501 (2021). https://doi.org/10.18280/ts.380229

28. Kumar, I., Mishra, R.K.: An efficient ICI mitigation technique for MIMO-OFDM system in time-varying channels. Math. Model. Eng. Probl. **7**(1), 79–86 (2020). https://doi.org/10.18280/mmep.070110

29. Jacobson, K., Murali, V., Newett, E., Whitman, B., Yon, R.: Music personalization at Spotify. In: Proceedings of the 10th ACM Conference on Recommender Systems, p. 373, September 2016

30. Harris, M., et al.: Analyzing the Spotify Top 200 Through a Point Process Lens. arXiv preprint arXiv:1910.01445 (2019)

31. Duman, D., Neto, P., Mavrolampados, A., Toiviainen, P., Luck, G.: Music we move to: Spotify audio features and reasons for listening. PLoS ONE **17**(9), e0275228 (2022)

32. Gupta, N., Kumar, I., Rathod, I., Sharma B, S.S.P.M.: Sustainable production systems with AI and emerging technologies: a moderator-mediation analysis. **12**, Special Issue 8, 2819–2832 (2023). https://doi.org/10.48047/ecb/2023.12.si8.200

33. Lozic, J.: Comparison of business models of the streaming platforms Spotify and Netflix. Economic and Social Development: Book of Proceedings, pp. 110–119 (2020)

34. South, T.: Network analysis of the Spotify artist collaboration graph. Aust. Math. Sci. Inst. 1–12 (2018)

35. Salameh, A., Bass, J.M.: An architecture governance approach for Agile development by tailoring the Spotify model. AI & Soc. **37**(2), 761–780 (2022)

36. Kim, J.: Music popularity prediction through data analysis of music's characteristics. Int. J. Sci. Technol. Soc. **9**(5), 239 (2021)

37. Sharma B, S.S.P.M., Ravishankar Kamath, H., Siva Brahmaiah Rama, V.: Modelling of cloud based online access system for solar charge controller. Int. J. Eng. Technol. **7**(2.21), 58–61 (2018)

38. Shalinee Gupta, Ms., Sharma B, S.S.P.M.: Design and development of an intelligent aqua monitoring system using cloud based online access control systems. Int. J. Rec. Technol. Eng. (IJRTE), **8**(4), November 2019. ISSN: 2277-3878

39. Dr. Ravishankar Kamath, H., Sharma B, S.S.P.M., Siva Brahmaiah Rama, V.: PWM based solar charge controller using IoT. Int. J. Eng. Technol. **7**(2.7), 284–288 (2018)

40. Dr. Ravishankar Kamath, H., Siva Brahmaiah Rama, V., Sharma B, S.S.P.M.: Street light monitoring using IOT. Int. J. Eng. Technol. **7**(2.7), 1008–1012 (2018)

41. Sharma B, S.S.P.M., Kumar, A., Meena, B.K.: An intelligent solar based farm monitoring using cloud based online access control systems. Int. J. Rec. Technol. Eng. (IJRTE), **8**(3), September 2019. ISSN: 2277-3878

Comparative Study of Predicting Stock Index Using Deep Learning Models

Harshal Patil[1,3], Bharath Kumar Bolla[2(✉)], E. Sabeesh[4],
and Dinesh Reddy Bhumireddy[5]

[1] Upgrad Education Pvt. Ltd., Mumbai, India
[2] Salesforce, Hyderabad, India
bolla111@gmail.com
[3] Liverpool John Moores University, London, UK
[4] IIT Jodhpur, Jodhpur, India
[5] Cardinal Health, Bengaluru, India

Abstract. Time series forecasting has seen many methods attempted over the past few decades, including traditional technical analysis, algorithmic statistical models, and more recent machine learning and artificial intelligence approaches. Recently, neural networks have been incorporated into the forecasting scenario, such as the LSTM and conventional RNN approaches, which utilize short-term and long-term dependencies. This study evaluates traditional forecasting methods, such as ARIMA, SARIMA, and SARIMAX, and newer neural network approaches, such as DF-RNN, DSSM, and Deep AR, built using RNNs. The standard NIFTY-50 dataset from Kaggle is used to assess these models using metrics such as MSE, RMSE, MAPE, POCID, and Theil's U. Results show that Deep AR outperformed all other conventional deep learning and traditional approaches, with the lowest MAPE of 0.01 and RMSE of 189. Additionally, the performance of Deep AR and GRU did not degrade when the amount of training data was reduced, suggesting that these models may not require a large amount of data to achieve consistent and reliable performance. The study demonstrates that incorporating deep learning approaches in a forecasting scenario significantly outperforms conventional approaches and can handle complex datasets, with potential applications in various domains, such as weather predictions and other time series applications in a real-world scenario.

Keywords: ARIMA · SARIMA · SARIMAX · RNN · CNN · LSTM · GRU · DeepAR · DSSM · DF-RNN · Deep Renewal · POCID · Thiels'U

Time series forecasting has been implemented traditionally using standard methods such as ARIMA, SARIMA, and SARIMAX [1]. A significant drawback of these methods has been their inability to handle multivariate datasets where exogenous variables significantly affect the forecasting predictions [2]. Furthermore, their accuracies of predictions have not been satisfying enough in many complex real-world scenarios [2]. The advent of Deep Learning has helped bridge this gap. Neural networks and their ability to achieve universal approximation is a well-established theory, as seen in scenarios such as regression and classification [3]. In the last two decades, models based on recurrent neural

© ICST Institute for Computer Sciences, Social Informatics and Telecommunications Engineering 2024
Published by Springer Nature Switzerland AG 2024. All Rights Reserved
P. Pareek et al. (Eds.): IC4S 2023, LNICST 536, pp. 45–57, 2024.
https://doi.org/10.1007/978-3-031-48888-7_4

networks (RNNs) and LSTMs (Long-short-term memory) have been widely used in the forecasting scenario with promising results, and their ability to process sequential data has been exploited to solve complex time series scenarios [4]. However, in the last decade, newer architectures such as Deep Factor RNN(DF-RNN) [5], DSSM [6], Deep AR [7], and Deep Renewal [8] have been shown to outperform classical RNN and LSTM-based deep learning models in various scenarios. Very little experimentation has been done using these approaches on the Stock Market data. Hence, a comparative study of these models on a widely established dataset such as the NIFTY 50 index would help establish the superiority of deep learning models over traditional approaches and evaluate the effectiveness of recent deep learning models. The research objectives are as follows.

– To evaluate the superiority of neural networks over traditional approaches in forecasting augmentation.
– To evaluate the performance of the models, varying levels of train data (50% and 25%), keeping the test data the same to assess the effect of lesser data on model's performance.
– To evaluate if the better models are performing consistently on all metrics (MSE, RMSE, POCID, Thelis'U, MAPE.

1   Literature Review

Various research has been done in the recent past to increase the efficiency of time series forecasting by incorporating deep learning methodologies. While traditional approaches have been used to solve time series problems in a univariate scenario, deep learning approaches have been used to approximate multivariate datasets with significantly higher efficiency.

1.1   Traditional Approaches

Time series forecasting has been an important research field since humans started to predict values associated with a time component. According to De Gooijer and Hyndman [9], the earliest statistical models for time series analysis, namely the Auto Regressive (AR) and Moving Average (MA) models, were developed in the 1940s.

These models aimed to describe time series autocorrelation and were limited to linear forecasting problems. As researchers delved deeper into the subject, they factored in parametric influences. In the early 1970s, Box et al. [10] developed the Box-Jenkins method, a three-step iterative process for determining time series, which became a popular approach for time-series modeling. With the advent of computers and increasing processing power, Autoregressive integrated Moving Average (ARIMA) models were used empirically for univariate and multivariate time series forecasting [11]. In the 1980s and 1990s, researchers started incorporating seasonality in time series modeling. Various methods, including X-11, X-12-ARIMA, etc., used decomposition to obtain seasonality and apply it in time-series forecasting [12].

In the past few decades, many methods have been implemented to forecast various domains. These methods include simple traditional technical analysis (also known as "charting") of price charts [13], algorithmic statistical models [14], and more recent Machine Learning and Artificial Intelligent approaches [15]. Computational time series forecasting has applications in various fields, from weather and sales forecasting to finance-related forecasting (budget analysis, stock market price forecasting). It is an indispensable tool for all fields that rely on time factors. Methods including Autoregression, Box Jenkins, and Holt-Winters were used to yield generally acceptable results.

1.2 Deep Learning Approaches

Recently, novel techniques and models have emerged utilizing deep learning methodologies. For instance, the Long- and Short-Term Time-Series Network (LSTNet) incorporates Convolutional Neural Network (CNN) and Recurrent Neural Network (RNN) to capture both short-term and long-term dependencies in time-series data [16]. Another approach proposes using the Gaussian Copula process (GP-Copula) in conjunction with RNN [17]. The Neural Basis Expansion Analysis for Interpretable Time Series forecasting (NBEATS) achieved state-of-the-art performance in the recent M4 time-series prediction competition [18]. It has been observed that deep learning methods possess an edge over traditional techniques with regard to overfitting, as evidenced in previous research by [19].

Many studies have shown that classical deep learning and machine learning models outperform ARIMA models in time-series forecasting. Various complex models, including Multi-Layer Perceptron, CNN, and LSTM, have been implemented and analyzed for time-series forecasting. These models can handle multiple input features, leading to higher accuracy than conventional methods. Feature extraction is a critical step in improving the performance of predictive models, even when simple features are used. Some studies have used modified deep networks to extract frequency-related features from time-series data using EMD and Complete Ensemble Empirical Mode Decomposition with Adaptive Noise (CEEDMAN). The extracted features were then fed to LSTM to predict one-step-ahead forecasting [20, 21]. Other studies have used image data features by decomposing raw time-series data into IMFs using IF and providing CNN to learn features automatically [22]. Data augmentation approaches such as adding external text-based sentiment data to the model-generated features, were also used [23].

Additionally, auto-regressive models have been proposed, such as DeepAR, which uses high-dimensional related time-series features to train Autoregressive Recurrent Neural Networks and has demonstrated superior performance compared to other competitive models [7]. Another study proposed a Multi-Step Time-Series Forecaster that uses various related time-series features to forecast demand on Amazon.com [24]. Furthermore, several state-of-the-art methods have been developed and proven to be highly promising in generalized competitions like M4 [25]. Finally, it has been shown that an ensemble of models consistently performs better than any single model [26].

2 Research Methodology

A novel python-based library, GluonTS, has been introduced to provide models, tools, and components for time-series forecasting [27]. Techniques such as DF-RNN, DSSM, Deep AR, and LSTNet have been implemented using relevant libraries from the GluonTS framework. The models used as baselines are ARIMA, SARIMA, SARIMAX, and Facebook's Prophet. Facebook's Prophet API is used for implementing the Prophet model.

2.1 Dataset Exploration

The dataset used in this research consists of the NIFTY 50 index consisting of the closing value of the stock indices from Jan 2011 till Feb 2022 (Fig. 1) [28]. A rolling window of length ten has been taken with context and prediction length of five each. Mean, and Standard deviation is calculated for ten window period to test for the stationarity of the time series. As seen in Fig. 2, where there is a variation in the mean and Standard deviation across the ten-window time frame, it is evident that the time series is not stationary. This is further confirmed by the ADF test, as seen in Table 1. Indices from Jan 2011 to 2020 have been used as Train, and the subsequent series have been used as Test data, as seen in Fig. 1.

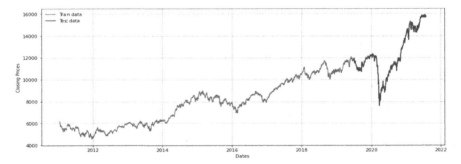

Fig. 1. Test-Train Split – NIFTY-50 Index.

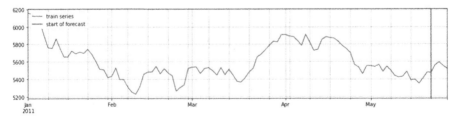

Fig. 2. Ten period Rolling Window – Non-uniform mean & STD – Non-stationary series.

2.2 Forecasting Models

Various models used in the experiments have been elaborated in the succeeding sections. Baseline forecasting models have been built using machine learning models such as ARIMA, SARIMA, and SARIMAX, while deep learning models have been built using DF-RNN, DeepAR, DSSM, and LSTNet.

ARIMA

ARIMA model has been created using the PMDARIMA library and hyper-parameters such as the number of auto-regressive terms (p), number of non-seasonal differences needed for stationarity (d), and number of lagged forecast errors in the prediction equation (q). The hyper parameters passed are shown below.

- $p - 0, 1, 2$
- $q - 0, 1, 2$
- Test to determine 'd' – Augmented Dickey-Fuller Test

The best model hyperparameters for ARIMA are (0,1,1) (Fig. 3).

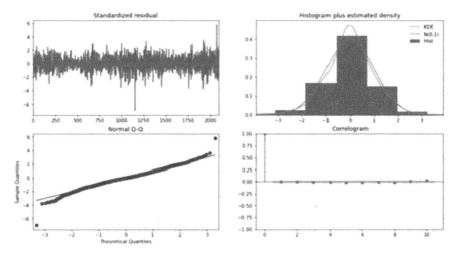

Fig. 3. ARIMA.

SARIMA

The model architecture is like ARIMA, except that SARIMA has an additional seasonal component. In addition to the existing hyperparameters of p, q, and d, seasonal hyperparameters, as mentioned below, are passed into the model. The best model hyperparameters for SARIMA are (0,1,1) (0,0,1) [30], as discussed below.

- P – Seasonal Autoregressive order = 0.
- D – Seasonal Difference order = 0.
- Q – Seasonal Moving Average order = 1.

– m – Number of timesteps for a single seasonal period.

SARIMAX

SARIMAX is similar to SARIMA except for the addition of an exogenous variable which is used as an additional feature in the learning process.

Deep Factor RNN (DF-RNN)

The Deep Factor RNN is a model that incorporates two significant factors, namely global and local fluctuations, to govern the progression of a given time series. These factors are learned using separate RNN models, each controlled by specific hyperparameters that determine the model's architecture, as illustrated in Table 1.

Table 1. Global vs. Local RNN Hyperparameters of DF-RNN.

Hyper-Parameters	Global RNN Model	Local RNN Model
Number of Hidden Layers	1	5
Number of Neurons in the Hidden Layer	50	5
Type of Cell	LSTM	LSTM

Deep AR

Deep AR, a probabilistic forecasting model, utilizes a global model to learn jointly from multiple related time series using negative binomial likelihood. The model is constructed on a recurrent neural network based on Long Short-Term Memory (LSTM) architecture. Hyperparameters, depicted in Table 2, are used in model building.

Table 2. Deep AR Hyperparameters.

Hyper-Parameters	Deep AR Model
Number of LSTM Layers	3
Number of LSTM Cells	40
Scaling	Enabled
Learning Rate	0.001

Deep State Space Model

The Deep State Space model [29] works on the principles of parametrizing linear state space of individual time series with a jointly learned recurrent neural network.

Deep Renewal

Deep Renewal Processes are a probabilistic intermittent demand forecasting method. This method builds upon Croston's framework and molds its variants into a renewal process. The random variables, M (The demand size at non-zero demand point) and Q (The inter-demand interval) are estimated using a separate RNN (Table 3).

Table 3. Deep AR Hyperparameters.

Hyper-Parameters	Deep Renewal Model
Skip size for skip RNN layer	3
Auto regressive window size – Linear	40
Learning Rate	0.001

3 Performance Measurement

Performance measurement of the forecasting model has been done using evaluation metrics such as MSE, RMSE, MAE, MAPE and custom metrics such as POCID and Theil's U.

3.1 Mean Square Error/Root Mean Square Error

Mean squared error is the mean of the squared error between the target variable (original observation) and the output variable (the predicted variable) in a given time series. Root mean squared error applies a square root to the MSE. The mathematical representation is shown in Eq. 1.

$$MSE = 1/N \sum_{i=1}^{N} \left(target_i - output_i \right)^2 \tag{1}$$

3.2 Mean Absolute Percentage Error

Mean Absolute percentage error defines the percentage difference between the target variable and the output variable w.r.t the output variable. The mathematical representation is shown in Eq. 2. Lower the value of MAPE better the model performance.

$$MAPE = 100/N \sum_{i=1}^{N} (target_i - output_i)/output_i \tag{2}$$

3.3 Mean Absolute Error

Mean absolute error is the difference between the target and output variables. The lower the MAE better is the model performance. The mathematical representation is shown in Eq. 3.

$$MAE = 1/N \sum_{i=1}^{N} \left(target_i - output_i \right) \tag{3}$$

3.4 POCID – Prediction of Change in Direction

The prediction of change in direction is the percentage of the number of correct decisions in predicting whether the time series in the next time interval will increase or decrease. The mathematical representation is shown in Eq. 4. The higher the value of POCID, the better the model performance.

$$POCID = 100 * \frac{\sum_{t=1}^{N} D_t}{N}$$

$$\text{where } D_t = \begin{cases} 1, & if \ (target_t - target_{t-1})(output_t - output_{t-1}) > 0 \\ 0, & otherwise \end{cases} \tag{4}$$

3.5 Theil's U

Theil's U is similar to the mean squared error except that the error is normalized w.r.t output variable of the previous time interval. U lesser than 1 indicates a better performance of the model. A value equal to 1 indicates a random model and a value greater than 1 indicates a model worse than a random model. Hence it is ideal for achieving a U of value 0. The mathematical representation is shown in Eq. 5.

$$\text{Theil's U(Normalized mean squared error)} = \frac{\sum_{t=1}^{N} \left(target_t - output_t \right)^2}{\sum_{t=1}^{N} \left(output_t - output_{t-1} \right)^2} \tag{5}$$

4 Results

As explained in the preceding sections, models have been evaluated using the metrics mentioned above on both machine learning and deep learning models.

4.1 Forecasting Using ARIMA, SARIMA and SARIMAX

Forecasting has been done over a 36-day horizon. From Table 4, it is evident that there is no significant difference in the model performance among the three baseline models. Furthermore, on analysis of standardized residuals for each forecasted point by all three models, they are uniformly distributed around the mean of zero.

4.2 Forecasting Using Facebook Prophet

The performance of the Facebook Prophet was assessed using cross-validation on the provided time series for various forecast horizons between 36 and 364 days. It was noted that the MSE and MAPE values increase linearly as the forecast horizon increases. This suggests that an increase in forecast horizon leads to a rise in prediction error. Moreover, when compared to the baseline models, namely ARIMA, SARIMA and SARIMAX, Prophet exhibited a relatively lower performance in terms of RMSE and MAPE (Table 5).

Table 4. ARIMA, SARIMA and SARIMAX Metrics.

Models	MSE	MAE	RMSE	MAPE
ARIMA	3003730.628	1387.5576	1733.1274	0.118263
SARIMA	3009965	1391.598	1734.925	0.118803
SARIMAX	3002241	1387.658	1732.698	0.118351

Table 5. Evaluation metrics for Facebook Prophet.

Error	36 Days	364 Days	Average
MSE	203,144	5,000,000	2,522,746
RMSE	Not Done	Not Done	1588.31
MAPE	0.05	0.19	**0.12**
POCID	Not Done	Not Done	**80.48**
Theil's U	Not Done	Not Done	494.96

4.3 Forecasting using Deep Learning Models on 100% Train Data

The Table 6 presents the performance evaluation of various deep learning models for time-series forecasting based on five evaluation metrics: MSE, RMSE, POCID, Theil's U, and MAPE. The models include RNN, GRU, LSTM, DF-RNN, DeepAR, DSSM, and Deep Renewal. Among these models, DeepAR performed the best with the lowest values for all evaluation metrics, indicating its high accuracy in forecasting. The LSTM and GRU models also performed well with relatively low values for all metrics. The DF-RNN model performs better than the DSSM and Deep Renewal models but not as well as the others. Therefore, the ranking of the models based on better performance is DeepAR, GRU, LSTM, DF-RNN, RNN, DSSM, and Deep Renewal.

Table 6. Deep Learning Models Performance on 100% train data.

Model	MSE	RMSE	POCID	Theil's U	MAPE
RNN	21341000	4619.6	49	30790	36.1
GRU	53167	230.6	52	892	5.5
LSTM	171396	414	52	898	17.4
DF-RNN	3754898	1937.8	25	1168	0.12
Deep AR	**35600**	**188.7**	75	**12**	**0.01**
DSSM	25041182	5004	75	7767	0.29
Deep Renewal	246363673	15696	75	77148	1.0

4.4 Forecasting using Deep Learning Models on 50% Train Data

Based on the metrics in the table, the Deep AR model seems to be the best performer (as depicted in Table 7), followed by the LSTM, GRU, and RNN models. The DF-RNN, DSSM, and Deep Renewal models performed poorly than the others.

Table 7. Deep Learning Models Performance on 50% train data.

Model	MSE	RMSE	POCID	Theil's U	MAPE
RNN	1779572	1334	52	883	16.5
GRU	330895	575	52	910	17.3
LSTM	438402	662	51	889	17.1
DF-RNN	56715481	7531	75	1168	0.12
DeepAR	**2823**	**53.1**	50	**0.83**	**0.003**
DSSM	303172659	17411	75	94228	0.99
Deep Renewal	245658060	15673	50	77269	0.99

4.5 Forecasting using Deep Learning Models on 25% Train Data

The table shows that the DeepAR model performs best with the lowest MSE, RMSE, Theil's U, and MAPE values, even with 25% of the actual train data (as depicted in Table 8). GRU and LSTM models perform similarly with slightly higher values of the evaluation metrics.

Table 8. Deep Learning Models Performance on 25% train data.

Model	MSE	RMSE	POCID	Theil's U	MAPE
RNN	3970916	1993	52	1114	18.2
GRU	252760	503	53	893	17.3
LSTM	2718187	1649	52	848	17.2
DF-RNN	56715481	4985	25	7719	0.32
DeepAR	**1886**	**43**	**50**	**0.76**	**0.002**
DSSM	2442782	1563	50	760	0.09
Deep Renewal	245151974	15657	50	77243	0.99

Overall, Deep AR and GRU consistently performed the best across all tables, regardless of the percentage of training data used. Surprisingly these models' performance was consistent despite lowering the train data (Figs. 4 and 5).

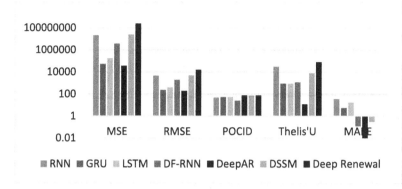

Fig. 4. Evaluation metrics of all models on 100% training data

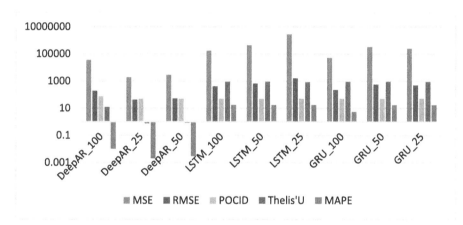

Fig. 5. Evaluation metrics of all models on 100% training data

5   Conclusion

This work compares traditional machine learning models with cutting-edge deep learning architectures for time series forecasting on stock market indices. The study employs several metrics, including MAE, MSE, RMSE, MAPE, POCID, and Theil's U, to evaluate the performance of the models. The results of the experiments demonstrate that state-of-the-art deep neural networks such as DeepAR and GRU outperform traditional forecasting models such as ARIMA, SARIMA, and SARIMAX. Moreover, DeepAR is stable across varying training data sizes and is consistent on all metrics. Furthermore, the study highlights the superiority of recurrent neural networks, their variants, such as LSTMs, for handling stock indices datasets, and their ability to outperform conventional machine learning and statistical-based algorithms. This makes them suitable for deployment in real-world scenarios. However, the study is limited by the use of univariate Stock market datasets. Future research on multivariate datasets could be explored to further establish the superiority of deep learning networks in time series forecasting.

References

1. Nokeri, T.C.: Forecasting using ARIMA, SARIMA, and the additive model. In: Implementing Machine Learning for Finance: A Systematic Approach to Predictive Risk and Performance Analysis for Investment Portfolios, pp. 21–50. Apress, Berkeley (2021)
2. Wang, S., Li, C., Lim, A.: Why are the ARIMA and SARIMA not sufficient. In: Proceedings of arXiv (2021). http://arxiv.org/abs/1904.07632. Accessed 06 June 2023
3. Sirisha, U.M., Belavagi, M.C., Attigeri, G.: Profit prediction using ARIMA, SARIMA and LSTM models in time series forecasting: a comparison. IEEE Access **10**, 124715–124727 (2022)
4. Sen, J., Mehtab, S.: Long-and-short-term memory (LSTM) networks architectures and applications in stock price prediction. In: Emerging Computing Paradigms, pp. 143–160. Wiley (2022)
5. Wang, Y., Smola, A., Maddix, D., Gasthaus, J., Foster, D., Januschowski, T.: Deep factors for forecasting. In: Proceedings of the 36th International Conference on Machine Learning, pp. 6607–6617. PMLR (2019). http://proceedings.mlr.press/v97/wang19k.html. Accessed 09 June 2023
6. Rangapuram, S.S., Seeger, M.W., Gasthaus, J., Stella, L., Wang, Y., Januschowski, T.: Deep state space models for time series forecasting. In: Advances in Neural Information Processing Systems, pp. 7785–7794 (2018)
7. Salinas, D., Flunkert, V., Gasthaus, J., Januschowski, T.: DeepAR: probabilistic forecasting with autoregressive recurrent networks. Int. J. Forecast. **36**(3), 1181–1191 (2020)
8. Snyder, R.D., Ord, J.K., Beaumont, A.: Forecasting the intermittent demand for slow-moving inventories: a modelling approach. Int. J. Forecast. **28**(2), 485–496 (2012)
9. De Gooijer, J.G., Hyndman, R.J.: 25 years of time series forecasting. Int. J. Forecast. **22**(3), 443–473 (2006)
10. Newbold, P.: The principles of the Box-Jenkins approach. Oper. Res. Q. (1970–1977) **26**(2), 397–412 (1975)
11. Kovvuri, A.R., Uppalapati, P.J., Bonthu, S., Kandula, N.R.: Water level forecasting in reservoirs using time series analysis – auto ARIMA model. In: Gupta, N., Pareek, P., Reis, M. (eds.) IC4S 2022. LNICST, vol. 472, pp. 192–200. Springer, Cham (2023). https://doi.org/10.1007/978-3-031-28975-0_16
12. Findley, D.F., Monsell, B.C., Bell, W.R., Otto, M.C., Chen, B.-C.: New capabilities and methods of the X-12-ARIMA seasonal-adjustment program. J. Bus. Econ. Stat. **16**(2), 127 (1998)
13. Lo, A.W., Mamaysky, H., Wang, J.: Foundations of Technical Analysis: Computational Algorithms, Statistical Inference, and Empirical Implementation (2000). http://papers.ssrn.com/abstract=228099. Accessed 21 June 2022
14. Webby, R., O'Connor, M.: Judgemental and statistical time series forecasting: a review of the literature. Int. J. Forecast. **12**(1), 91–118 (1996)
15. Sezer, O.B., Gudelek, M.U., Ozbayoglu, A.M.: Financial time series forecasting with deep learning: a systematic literature review: 2005–2019. Appl. Soft Comput. **90**, 106181 (2020)
16. Lai, G., Chang, W.-C., Yang, Y., Liu, H.: Modeling long- and short-term temporal patterns with deep neural networks (2018). http://arxiv.org/abs/1703.07015. Accessed 21 June 2022
17. Salinas, D., Bohlke-Schneider, M., Callot, L., Medico, R., Gasthaus, J.: High-dimensional multivariate forecasting with low-rank Gaussian Copula Processes (2019). https://proceedings.neurips.cc/paper/2019/hash/0b105cf1504c4e241fcc6d519ea962fb-Abstract.html. Accessed 21 June 2022
18. Oreshkin, B.N., Carpov, D., Chapados, N., Bengio, Y.: N-BEATS: neural basis expansion analysis for interpretable time series forecasting (2020). http://arxiv.org/abs/1905.10437. Accessed 21 June 2022

19. Khare, K., Darekar, O., Gupta, P., Attar, V.Z.: Short term stock price prediction using deep learning. In: 2017 2nd IEEE International Conference on Recent Trends in Electronics, Information & Communication Technology (RTEICT), pp. 482–486 (2017)
20. Guo, C., Kang, X., Xiong, J., Wu, J.: A new time series forecasting model based on complete ensemble empirical mode decomposition with adaptive noise and temporal convolutional network. Neural. Process. Lett. **55**, 4397–4417 (2022). https://doi.org/10.1007/s11063-022-11046-7
21. Cao, J., Li, Z., Li, J.: Financial time series forecasting model based on CEEMDAN and LSTM. Physica A **519**, 127–139 (2019)
22. Zhou, F., Zhou, H., Yang, Z., Gu, L.: IF2CNN: towards non-stationary time series feature extraction by integrating iterative filtering and convolutional neural networks. Expert Syst. Appl. **170**, 114527 (2021). https://doi.org/10.1016/j.eswa.2020.114527
23. Atha, S., Bolla, B.K.: Do deep learning models and news headlines outperform conventional prediction techniques on forex data? In: Rout, R.R., Ghosh, S.K., Jana, P.K., Tripathy, A.K., Sahoo, J.P., Li, K.C. (eds.) Advances in Distributed Computing and Machine Learning. LNNS, vol. 427, pp. 413–423. Springer, Singapore (2022). https://doi.org/10.1007/978-981-19-1018-0_35
24. Wen, R., Torkkola, K., Narayanaswamy, B., Madeka, D.: A multi-horizon quantile recurrent forecaster. arXiv (2018). https://doi.org/10.48550/arXiv.1711.11053
25. Makridakis, S.S., Assimakopoulos, V.: The M4 competition: results, findings, conclusion and way forward. Int. J. Forecast. **34**(4), 802–808 (2018). https://doi.org/10.1016/j.ijforecast.2018.06.001
26. Dietterich, T.G.: Ensemble methods in machine learning. In: Dietterich, T.G. (ed.) MCS 2000. LNCS, vol. 1857, pp. 1–15. Springer, Heidelberg (2000). https://doi.org/10.1007/3-540-45014-9_1
27. Alexandrov, A., et al.: GluonTS: probabilistic time series models in Python. arXiv (2019). http://arxiv.org/abs/1906.05264. Accessed 21 June 2022
28. A Novel Machine Learning Approach for Predicting the NIFTY50 Index in India | SpringerLink. https://link-springer-com.ezproxy2.library.arizona.edu/article/10.1007/s11294-022-09861-8. Accessed. 24 July 2023
29. Rangapuram, S.S., Seeger, M.W., Gasthaus, J., Stella, L., Wang, Y., Januschowski, T.: Deep State Space Models for Time Series Forecasting (2018)

Enhancing Heart Disease Prediction Through a Heterogeneous Ensemble DL Models

J. N. S. S. Janardhana Naidu[1]([✉]) [iD], Mudunuri Aniketh Varma[1],
P. Shyamala Madhuri[1] [iD], D. Shankar[1] [iD], Durga Satish Matta[1] [iD],
and Singaraju Ramya[2] [iD]

[1] Department of Computer Science and Engineering, Vishnu Institute of Technology,
Bhimavaram 534202, Andhra Pradesh, India
janardhana.j@vishnu.edu.in
[2] Department of Computing Technologies, SRM Institute of Science and Technology, Chennai,
India

Abstract. Accurate and highly effective forecasting approaches for timely detection and management are imperative given that cardiovascular illness is one of the biggest causes of death world wide. Machine learning (ML) and Deep learning (DL) methodologies have produced promising outcomes. In particular, ensemble DL models have gained attention for their ability in order to capitalize on the advantages of many models to enhance predictive performance. This study focuses on applying a cardiovascular disease prognosis with a DL ensemble. The model combines the predictions of multiple DL models, networks of neurons, such as supervised DL Models (CNN and RNN), to magnify accuracy and robustness. In this research, the effectiveness of an ensemble model is measured using an ample dataset, comparing it with individual DL models. The model's prediction skills are evaluated by utilizing a verity of Evaluation metrics. The findings highlight the effectiveness of cardiovascular disease ensemble DL models prediction, showcasing their potential for enhancing diagnostic accuracy in clinical settings and aiding healthcare professionals in making informed decisions for patient care.

Keywords: Heart disease prediction · Ensemble DL · ML · CNN · RNN · healthcare decision-making

1 Introduction

According to mortality rates, heart disease is the deadliest ailment afflicting humans today. Heart disease happens when the heart is unable to Diagnosing cardiac disease might be complicated by coexisting disorders such high blood pressure, diabetes, and abnormal cholesterol levels. These additional symptoms can make it more difficult to pin down the root cause of the cardiac problem, which can then make it more difficult to treat. In order to recognize cardiac illness early and accurately is crucial for preventing and treating heart failure [1].

© ICST Institute for Computer Sciences, Social Informatics and Telecommunications Engineering 2024
Published by Springer Nature Switzerland AG 2024. All Rights Reserved
P. Pareek et al. (Eds.): IC4S 2023, LNICST 536, pp. 58–73, 2024.
https://doi.org/10.1007/978-3-031-48888-7_5

Diagnostic criteria for coronary heart disease according to antiquity have been discredited for various reasons. A vast volume of intricate medical records and a selection of ML and DL methodologies, support forecast heart illness with less involvement from clinicians. Mainly, we want to save people's lives by recognizing irregularities in heart conditions, which would be accomplished by finding and analyzing raw data derived from heart disease data. Non-invasive measures and also ML and DL techniques are effective in differentiating between healthy individuals and those who have cardiac illness.

ML is both a type of AI and a method for accomplishing AI tasks for developing algorithms that take providing inputs such as historical data and then employing statistical analysis to make predictions about future output. ML is an effective method for developing complex algorithms for analyzing high-dimensional and biomedical data [2]. DL is a subset of ML that uses numerous layers of expert systems to process and calculate a plethora of data. The human brain's activity and operation are the basis for the DL algorithm. The DL algorithm is capable of erudition without the need for human intervention and can handle both organized and unstructured information. It is useful for extracting valuable information from a huge clinical data set [3] and helping the health care professionals to make decisions quickly. Deep learning is currently implemented in various other domains, including but not limited to finance, banking, and e-commerce. Information extraction, representation learning, and outcome prediction, are some of the DL techniques and frameworks used in healthcare applications [4]. DL algorithms are based on expert systems, just as how the human brain uses millions of neurons to process information.

Heart disease is a common and dangerous disorder that needs precise and effective prediction models for quick diagnosis and treatment. ML and DL techniques emergently become formidable tools in healthcare, showing promise in areas related to heart disease prediction. Ensemble DL models, in particular, have gained attention for their ability to harness the strengths of multiple models, leading to improved predictive performance. This study focuses on utilizing an ensemble DL model for cardiac disease prognosis. The model combines predictions from various DL models, such as CNN and RNN to ameliorate both accuracy and robustness. A substantial data set serves as a gauge for the ensemble model's efficacy, and comparisons are formed against individual DL models. When evaluating the capacity for prediction of an ensemble model, evaluation measures such as accuracy, precision, recall, and F1-score are used. The study's findings demonstrate the effectiveness of ensemble DL models for heart disease prediction, showcasing their potential to enhance diagnostic accuracy in clinical settings and provide valuable support for healthcare professionals in making informed decisions for patient care.

In order to acquire reliable results from ML and DL methods, the heart disease dataset undergoes a number of adjustments. First, the basic data on heart disease is analyzed and inserted into either the machine or DL model. In the second step, preprocessing techniques are applied to remove irrelevant, noisy and inconsistent data. Then, applying the feature selection method, select the significant features for analysis. Finally, categorization methods are used to the extracted data and the prediction is made. These steps are shown in below Fig. 1.

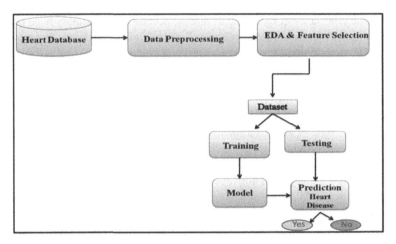

Fig. 1. The Heart Disease Prediction Model

2 Background

Different types of ML and DL approaches are covered in this section along with a brief discussion of each.

2.1 Machine Learning Techniques

All three of these types of learning—supervised, unsupervised, and reinforcement has their uses add these three categories used to classify ML algorithms. In a supervised learning method, a target or dependent variable must be predicted from a collection of independent factors. A model that maps inputs to target outputs is developed using a set of variables. The model is trained on the training data until it gets the required level of accuracy. Regression, Decision Tree (DT), Random Forest (RF), KNN, Logistic Regression (LR) etc. are few examples of supervised learning. In unsupervised learning, there is no desired outcome or outcome variable for which predictions or estimates must be made. Unsupervised learning is used to find patterns from data sets that have neither been classified nor labelled. Unsupervised Learning examples are K-means and Apriori algorithms. The machine is trained to make certain decisions using the Reinforcement Learning method. To make accurate decisions, the computer learns from its past experiences and works to acquire as much information as possible. Markov Decision Process and Q learning are examples of this type of algorithm.

2.2 Deep Learning Techniques

DL techniques employ neural networks to compute information in the same way that the human brain uses millions of neurons to do so. It uses multiple layers of neural networks to do data processing and computations on a large amount of data [5]. Because so much computing occurs between the input and output layers, this learning process is referred to as DL. It requires more time to train a model on a huge amount of data, but it requires much less time to execute than other ML methods [6]. This is the main reason for DL becoming more popular day by day than ML. It comprises both supervised and unsupervised learning algorithms because it is a subset of a ML algorithm. Supervised learning uses ANN, CNN, and RNN as examples. AutoEncoders and Boltxmann machines are used in unsupervised learning.

Ensembling is a well-established algorithm that has been shown in Fig. 2. to increase the precision of machine learning models, reduce overfitting, and increase the robustness of the system.

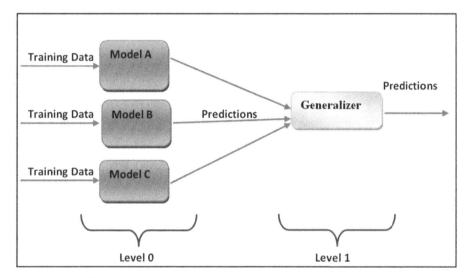

Fig. 2. The organization and mechanisms of ensemble learning

There are several ways to combine multiple deep learning models for ensemble learning. Here are a few examples:

Bagging: In bagging, multiple deep learning models are trained independently on various subset of the training data. The outputs of various models are averaged to get the final projection. Bagging can improve the robustness of the system by reducing the variance of the models.

$$\hat{y} = (\hat{y}_1 + \hat{y}_2 + \ldots + \hat{y}_n)/n \tag{1}$$

where $\hat{y}$ is the final ensemble prediction, $\hat{y}_1, \hat{y}_2, \ldots, \hat{y}_n$ are the predictions from individual models, and n is the number of models.

Boosting: The process of boosting includes training many DL models in succession, with each subsequent system concentrating on the errors generated by the models that came before it. The ultimate forecast is arrived at by adding up the results of all of the models. By decreasing the inherent bias of the models, boosting has the potential to make the methodology more accurate.

$$\hat{y} = \alpha_1 \hat{y}_1 + \alpha_2 \hat{y}_2 + \ldots + \alpha_n \hat{y}_n \tag{2}$$

Here $\hat{y}$ is the final ensemble prediction, $\hat{y}_1, \hat{y}_2, \ldots, \hat{y}_n$ are the predictions from specific models, and $\alpha_1, \alpha_2, \ldots \alpha_n$ are the weights provided to individual predicted model.

Stacking: Multiple deep learning models are trained independently using training data in stacking, and the results are then used as features in a meta model that generates the final prediction. Stacking can improve the accuracy of the system by integrating the benefits of various models. The mode of ensembling is shown in Fig. 3 (Table 1).

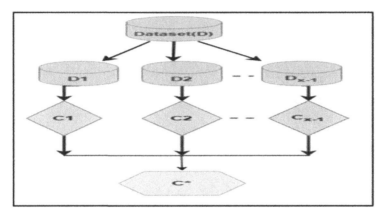

Fig. 3. Ensemble Stacking

3 Literature Review

The proper diagnosis of patients and the identification of potential dangers are essential components of heart disease study. In such a context, it is essential to consider different evaluation metrics to assess the performance of the models accurately.

While precision, recall, and F1 score are important metrics to consider, accuracy is also a crucial measure that should not be overlooked. Accuracy is a metric that quantifies the proportion of correctly predicted instances in the dataset, relative to the total number of instances. It serves as an indicator of the model's capability to correctly classify both positive and negative instances.

In the case of heart disease analysis, accuracy is important because misdiagnosis or misclassification can have serious consequences. False positives and false negatives can lead to unnecessary procedures, missed diagnoses, or wrong treatment plans, which can affect patient outcomes.

Despite their significance, precision and recall do not account for a model's total accuracy. A high recall score indicates a low false-negative rate for the model, but it may classify some negative cases as positive.

In summary, the choice of metric in heart disease analysis should depend on the specific goals and priorities of the analysis. While accuracy is a valuable metric for overall performance, precision and recall are important in situations where false positives or false negatives can have serious consequences. The F1 score can be a overall measure of model performance that combines both precision and recall.

M. Kavitha et al. [7] developed a mixed-method model by fusing RF and DT techniques. In comparison to the RF tree model and the DT model, the proposed hybrid model demonstrates superior accuracy. The experimental results indicate that the hybrid model achieves an accuracy of 88.7%, outperforming the other models.

Hager Ahmed et al. [8] compared the accuracy of various Supervised ML classification algorithms, including DT, RF, SVM, and LR Classifier, was measured using both a selection of features and the entire set of features. The suggested approach exhibits a notable benefit in its capacity to effectively handle Twitter updates that encompass patient information. The framework's fundamental design involves the integration of Apache Kafka and Apache Spark. In comparison to the other designs, the RF classifier received the best with 94.9% accuracy. To extract essential features from the dataset, both the univariate feature selection and Relief methods were employed.

Pooja Anbuselvan [9] analyzed a range of various supervised ML models such as LR, NB, SVM, K-NN, DT, RF and the ensemble technique of XGBoost. Among these models, the proposed approach utilizing RF achieved the highest accuracy of 86.89%, surpassing the performance of other models.

Mohan et al. [10] created a novel approach to identifying and grouping critical characteristics to enhance the precision of cardiovascular illness prognosis. This suggested approach, a Hybrid RF with a Linear Model (HRFLM), for predicting cardiovascular disease showed an accuracy of 88.7%.

Considering the reason of making accurate predictions about cardiovascular illness, Budholiya et al. [11] created an XGBoost Classifier using a Bayesian optimization approach. With an impressive 91.8% accuracy for predictions, this approach outperforms the two most popular current tree-based concepts, RF and Extra Tree.

Spencer and colleagues [12] employed multiple ML techniques and feature selection algorithms to generate diverse tasks. The overall performance of the chi-squared feature choosing process and the BayesNet classification method for the model that was recommended was 85.0%.

The CHI-PCA approach, built by Escamilla et al. [13], when combined with RF, demonstrated remarkable precision. It was determined that the Cleveland database had a precision of 98.7%, the Hungarian database had an accuracy of 99.0%, whereas the Cleveland-Hungarian (CH) database had a reliability of 99.4%.

In [14], ApurbRajdhan et al., utilizing a RF method, introduced a novel model. When compared to other ML techniques, the suggested model's preciseness of 90.16% comes out as very excellent.

When compared with various categorization methods, such as K-NN, SVM, Naive Bayes, and RF classifier, Youness et al. [15] proposed an ANN model with impressive accuracy of 99.65% was developed by incorporating Particle Swarm Optimization (PSO) technique and Ant Colony Optimization (ACO) approach.

Haq AU et al. [16] developed a hybrid intelligent ML-based predictive system was evaluated using complete and reduced feature sets. The utilization of the Relief feature selection technique in conjunction with 10-fold cross validation and LR resulted in the attainment of an overall accuracy rate of 89%.

Awais Mehmood et al. [17] used deep learning to present a novel heart disease prediction model. Convolutional Neural Networks (CNN) algorithm, giving an accuracy of 97%.

Kazeem et al. [18] developed the models using hybrid algorithms like Boruta Algorithm and Deep Neural Network Algorithm (BADNN), Genetic Algorithm and Deep Neural Network (GADNN), and Boruta Algorithm and Neural Network Algorithm(BANN) all achieved two-way hybrid accuracy of 97%, 87%, and 100%, respectively.

FarmanAli et al. [19] developed a model that has an accuracy of 98.5% in predicting heart disease and proved that proposed system accuracy is higher than the existing traditional classifier model. The comparison is carried out based on metrics such as weighting techniques, feature selection and feature fusion.

To produce predictions based on learnt records, Mienye et al. [20] suggested an improved sparse auto encoder-based ANN framework. When using Adam's improved approach and batch normalization, the model's accuracy on processed data was 90%. The proposed model is more accurate than ANN and other standard methods.

Sumit Sharma et al. [21] used Talos optimization, a new DNN optimization technique, to implement a model using deep learning neural networks (DNN). Talos provides better accuracy of 90.76% to other optimizations.

Simanta et.al. [22] Proposed a Deep Learning Modified Neural Network (DLMNN)-based IoT-centric prediction model for heart disease classifier aimed to identify the heart disease of the patient more accurately. When results from this model are contrasted with those from other models, it is discovered that it has a 95.87% accuracy level.

P. Ramprakash et al. [23] proposed a new heart disease prediction model constructed by applying Deep Neural Network and chi-square statistical model. The accuracy of the developed model is compared with the accuracy of the existing models using DNN and ANN and stated that the proposed model more efficiently predicts the presence of heart disease than the other models.

The summary of prediction accuracy of both Machine and Deep Learning models are shown in Table 2.

Table 1. Classifying Ensemble approaches

Techniques Used	Fusion methods applied	Model Dependent	Type of Heterogeneity
Boosting	Weight Voting	Sequential	Homogenous
Gradient Boosting	Weight Voting	Sequential	Homogenous
AdaBoost	Weight Voting	Sequential	Homogenous
Bagging	Weight Voting	Parallel	Homogenous
Random Forest	Weight Voting	Parallel	Homogenous

Table 2. Machine and Deep learning models with accuracy

S.No	Authors	Methods	Accuracy
1	PoojaAnbuselvan	Random Forest	86.89%
2	Arabasadi	NN-Genetic algorithm	89.04%
3	Sumit Sharma	DNN optimization technique	90.76%
4	Doppala	Hybrid Machine Learning	94.20%
5	Hager Ahmed	DT, SVM and Logistic Regression	94.90%
6	Simanta Shekhar	Deep Learning Modified Neural Network (DLMNN)	95.87%

Table 3. Implementation of CNN and RNN Model with accuracy

Model	Input	Units/Layer1	Units/Layer2	Units/Layer3	Accuracy
Model-1	1025×7	64 units	32 units	1 unit	81.46
Model-2	1025×7	32 units	16 units	1 unit	80.48
Model-3	1025×7	128 units	64 units	1 unit	83.41
Model-4	1025×7	64 units	1 unit	-	96.58

4   Methodology

In this section a brief description about the feature selection is discussed. In heart disease analysis, the feature selection phase plays a crucial role as it allows users to identify the vital variables or features that have a substantial impact on predicting heart disease. The primary objective of feature selection is to enhance the accuracy and efficiency of the prediction model by eliminating irrelevant or redundant features, thereby improving the overall performance of the analysis (Fig. 4).

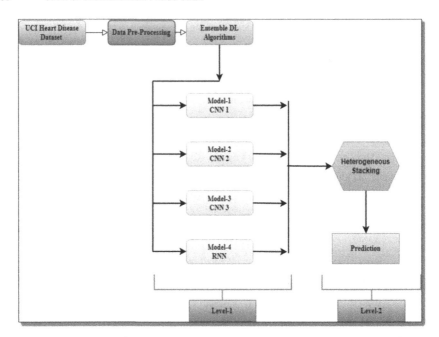

Fig. 4. Heterogeneous Ensemble DL Model

4.1   Information About the Dataset

Public Health Dataset was the dataset used in this study and exploratory data analysis is shown in Fig. 7.

1. **Span(age)** - the patient's age, quantified in years, and sex (one for males and zero for females).
2. **Chest Pain(cp)** - chest pain type.
3. **RBP(Trestbps)** - arterial pressure at rest (in millimeters of mercury (Hg)) at the time of patient admission to the clinic. The typical range for blood pressure is 120/80; if your reading is within that range, everything is great; however, if it is slightly higher than expected, we need to try to bring it down. Alter your way of living in a healthy way.
4. **Cholesterol(chol)-**A person's serum cholesterol will reveal their triglyceride level.
5. **Fasting glucose levels (fbs)–**Fasting blood sugar (FBS) refers to the measurement of blood sugar levels after a period of fasting. A value of 120 mg/dL or higher is considered indicative of elevated blood sugar (1 true). Normal blood sugar levels typically range from below 100 mg/dL (5.6 mmol/L) to 125 mg/dL (5.6 to 6.9 mmol/L), which serves as the threshold for prediabetes.
6. **Resting(Restecg)-**electrocardiogram findings while at rest.
7. **MaxHeartRate(thalach)-**The maximum achievable heart rate is determined by subtracting your age from 220.
8. **Exercise-**angina pectoris:exercise-induced angina (1yes).
9. **ST depression(Oldpeak)** - During exercise, there is a notable occurrence of ST depression in comparison to the resting state.

10. **Slope** - the angle of the exercise's ST segment peak.
11. **Ca** - Fluoroscopy-colored main vascular count (0–3).
12. **Thalassemia(Thal)-**No reason was given, although thalassemia categorization includes 3 instances of normal findings, 6 instances of fixed defects, and 7 instances of reversible defects is most likely the cause
13. **Target(T)** - (Angiographic disease status) No Heart Disease = 0, Heart Disease = 1.

4.2 Data Preparation

Once data is collected, it should be prepared for analysis. This includes cleaning and organizing the data, filling in any missing values, and transforming any categorical variables into numerical ones.

The ensemble DL model is made from the multiple models, containing multiple structures of layers. The heart analysis is made from the data containing multiple features co-related to rate of heart analysis. Initially, the data is pre-processed by handling null values of the dataset chosen which results to get much accurate results. Exploratory data analysis (EDA) is done on the data to select the features that are contributing much to the chance of heart attack. In the model, the correlation value > = 0.3 is termed to be preferably chosen. So, the feature selection is made based on the moderately correlated features. The training and testing data are splitted with a test size of 20% of the entire dataset. To avoid outliers of the data, robust scaling is used for the data.

4.2.1 Data Pre-Processing and Feature Selection

By using the robust scalar during data pre-processing, EDL for heart disease prediction benefit from improved performance and stability. The scalar normalizes input data, making it less sensitive to outliers and non-normal distributions. This robust normalization enhances the models' ability to learn and generalize from features, resulting in increased

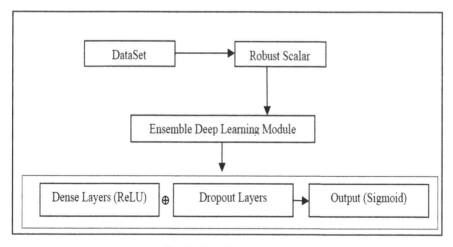

Fig. 5. Data Preprocessing

accuracy and reliability in heart disease prediction shown in the Fig. 5. In order to prevent the model's generalizability from being compromised, it is important that the duplicates be safely eliminated.

4.3 Hypothesis Testing

The null hypothesis that there is no correlation between a certain variable and the occurrence of heart disease can be tested using the chi-square test. If the estimated chi-square statistic is larger than the crucial value, then you can conclude that cardiac disease is occurring and reject the null hypothesis.

4.4 Model Building

After performing hypothesis testing, we can use the significant variables to model cardiac disease. One common method is logistic regression, which estimates the important predictors; estimate the likelihood of getting heart disease.

In ensembling multiple models, the stacking method is used in which entire training data is used to train each and every model (Table 3).

Performance Analysis
See Fig. 6.

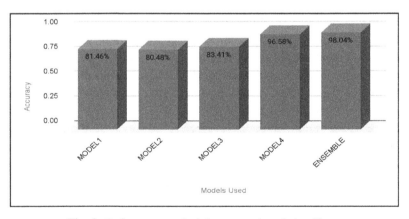

Fig. 6. Performance analysis between selected algorithms.

Evolution Metrics

1. Accuracy Processing

Accuracy is a term used in evaluation metric for classification problems in machine learning. It measures the percentage of the test dataset that was correctly labeled.

Mathematically, accuracy is defined as:

Accuracy $=$ (No of Positive Prediction)/(Total no of Prediction)

2. Weighted Average Ensemble:

Weighted average ensemble combines the outputs of multiple models in ensemble deep learning. It assigns weights based on model performance, allowing accurate models to have a larger influence. This improves performance compared to other methods, especially in heart disease prediction for image recognition. Some of the Metrics or Activation functions used in Convolutional Networks are as follows.

Evaluation Process Used
Evaluation metrics include the confusion matrix, precision, accuracy score, recall, sensitivity, and F1 score. The confusion matrix consists of TP, TN, FN, and FP representing true positives, true negatives, false negatives, and false positives, respectively.

Table 4. Binary Classification Confusion Matrix

	Predicted Value 0	Predicted Value 1
Actual Value 0	TN	FP
Actual Value 1	FN	TP

In Table 4. P $=$ positive, N $=$ negative. The accuracy score evaluates model performance by considering the ratio of correct predictions to total predictions. It is calculated as

$$(TP + TN)/(TP + TN + FP + FN) \tag{3}$$

Specificity, or the true negative rate, evaluates the classifier's performance by identify negative cases accurately and is given by

$$TN/(TN + FP) \tag{4}$$

According to the research on ML and DL techniques, rather than using just one strategy, assembling a number of techniques results in increased accuracy, which in turn results in better performance for the model (Figs. 8 and 9).

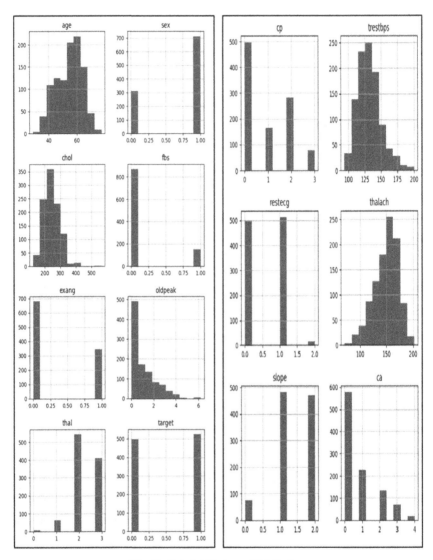

Fig. 7. DataSet Features Distribution

By employing the proposed ML method outlined in this study, the analysis of real time patient data can be significantly improved. Based on CNN's 97% accuracy, Generic and Deep neural network's 87%, Sparse auto encoder based ANN's 90%, DNN and Talos optimization's 90.76%, and IoT centered DLMNN's 95.87% accuracy, Ensembled deep learning model is the most accurate mode. Ensembling of multiple deep learning models achieves higher accuracy (98.04%) with multiple layers in each of the models.

Some of the limitations of ensemble learning can be overcome or mitigated in the future with advancements in technology and research. Researchers can develop more

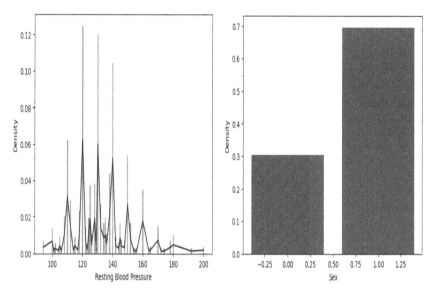

Fig. 8. Distribution of BP and Sex

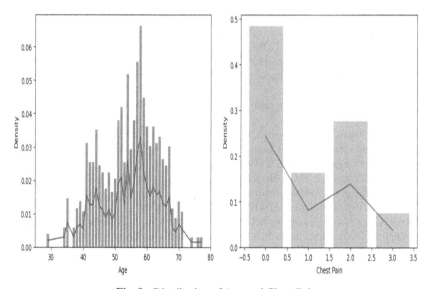

Fig. 9. Distribution of Age and Chest Pain

computationally efficient ensemble methods that require less training time and computational resources. One approach could be to use approximate methods such as random projections or sketching to decrease the amount of input data and the quantity of base models.

5 Conclusion

In this paper, we looked at the fundamental ideas behind the DL and ML models to prognosticate the onset of cardiovascular problems. Analysis of studies aimed at predicting cardiovascular disease by employing a selection of ML and DP techniques. Different models' performance was discussed and reviewed. Obstructive on the tools, dataset and the techniques, the models have varied accuracy. In recent years the usage of DL algorithms on huge datasets yields better accuracy in heart disease prediction models. This article will be beneficial for the researchers to get an idea of the present and existing models and work to design future models that are more accurate.

References

1. Shah, D., Patel, S., Bharti, S.K.: Heart disease prediction using machine learning techniques. SN Comput. Sci. **1**(6), 1–6 (2020). https://doi.org/10.1007/s42979-020-00365-y
2. Fatima, M., Pasha, M.: Survey of machine learning algorithms for disease diagnostic. J. Intell. Learn. Syst. Appl. **09**(01), 1–16 (2017). https://doi.org/10.4236/jilsa.2017.91001
3. Jaya Lakshmi, A.: A review on deep learning algorithms in healthcare. Turkish J. Comput. Math. Educ. (TURCOMAT). **12**, 5682–5686 (2021). https://doi.org/10.17762/turcomat.v12 i10.5379
4. Shickel, B., Tighe, P.J., Azra Bihorac, Rashidi, P.: Deep EHR: a survey of recent advances in deep learning techniques for electronic health record (EHR) analysis. IEEE J. Biomed. Health Inform. **22**, 1589–1604 (2018). https://doi.org/10.1109/jbhi.2017.2767063
5. Sarker, I.H.: Deep learning: a comprehensive overview on techniques, taxonomy, applications and research directions. SN Comput. Sci. **2**(6), 1–20 (2021). https://doi.org/10.1007/s42979-021-00815-1
6. Xin, Y., et al.: Machine learning and deep learning methods for cybersecurity. IEEE Access. **6**, 35365–35381 (2018). https://doi.org/10.1109/access.2018.2836950
7. Kavitha, M., Gnaneswar, G., Dinesh, R., Sai, Y.R., Suraj, R.S.: Heart disease prediction using hybrid machine learning model. https://doi.org/10.1109/ICICT50816.2021.9358597. https://ieeexplore.ieee.org/document/9358597
8. Ahmed, H., Younis, E.M.G., Hendawi, A., Ali, A.A.: Heart disease identification from patients' social posts, machine learning solution on Spark. Future Gener. Comput. Syst. **111**, 714–722 (2019). https://doi.org/10.1016/j.future.2019.09.056
9. Anbuselvan, P.: Heart disease prediction using machine learning techniques. Int. J. Eng. Res. Technol. **9** (2020). https://doi.org/10.17577/IJERTV9IS110259
10. Mohan, S., Thirumalai, C., Srivastava, G.: Effective heart disease prediction using hybrid machine learning techniques. IEEE Access. **7**, 81542–81554 (2019). https://doi.org/10.1109/access.2019.2923707
11. Budholiya, K., Shrivastava, S.K., Sharma, V.: An optimized XGBoost based diagnostic system for effective prediction of heart disease. J. King Saud Univ. Comput. Inf. Sci. **34**, 4514–4523 (2020). https://doi.org/10.1016/j.jksuci.2020.10.013
12. Spencer, R., Thabtah, F., Abdelhamid, N., Thompson, M.: Exploring feature selection and classification methods for predicting heart disease. Digit. Health **6**, 205520762091477 (2020). https://doi.org/10.1177/2055207620914777
13. Gárate-Escamila, A.K., Hajjam El Hassani, A., Andrès, E.: Classification models for heart disease prediction using feature selection and PCA. Inform. Med. Unlocked. **19**, 100330 (2020). https://doi.org/10.1016/j.imu.2020.100330

14. Rajdhan, A., Agarwal, A., Sai, M., Ravi, D.: Heart disease prediction using machine learning. Int. J. Eng. Res. Technol. **9** (2020). https://doi.org/10.17577/ijertv9is040614

15. Khourdifi, Y., Bahaj, M.: Heart disease prediction and classification using machine learning algorithms optimized by particle swarm optimization and ant colony optimization. Int. J. Intell. Eng. Syst. **12**, 242–252 (2019). https://doi.org/10.22266/ijies2019.0228.24

16. Haq, A.U., Li, J.P., Memon, M.H., Nazir, S., Sun, R.: A hybrid intelligent system framework for the prediction of heart disease using machine learning algorithms. Mob. Inf. Syst. **2018**, 1–21 (2018). https://doi.org/10.1155/2018/3860146

17. Mehmood, A., et al.: Prediction of heart disease using deep convolutional neural networks. Arab. J. Sci. Eng. **46**, 3409–3422 (2021). https://doi.org/10.1007/s13369-020-05105-1

18. Dauda, K.A., Olorede, K.O., Aderoju, S.A.: A novel hybrid dimension reduction technique for efficient selection of bio-marker genes and prediction of heart failure status of patients. Sci. Afr. **12**, e00778–e00778 (2021). https://doi.org/10.1016/j.sciaf.2021.e00778

19. Ali, F., et al.: A smart healthcare monitoring system for heart disease prediction based on ensemble deep learning and feature fusion. Inf. Fusion **63**, 208–222 (2020). https://doi.org/10.1016/j.inffus.2020.06.008

20. Mienye, I.D., Sun, Y., Wang, Z.: Improved sparse autoencoder based artificial neural network approach for prediction of heart disease. Inform. Med. Unlocked **18**, 100307 (2020). https://doi.org/10.1016/j.imu.2020.100307

21. Sharma, S., Parmar, M.: Heart diseases prediction using deep learning neural network model. Int. J. Innov. Technol. Exploring Eng. **9**, 2244–2248 (2020). https://doi.org/10.35940/ijitee.c9009.019320

22. Sarmah, S.S.: An efficient IoT-based patient monitoring and heart disease prediction system using deep learning modified neural network. IEEE Access. **8**, 135784–135797 (2020). https://doi.org/10.1109/access.2020.3007561

23. Ramprakash, P., Sarumathi, R., Mowriya, R., Nithyavishnupriya, S.: Heart disease prediction using deep neural network. https://doi.org/10.1109/ICICT48043.2020.9112443. https://ieeexplore.ieee.org/abstract/document/9112443. Accessed 08 Sept 2022

24. Doppala, B.P., Bhattacharyya, D., Chakkravarthy, M., Kim, T.: A hybrid machine learning approach to identify coronary diseases using feature selection mechanism on heart disease dataset. Distrib. Parallel Databases **7**, (2021). https://doi.org/10.1007/s10619-021-07329-y

IoT-Based Pesticide Detection in Fruits and Vegetables Using Hyperspectral Imaging and Deep Learning

Anju Augustin[✉] and Cinu C. Kiliroor

Department of Computer Science and Engineering, Indian Institute of Information Technology, Kottayam, Kottayam, India
{anjuaugustin.23phd21006,cinu}@iiitkottayam.ac.in

Abstract. Fruits and vegetables contain rich nutrients and vitamins. So that they are part of our daily diet. For proper cell growth and health, we need these nutrients. Today in most crops during their growth and post-harvesting preservation different kinds of pesticides were used. Normal usage of such pesticides not that much affects health. But the actual situation is beyond our control. From soil preparation to the post-harvesting stage, pesticides are being added at alarming rates. It affects our health in a harmful way and leads to major health issues. Various studies exist to detect the pesticide levels in fruits and vegetables. This article analyses different existing methods of pesticide detection and examines their features and problems. Through this study, it is understood that Hyperspectral Imaging (HSI) is a very good method, and with it, more accurate results can be obtained by Transfer Learning (few-shot learning). This paper proposes an architecture and algorithm based on HSI and few-shot learning. Future studies are needed in this area to convert an RGB image to a spectral image because the HSI device is very expensive.

Keywords: Pesticide detection · Hyperspectral imaging · Deep learning

1 Introduction

Fruits and vegetables are a major part of our daily diet. We can consume almost all fruits and vegetables without cooking. Due to the huge demand of fruits and vegetables, it is necessary to protect them from pests and insects. Different kinds of pesticides and insecticides are used to protect these crops. Recent studies show that there is no control to limit the usage of these pesticides. But what happened is that these fruits and vegetables themselves contain chemicals and pesticides. Today in most crops during their growth and post-harvesting preservation different kinds of pesticides were used. Normal usage of such pesticides not that much affects health. But the actual situation is beyond our control. From soil preparation to the post-harvesting stage, pesticides are being added at alarming rates. When we are consuming this, it leads to several dangerous diseases like

© ICST Institute for Computer Sciences, Social Informatics and Telecommunications Engineering 2024
Published by Springer Nature Switzerland AG 2024. All Rights Reserved
P. Pareek et al. (Eds.): IC4S 2023, LNICST 536, pp. 74–83, 2024.
https://doi.org/10.1007/978-3-031-48888-7_6

cancer, liver problems, etc. Fruits and vegetables with these chemicals affected our full body functioning and health. The pesticides present in it cannot be detected by the naked eye, smell, or imaging. Because they are present on the outer surface as well as the inner parts of fruits and vegetables and are able to change the chemical structure also. Therefore, detecting the presence and levels of these pesticides is a challenging task and requires tools and methods.

Hyperspectral imaging (HSI) is one of the growing fields today. In this imaging, we get a typical image spectrum of objects. What is special about this imaging method is that we can detect the presence of particles beyond our naked eyes through this image. It gives us multiple levels of the spectrum when normal images only give three like RGB (Red-Green-Blue). The HSI images wavelength ranges from 400 to 1100 nm, and our naked eye can only be visible from 380 to 700 nm. But the spectral camera is very expensive.

Compared with traditional machine learning methods deep learning (DL) models are very efficient and performance-wise also good. DL models can extract high-level features automatically using multiple layers. But it has some problems that they lead to overfitting if the dataset contains less number of samples. Transfer Learning (TL) is one of the solutions for that. This paper investigates and analyzes different methods and techniques that are used in the field of pesticide detection in fruits and vegetables by deep neural networks.

2   Related Works

A lot of work is going on in this area today. Devi et al. [1] suggested an IoT-based pesticide detection using different types of sensors. Here four sensors were used to detect gases, moisture, pH, and temperature. Support Vector Machine (SVM) with Radial Basis Function (RBF) kernel and Convolutional Neural Network (CNN) with GoogleNet architecture are the models used here. This system outputs real-time values but there have some problems SVM gave fast results but the affected percentage is not mentioned and is less accurate. CNN performance is better than SVM but the training time is 6.5 min more.

Kandasamy Sellamuthu et al. [2] proposed a system that mainly uses a Q-learning-based Recurrent Neural Network (RNN) so that it can handle complex, high-volume, and fast-happening data. There is a Q-table associated with it, by using this table the system updates the next action. This model analyses the pesticide based on the data drawn from three sources, soil, vegetables, and fruits. From these datasets, three components of the pesticides were extracted. But the problem with this is that a full understanding of pesticide utilization is not possible. Also, it is not supported the real-time pesticide detection.

Bo Jiang et al. [3] implemented a CNN-based hyperspectral imaging system. Here a sample of Apple is considered for the detection, with AlexNet CNN architecture. The neural network had eight convolution and max-pooling layers for the feature extraction. Anise t al. [4] explained a deep learning-based plant disease identification method. Five types of Deep Neural Network (DNN) classification models and five types of corn leaves datasets were used here. The models are Inception V3, ResNet-50, VGG- 16, DenseNet-169, and Xception. The main aim of this work is to find the best model and best dataset based on accuracy. The result shows that the DenseNet-169 is the best DNN model with 80.33 % accuracy.

A Visible-Near-infrared (Vis-NIR) and Near-infrared (NIR) hyperspectral imaging system is suggested by Weixin Ye [5]. This method takes three grapes varieties and sprayed them with three types of pesticides. Then analyze the result with SVM and ResNet models. Here also ResNet gives better results compared with SVM. Identification of pesticide residue in black tea leaves by fluorescence hyperspectral technology is proposed by Jie SU [6]. A 1D- CNN and Random forest ML models are used here. Tiago Domingues et al. [7] presented a survey of different feature extraction methods and ML models that are used in the field of disease and pest detection in tomato leaves. Commonly used models are SVM, Random Forest (RF), and Artificial Neural Network (ANN) models. This paper suggested new and emerging transfer learning methods to avoid the problems like overfitting.

Xiaoyan Tang et al. [8] proposed a sensor-based pesticide contamination detection called an electronic nose, in tea leaves. For that, three Portable Electronic Nose-3 (PEN -3) electronic metal oxide sensors and Back Propagation algorithms with three hidden layers were used. Applications of different nanosensors explained by Rabisa Zia [9]. In this paper, different nanosensors that are mainly used in the pesticide detection area and their characteristics are also described. A review paper by Lili Li [10] explained different datasets and DL models used for leaf image processing.

A system for the freshness of the chicken checking is proposed by Rajina R Mohamed [11]. An electronic nose with Back Propagation (BPP) algorithm and SVM is suggested here. When comparing the existing works some of them need laboratory situations to measure pesticide concentration and some can detect the presence real-timely. The laboratory measurements are time-consuming as well as need costly equipment. There does not exist a good and accurate method for real-time detection.

3   Technologies Used for Detecting the Presence of Pesticides

Most countries have crossed the limit of minimum usage of pesticides and chemicals in agricultural crops. There has a Maximum Residue Level (MRL) to control the usage of pesticides in crops. It is the highest residue level that is tolerated in food or food crops. But today it crosses the limit, for preventing from insects or to get a high yield. It leads to dangerous situations of health problems. There have different technologies and ML/DL models for real-time pesticide detection. The Table 1 shows the existing technologies and architectures used in this field with the available datasets.

In these related works, the experiments were done with varieties of fruits or agricultural items like Corn, Black tea, tomato leaves, etc. Some of the datasets are already available in data repositories. But some datasets were created in the laboratory itself, by collecting different pesticides and crops.

Table 1. Summary of different technologies, DL-models, and dataset used for pesticide detection

Ref.	Fruit/Veg.	Technology	Architecture	Dataset
[1]	Any fruit	IoT with 4 sensors	SVM (RBF), CNN	Fruit dataset
[2]	Any fruit	Q-learning	RNN	soil, fertilizer plant village
[3]	Apple	HSI	AlexNet with 8 layers	Created dataset
[4]	Corn		Inception V3, Resnet 50, DenseNet-169	Plant Village
[5]	Grape	Vis-NIR & NIR HSIs	CNN, ResNet	Created dataset
[6]	Black tea	Fluorescent HSI	1D CNN, RF	Plant Village, Plant Doc
[7]	Tomato leaves		SVM, RNN, TL	Flavia & Malayekew Leaf
[8]	Tea leaves	Electronic nose (PEN-3)	Back propagation	Created data
[10]	Leaves	HSI	GoogleNet, GAN, VGG-16	Plant village, Plant pathology challenge

For generating the dataset in a laboratory, first made pesticide solutions with different concentration levels. Then dip each of the crops separately. These data were used for the analysis with the above mentioned technologies. Sensor-based methods directly sense the presence of pesticides in fruits and vegetables and image-based methods detect chemicals by processing the images of crops that are captured by cameras.

Analysis of existing methods in pesticide detection in fruits and vegetables shows that Hyperspectral Imaging (HSI) gives better accuracy compared with other methods. HSI is not affected by environmental conditions and it can detect the presence of chemicals that present inside the crop. Convolutional Neural Networks (CNN) require a large number of data for the operation. But for created data, it is not possible to make a dataset with considerable samples. It leads to overfitting problems during convolution operation. Transfer learning or few-shot learning is a good deep neural network to avoid overfitting with a small dataset. Here proposing an IoT-based system that uses HSI and transfer learning for detecting the presence of pesticides.

3.1   Hyperspectral Imaging (HSI)

Hyperspectral imaging is an imaging method that is used to produce the spectrum of the image. Normally images that are taken by Digital Camera are in RGB format. But in HSI, different spectra of images are generated. Mainly two types of spectrum are available that are Near-infrared (NIR) and Visible-Near-infrared (Vis-NIR). NIR is a destructive type which means it makes defects in the material surface in which the image is taken. Vis-NIR is not a destructive method. The HSI is mainly concentrated on visible and near-infrared (400 nm–1000 nm) [10]. The output of HSI is not similar to the object, but it generates the color spectrum of the object.

3.2   Transfer Learning/Few-Shot Learning

Few-shot learning or transfer learning is a pre-trained model. The main advantage of this model is that it can avoid overfitting when working with less number of samples. It creates or transfers knowledge from a large data set, this is called pre-training. Then another model (a few-shot model) is trained with this knowledge and is used with a smaller dataset.

Figure 1 shows a typical example of transfer learning. DL model 1 is trained by using a large dataset. This knowledge is passed to DL model 2. In model 2 a small dataset and this pre-trained knowledge is used for its training. The overfitting problem is avoided in this way.

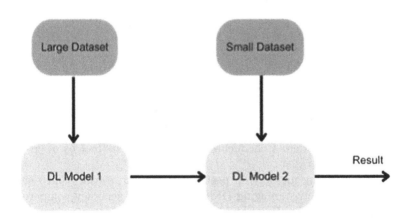

Fig. 1. Transfer Learning workflow diagram

3.3   IoT-Based Pesticide Detection System Using HSI and Deep Learning

Real-time detection of pesticides is very essential in the current situation. IoT-based hyperspectral imaging helps a lot in the real-time detection of pesticides and the conditions of the atmosphere and natural aroma do not affect it. A solution for overfitting with less number of samples is transfer learning or few-shot learning.

For example, markets and fruit processing industries handle different types of fruits and vegetables daily. Large amounts of pesticides are present in it, and consuming these fruits and vegetables leads to dangerous diseases. Only real-time as well as fast methods can detect and thereby control it. The proposed system can be used by Government authorities, consumers, or fruit-processing industries for the effective detection of pesticides.

During the training phase, we train the model using the dataset. Here the model that is used is few-shot learning. It has pre-trained knowledge by training with a large dataset. The dataset can download from any of the data repositories or by creating it in a laboratory environment. Figure 2 shows the block diagram of the proposed system. It has two modules IoT module and a deep learning module. In the IoT module, an image is captured by an HSI camera and this image is sent to the deep learning module.

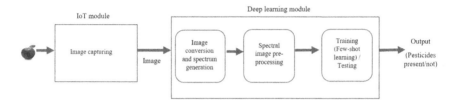

Fig. 2. Block diagram of the proposed model

In the deep learning module, from the HSI image, the spectral vector or spectrum of the image is generated. After the spectral vector generation preprocessing of the image is carried out. In preprocessing remove noise and unwanted spectrum values. This processed image or spectral output is classified with the help of a CNN model.

Algorithm 1 takes two inputs, spectral images of the crops and a dataset that was created in the field. The output of the algorithm shows whether the pesticide is present in the particular crop or not. Let's see how this algorithm works. Processing of the dataset is done first. During the preprocessing step noise and unwanted values were removed. Training of this processed dataset is done by a transfer learning model. Using an HSI camera image of the particular crop is taken. Classification of this spectral image or vector is performed using a CNN model.

Algorithm 1
Input: *Spectral images of crops*
Dataset: Dataset contains HSI images
Output: *Pesticide present/ not*
Procedure

1. *Dataset preprocessing*
2. *Train DL model(few-shot learning) with dataset.*
3. *Spectral vector generation from the input image.*
4. *Fit the model with test data.*
5. *Classification with CNN model.*

4   Result and Discussion

Pesticide detection in fruits and vegetables is one of the important areas that is most needed in today's society. By analyzing the existing method, these systems have different features as well as problems. Table 2 explains the problems of existing methods or their future needs. From the table, it is understood that most of the sensors are affected by environmental situations and have some precision issues. Hyperspectral imaging is a good method, but the device is very costly, more than four lakhs. This is a complex system so we need an expert person to handle this equipment. To avoid these issues future research is needed in this area, to create an algorithm that can convert the normal RGB image to spectral image with high accuracy.

Table 2. Limitations of existing methods

Ref.	Problem/Features
[1]	CNN has better performance than SVM but takes more training and testing time
[2]	A complete understanding of pesticide utilization is not possible
[3]	Overfitting problem due to small dataset and need to improve extraction ability of the network
[5]	Need a large number of samples and deep transfer learning
[6]	Increase the number of samples and pesticide species. A real-time non-destructive identification system is essential
[7]	Lack of data leads to overfitting
[8]	The natural aroma of tea leaves interferes with containing pesticides. The system was susceptible to temperature, humidity & atmospheric pressure. The electronic nose test must perform immediately after spraying
[9]	Nanosensors have high sensitivity, specificity, and minimum response time but they are affected by the environmental condition
[10]	Real-time detection is difficult because of the complex background and small lesions
[11]	Odour must be cleared up each time and do not mix with previous samples. Need more samples

Each of the related works uses different kinds of image datasets for the detection of pesticides. Several systems are working with the PlantVillage dataset. It is a publicly available dataset with 38 categories of species. Figure 3 shows the comparative results of the PlantVillage dataset with different deep-learning models in pesticide detection. All models show accuracy and performance above 95 percent with this dataset. DenseNet-169 gives better results than other models in terms of accuracy.

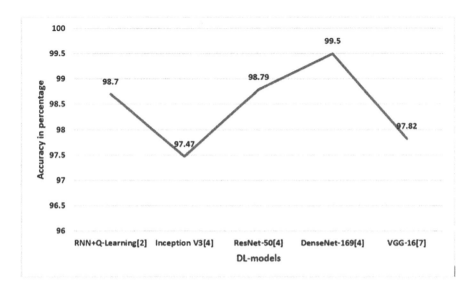

Fig. 3. Performance analysis of PlantVillage dataset with different DL-models

When comparing the architectures and technologies used in currently available pesticide detection systems understood that sensor-based methods and image-based systems have their own characteristics. Figure 4 shows a comparison graph of accuracies that are obtained with different existing systems that are based on image analysis. The graph easily infers that deep learning methods with more hidden layers give better accuracies compared with traditional machine learning as well as low-level deep learning models. Also, HSI images or spectral imaging methods give more results than normal RGB images.

Figure 5 shows a graphical analysis of the performance of the pesticide detection systems that are mainly based on sensors and IoT with various ML or DL models. It is an emerging area in the field of real-time pesticide detection. The sensors are easy to handle, deploy and analyze but it has some problems like being easily susceptible to environmental situations.

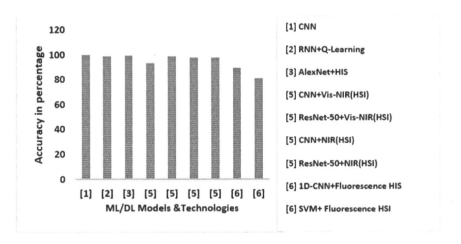

Fig. 4. Performance analysis of image-based systems

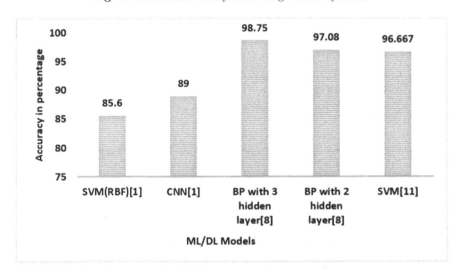

Fig. 5. Performance analysis of sensor-based systems

5   Conclusion

In this paper, we have presented different literature and existing works that address pesticide detection in fruits and vegetables using deep learning. DL models such as CNN, backpropagation, and transfer learning can be employed to accurately predict or detect the presence of dangerous pesticides that are present in the crops. These models have better performance than older machine-learning models based on manual feature extraction. However, deep-learning models need a large volume of data, which leads to the overfitting problem. Transfer learning (few-shot learning) has proven useful to tackle this issue. Hyperspectral imaging

(HSI) is a good solution for pesticide detection in terms of performance compared with standard image data.

This article aimed to provide a general overview of the different deep learning models, techniques, and dataset preparation in this area of research. And proposes a system that uses HSI and transfer learning to solve problems in this pesticide detection field. Further development in this area may help fill the current issues gap.

References

1. Devi, D., Anand, A., Sophia, S., Karpagam, M.: IoT deep learning based prediction of amount of pesticides and diseases in fruits. In: Conference Proceeding of the International Conference on Smart Electronics and Communication (ICOSEC 2020), pp. 848–853. IEEE Xplore (2020)
2. Sellamuthu, K., Kaliappan, V.K.: Q-learning based pesticide contamination prediction in vegetables and fruits. Comput. Sci. Eng. **45**(1), 715–736 (2023)
3. Jiang, B., He, J., Yang, S., Fu, H.: Fusion of machine vision technology and AlexNet- CNN deep learning networks for detection of post-harvest apple pesticide. Artif. Intell. Agric. **1**, 1–8 (2019)
4. Ahmad, A., Gamal, A.E., Saraswat, D.: Towards generalization of deep learning based plant disease identification under controlled and field condition. IEEE Access **11**, 9042–9057 (2023)
5. Ye, W., Yan, T., Zhang, C., Duan, L.: Detection of pesticide residue level in grape using hyperspectral imaging with machine learning. Foods **11**(11), 1609 (2022)
6. Su, J., Hu, Y., Zou, Y., Geng, J., Wu, Y.: Identification of pesticide residue in black tea by fluorescence hyperspectral technology combined with machine learning. Food Sci. Technol. **42**, e55822 (2022)
7. Domingues, T., Brandão, T., Ferreira, J.C.: Machine learning for detection and prediction of crop diseases and pests: a comprehensive survey. Agriculture **12**(9), 1350 (2022)
8. Tang, X., Xiao, W., Shang, T., Zhang, S.: An electronic nose technology to quantify pyrethroid pesticide contamination in tea. Chemosensors **8**(2), 30 (2020)
9. Zia, R., Taj, A., Younis, S., Bukhari, S.Z.: Application of nanosensors for pesticide detection, pp. 259–302. Elsevier (2022)
10. Li, L., Zhang, S., Wang, B.: Plant disease detection and classification by deep learning: a review. IEEE Access **9**, 56683–56698 (2021)
11. Mohamed, R.R., Hashim, W., Azahar, T.M., Yaacob, R.: Food freshness detection using smart machine learning classification. J. Pharm. Negative Results **13**, 7410–7426 (2022)
12. Aherwadi, N., Mittal, U., Singla, J.: Prediction of fruit maturity, quality, and its life using deep learning algorithms. Electronics **11**(24), 4100 (2022)

Statistical Analysis of Hematological Parameters for Prediction of Sickle Cell Disease

Bhawna Dash[✉], Soumyalatha Naveen, and UM Ashwinkumar

School of CSE, REVA University, Bangalore, India
dashbhawna2000@gmail.com

Abstract. About 30 million people worldwide are affected by the monogenic recessive -globin gene abnormality known as sickle cell disease (SCD), which is a significant public health issue. From asymptomatic to severely symptomatic illnesses that might cause patient mortality, pathological features range. The most common presenting symptom of SCD is vasooclussive crisis (VOC). The red cell membrane of the Sickle Red Blood Cells (SRBCs) is damaged by repeated cycles of sickling and desickling processes caused by the formation and aggregation of HbS (sickle hemoglobin) polymers. Cellular dehydration (reduction of ion and water content), increased viscosity (red cell density) and a transient increase in intracellular calcium are all associated with HbS polymerization. As a result, SRBCs become adhesive and inflexible (rigid), resulting in premature destruction. The decreased life span of SRBCs causes chronic hemolytic anemia, and capillary blockage causes tissue hypoxia and subsequent organ damage. So, it is important to monitor patients suffering from sickle cells.

Here we have used machine learning to visualize those patients and categorize them according to their hemoglobin level, percentage of reticulocyte count and serum Lactate dehydrogenase (LDH) level which is regarded as a marker of hemolysis. In this article we propose a framework which uses the statistical analysis using Linear Regression technique on a sickle cell patients dataset showing how hemoglobin is depleted in a body by the use of two parameters called LDH and Retics.

Keywords: Sickle Cell Disease · RBC · WBC · Hemoglobin · Reticulocyte · Machine learning

1 Introduction

Hemoglobinopathy, one of the most prevalent monogenic disorders affecting humans, is responsible for some of the serious genetic and social health difficulties in India, South Africa, Saudi Arabia, South America, and other South Asian and African nations (see Fig. 1). In the history of haemoglobinopathies, sickle cell disease (SCD) is one of the oldest recognized molecular disorder [HbS (HBB ^{Glu6Val})] whereas the HbE disease (HBB ^{Glu26Lys}) is the most widely reported hemoglobin disorders after HbS. HbE is due to a mutation in which lysine is exchanged for glutamic acid in the Beta chain of hemoglobin

P. Pareek et al. (Eds.): IC4S 2023, LNICST 536, pp. 84–94, 2024.
https://doi.org/10.1007/978-3-031-48888-7_7

at the 26th position. SCD affects a sizable portion of population in India residing in the area which is spread throughout the Central Part of India from Odisha to Maharashtra and Gujarat. Sickle Hemoglobin (HbS) in the central belt of India, Hemoglobin E (HbE) in West Bengal and Northeastern States and Hemoglobin D (HbD$_{Punjab}$) in Northwest Parts of India are the three main genotypes of this disorder that are frequently observed in our country [1].

The most prevalent blood condition anemia is brought on by a deficiency of RBCs which makes it difficult for the body to get adequate oxygen. Acute anemia is caused by a sharp fall in RBC, whereas chronic anemia is caused by a gradual decline in RBC, and it frequently co-occurs with inflammatory illnesses. RBCs aren't formed as they should be in people with sickle cell disease. RBCs resemble round or spherical discs in normal human beings whereas they resemble a crescent moon, or an old farming tool called a sickle in SCD. It is a hereditary hemoglobin disorder as represented in Fig. 2.

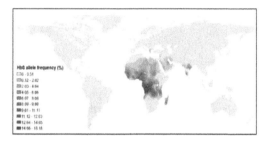

Fig. 1. Map depicting the global prevalence of the HbS allele [2]

Typically, SCA (Sickle Cell Anemia) symptoms and signs begin appearing around five months of age. Sickle cells quickly disintegrated and died, leaving just a small number of RBC in the circulation. The life of normal RBC typically lasts for about four months before they require replacement with new cells, while sickle RBCs often degrade in about two to three weeks, causing a lack of RBC.

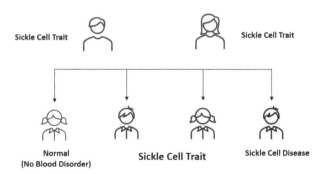

Fig. 2. Inheritance pattern of sickle cell hemoglobin gene from parents to the offspring

Individuals having sickle cell disease report edema, frequent infections, eye problems, slow growth and delayed puberty. As discussed, Hemoglobin S is an abnormal hemoglobin type that contributes to SCA. When both the parents pass the recessive sickle cell gene to their child, the child gets affected and develops the sickle cell homozygote phenotype.

SCD cannot be cured; however, it can be managed to lessen the symptoms and avoid complications. Hence, it is vital that these patients be monitored clinically and hemato-biochemical investigations on regular basis. Based on the steady state and crisis data patients can be monitored successfully [3, 4]. By analyzing the data using machine learning techniques the clinical and physiological state of the patients can be predicted well in advance to avoid any future complicacies. Keeping in view of this the present study was addressed. The data of SCD cases included in this study has been derived from a patient database of Odisha. Various tools of machine learning have been leveraged to predict and assess the health status of the patients those have been suffering from sickle cell anemia disease in Odisha.

2 Literature Survey

Sen et. al. (2021) took various microscopic blood samples and used techniques such as image processing and machine learning to make the process of detecting sickle cells automatic and have classified the RBC thus detected into three shape-based categories: circular, elongated sickle cell shaped and others, they are then preprocessed, and thresholding technique called Otsu is applied for segmentation [5].

Petrović et. al. (2020) used the smear from peripheral blood to observe the images of red blood cells and segmented the image by preprocessing and used machine learning techniques to classify their morphology [6].

In a case study of Nigeria Nkpordee (2022) have used different trend models of time series and statistics for a six-year projection of SCD in Nigeria and how it will decline in the year ahead [7].

Patel et. al. (2021) has shown how early detection of sickle cells can help patients to identify their symptoms and help the patients to take medications and can take regular blood transfusion sessions along with pain relieving medications. Sometimes manual assessment might lead to false classification. Therefore, using data mining techniques including classification algorithm they have sought to identify the sickle cells in human body with high accuracy [8].

Yang (2018) and Yeruva (2021) have employed machine learning algorithms to predictably understand the timing/situation of hospital re-admissions in SCD. In their research paper they have described how they partitioned their patients into groups for testing and training. The cases of unplanned treatment in the hospitals admissions were categorized for testing and training dataset where they applied machine learning algorithms. The prediction was then later assessed using various prediction algorithms such as specificity, sensitivity, and C-Statistic [9, 10].

Dean (2019) has used Multinomial Logistic Regression where they analyzed the pain scores of forty patients, and they devised a model of machine learning to predict the pain scores of a SCD patient with promising results [11]. Using proper optimization techniques, machine learning algorithms, and statistics, it can be predicted whether the number of patients suffering from SCD will decrease with the use of a proper data set and patients as input.

A low cost, cost effective easy to use sickle cell screening device is proposed to be used in developing countries as elucidated by Wing (2019) [12] can detect hemoglobin non-invasively.

Stone (2021) [13] in this case report demonstrated the severity of a delayed hemolytic transfusion reaction caused by anti-Fy3 in a SCD patient having red cell exchanging before hematological precursor cell harvest for gene therapy.

Ranjana (2020) [14] used a automatic categorization of the SCA system explored in this study. In the beginning, the original images are pre-processed using the median filter. The Grey Level Co-occurring Model (GLCM) and Haralick characteristics are then retrieved. Finally, for prediction, the random forest (RF) predictor is used. Using an RF classifier, the SCA system achieves a classification accuracy of 95%.

Patgiri [15] demonstrated a hybrid segmentation procedure that combines two segmentation approaches, notably fuzzy C-means segmentation with adaptive (local) thresholding. In this study, four distinct adaptive thresholding approaches are used with fuzzy C-means. The main axis, and secondary axis, aspect ratio, surface dimension, circumference, dimension factor (metric value), eccentricity, and solidity of each cell in the sample blood smear were retrieved for this analysis. These eight characteristics are used to train and test the classifiers. For categorization, two supervised classifiers, namely the Nave Bayes classifier and the K-nearest neighbor classifier, were exhibited on a dataset of ten image data samples, and the evaluated results for all of the hybrid combinations were compared.

Even though a lot of techniques have been used or the prediction or image segmentation in various machine learning dataset the complicated clinical symptoms of SCD have not been addressed fully till date. So, it is necessary to predict the outcome for the year ahead and come up with some solution that will prevent the patients going through the tedious process of regular blood transfusions and doctor visits. Although these methods yield the best results but considering the complex clinical manifestations of symptoms from patients to patients implementing those methods has been challenging so far.

3 Linear Regression

Linear Regression is a widely known and recognized algorithm and is categorized under supervised learning technique. When a set of independent variables is given, the logistic regression is used for carrying out the prediction of dependent variables categorically, such that output result can be a categorical or discrete value. As it is an analysis of independent variable, and it can be represented in Eq. 1:

$$y = c0 + c1x + e \tag{1}$$

wherein y has been assumed to be a dependent variable and x to be an independent variable, the variable c0 is a constant term and an intercept of the regression line on the vertical axis and c1 is the regression coefficient that lies on the slope of regression line and e can be a random error as shown in Fig. 3.

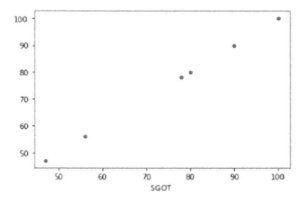

Fig. 3. Graph of linear regression

As specified earlier, the goal is to find out the best possible values for c0 and c1, the objective should be to minimize the error between the predicted value and the actual value as shown in Eqs. 2 and 3.

$$Minimize 1/n \sum_{i=1}^{n} (pred(i) - y(i))^2 \qquad (2)$$

$$Q 1/n \sum_{i=1}^{n} (pred(i) - y(i))^2 \qquad (3)$$

This function as mentioned above is aimed at minimizing the error values among the actual and the predicted values. Here the error difference is squared, added up across all the data points and then divided by the total number of data points. The result obtained (Q) is the average squared error across all data points. Hence the above cost function is also referred to as Mean Squared Error (MSE) function. With the MSE function the values of c0 and c1 are changed so that MSE settles at the minima.

Figure 4 depicts the process of fitting a linear regression model. Import data as an input, fit an optimization technique and a cost function for performance, verify its quality, change it to increase quality, and then find an output for the workflow.

Stochastic Gradient Descent (SGD) is a form of gradient descent variant used to optimize machine learning models. Only one random training example is used in this variant to calculate the gradient and update the parameters at each iteration. This algorithm is useful when the optimal points are not found by equating the slope of the function to zero (0). Linear regression on the other hand has the sum of squared residuals mentally mapped as the function "y" and the weight vector as "x" in the parabola above.

4   Proposed Work

The proposed approach extracts sickle cell data from a hospital in the western part of Odisha. The extracted dataset was then analyzed and usedS it as input for our machine learning model. So, first, we preprocessed and cleaned the data, and then we fed it into appropriate models for training and testing. Following the visualization and train-test

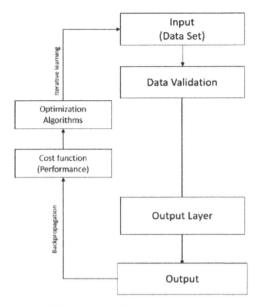

Fig. 4. Flowchart of the working of linear regression while using a dataset.

split, we selected an acceptable model (here, linear regression) for our planned work. We arrived at a proper conclusion after obtaining the accuracy and proper graphs (as indicated in the Graphs and Results section) (Fig. 5).

By using these statistical methods, we can classify, predict and find an optimal model that can help us identify the health / clinical status of people affected from sickle cell anemia or who have less amount of hemoglobin produced in their body.

Pseudocode
Input: *Patience dataset*
Pre-processing of data
Divide data into Train and Test with 80% and 20% respectively.
For each data in dataset
Linear regression (Train, Test data)
Perform gradient descent.
Predict test result.
Output: *predicted percentage, Correlation matrix*

5   Experimental Results

The proposed work as presented under section III has been analyzed using stochastic gradient descent analysis in conjunction with three different machine learning algorithm such as linear regression, decision tree classifier and support vector machine. The data

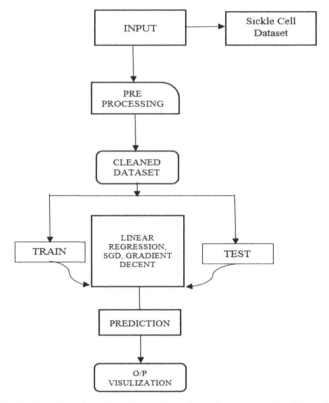

Fig. 5. Flowchart describing the methodology that was used in this study.

has been collected from the western part of Odisha state, highly affected with SCD. The data has been collected under six different categories such as WBC, RBC, HGB, BIT, LDH and RETICS%. All the data has been converted into it's per unit level except RETICS which is in percentage.

The data has been processed for redundancy analysis with a new value of 0.17. The scattered plot analysis has been carried out for two target data namely LDH, RETICS% with all the other four parameter as predictor. The effect of LDH and RETICS categorically analyzed with HGB. Out of the total dataset 80% data has been used for training and 20% has been used for testing the model.

Table 1 shows the statistical data of three different analyses of three different ML algorithms such as Decision Tree, Linear Regression and Support Vector Machine. It is found that the linear regression is having lowest RMSE of 3.60 and R^2 error is –0.39. However, the mean average error is 2.72; this could be due to similar type of data in the available training dataset. Further analysis has been carried out with Linear Regression with Gradient Descent.

Table 1. Statistical analysis of ML algorithms

	RMSE	MSE	R^2 Error	MAE
DT	4.32	18.70	–0.93	2.41
LR	3.60	13.03	–0.39	2.72
SVM	5.32	28.30	–1.92	2.56

Figure 6 shows the regression analysis of RETICS vs HGB where most of the data possesses negative slope characteristics which means that with increasing HGB content there is a decrease in the RETICS. In most of the cases the HGB content carries in between 8.5 to 11, which corresponds to a decrease in 17% of RETICS%

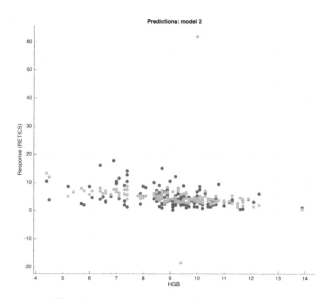

Fig. 6. Regression analysis of RETICS vs HGB

Similarly, Fig. 7 represents the statistical graphical analysis of LDH vs HGB. However, with the same range of HGB (refer Fig. 6) the LDH content varies between 385–460.

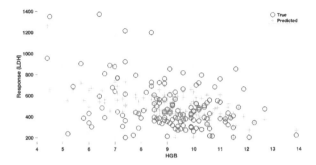

Fig. 7. Regression analysis of LDH vs HGB

Table 2 shows the analysis of different ML algorithms for LDH vs HGB. Figure 8 shows the heat map based on autocorrelation function where all the diagonal elements have a magnitude 1 per unit.

Table 2. Statistical analysis of ML algorithm

	RMSE	MSE	R^2 Error	MAE
DT	383.04	1.46×10^5	−1.08	180.6
LR	274.32	75251	−0.07	181.41
SVM	286.16	81885	−0.165	155.05

Table 3. Correlation statistical analysis with Hb

Sr. No	Parameter	Magnitude	Remarks
1	RETICS%	−0.39248	NEGATIVE
2	LDH	−0.37992	NEGATIVE
3	BIT	−0.137320	NEGATIVE
4	RBC	0.300613	POSITIVE
5	WBC	−0.09364	NEGATIVE

Figure 8 this correlation depicts that the patient does not have enough hemoglobin produced in this body as the formation of LDH is high in their body hence they cannot carry enough oxygen to supply throughout their body.

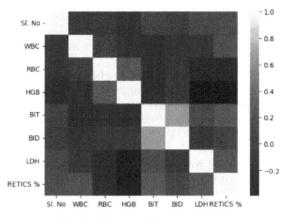

Fig. 8. Heatmap of the parameters.

6 Conclusion and Future Work

From the present study with the data of the patients who have sickle cell disorder (HbSS) and using the ML models it can be predicted that the patients having low hemoglobin might face many clinical symptoms due to the formation of high level of LDH, more numbers of WBC and reticulocyte counts. Due to the low hemoglobin in their body they are unable to meet the oxygen demand of the body and subjected to deoxygenated state and leads to high amount of lactic acid and higher count of reticulocyte or premature RBCs in their circulation. Hence patients must go through regular blood transfusions and hemoglobin tests. The traditional method of measurement of hemoglobin is accurate, infants and adults are hesitant to use it since it is painful, and regular blood extraction makes them uncomfortable. As a result, introducing a non-invasive way will be beneficial to determine their hemoglobin level any place and without any pain. Till then regular management of the patients with the clinical and hematological data set using machine learning techniques will be of great importance.

References

1. Chhotray, G.P., Dash, B.P., Ranjit, M.: Spectrum of Hemoglobinopathies in Orissa, India (2004). https://doi.org/10.1081/hem-120034244
2. Piel, F.B., et al.: Global distribution of the sickle cell gene and geographical confirmation of the malaria hypothesis (2010). https://doi.org/10.1038/ncomms1104
3. Serjeant, G.R.: One hundred years of sickle cell disease (2010). https://doi.org/10.1111/j.1365-2141.2010.08419.x
4. Meher, S., et al.: Haptoglobin Genotypes Associated with Vaso-Occlusive Crisis in Sickle Cell Anemia Patients of Eastern India (2021). https://doi.org/10.1080/03630269.2020.1801459
5. Sen, B., Ganesh, A., Bhan, A., Dixit, S., Goyal, A.: Machine learning based diagnosis and classification of sickle cell anemia in human RBC. In: 2021 Third International Conference on Intelligent Communication Technologies and Virtual Mobile Networks (ICICV), Tirunelveli, India, pp. 753–758 (2021). https://doi.org/10.1109/ICICV50876.2021.9388610

6. Petrović, N., Moyà-Alcover, G., Jaume-i-Capó, A., González-Hidalgo, M.: Sickle-cell disease diagnosis support selecting the most appropriate machine learning method: towards a general and interpretable approach for cell morphology analysis from microscopy images (2020). https://doi.org/10.1016/j.compbiomed.2020.104027

7. Nkpordee, L., Wonu, N.: Statistical modelling of genetic disorder in Nigeria: a study of sickle cell disease. Faculty Nat. Appl. Sci. J. Sci. Innov. 3(2), 10–19 (2022). https://www.fnasjourn als.com/index.php/FNAS-JSI/article/view/27

8. Patel, A., et al.: Machine-learning algorithms for predicting hospital re-admissions in sickle cell disease. Brit. J. Haematol. **192**(1), 158–170. Wiley (2020). https://doi.org/10.1111/bjh. 17107

9. Yang, F., Banerjee, T., Narine, K., Shah, N.: Improving pain management in patients with sickle cell disease from physiological measures using machine learning techniques. In: Smart Health, vols. 7–8, pp. 48–59. Elsevier BV. (2018). https://doi.org/10.1016/j.smhl.2018.01.002

10. Yeruva, S., Gowtham, B.P., Chandana, Y.H., Varalakshmi, M.S., Jain, S.: Prediction of anemia disease using classification methods. In: Machine Learning Technologies and Applications, pp. 1–11. Springer, Singapore (2021). https://doi.org/10.1007/978-981-33-4046-6_1

11. Dean, C.L., et al.: Challenges in the treatment and prevention of delayed hemolytic transfusion reactions with hyperhemolysis in sickle cell disease patients. Transfusion **59**(5), 1698–1705. Wiley (2021). https://doi.org/10.1111/trf.15227

12. Wing, J., et al.: A low-cost, point-of-care sickle cell anemia screening device for use in low and middle-income countries. In: 2019 IEEE Global Humanitarian Technology Conference (GHTC), Seattle, WA, USA, pp. 1–4 (2019). https://doi.org/10.1109/GHTC46095.2019.903 3017

13. Stone, E.F., et al.: Severe delayed hemolytic transfusion reaction due to anti-Fy3 in a patient with sickle cell disease undergoing red cell exchange prior to hematopoietic progenitor cell collection for gene therapy (2020). https://doi.org/10.3324/haematol.2020.253229

14. Ranjana, S., Manimegala, R., Priya, K.: Automatic classification of sickle cell anemia using random forest classifier. In: Proceedings of the European Conference on Medical Advances, LNCS, vol. 9999, p. 2020. Springer, Heidelberg (2020)

15. Patgiri, C., Ganguly, A.: Adaptive thresholding technique based classification of red blood cell and sickle cell using Naïve Bayes Classifier and K-nearest neighbor classifier (2021).https:// doi.org/10.1016/j.bspc.2021.102745

Optimized Route Planning and Precise Circle Detection in Unmanned Aerial Vehicle with Machine Learning

Ankit Garg, Priya Mishra, and Naveen Mishra[✉]

Department of Communication Engineering, School of Electronics Engineering, Vellore Institute of Technology, Vellore, Tamil Nadu, India
naveenmishra.ece@gmail.com

Abstract. This research paper presents a distributed architecture for optimized route planning and precise circle detection in unmanned aerial vehicle with machine learning. The architecture focuses on three key areas: motion planning, control, and the integration of a web application with machine learning (ML) for autonomous drones. By leveraging advanced planning and control algorithms, the architecture enables UAVs to navigate dynamic environments, execute complex maneuvers, and maintain stability. The ML-integrated web application enhances decision-making for detection, optimizing route planning. Extensive simulations and real-world experiments validate the effectiveness and scalability of the proposed architecture, making it a valuable tool for advancing research in autonomous UAV systems.

Keywords: Autonomous unmanned aerial vehicles · distributed architecture · motion planning · control · machine learning · web applications · experimentation

1 Introduction

In recent years, the field of intelligent unmanned aerial vehicles (UAVs) has grown quickly, and now it presents a wide range of distributed autonomous robotics research opportunities [1, 2]. With the aim of designing controllers that facilitate the autonomous flight of a UAV from one waypoint to another in comparison to prior research which has concentrated on low level control capabilities. The most frequent mission scenario entails positioning sensor pay-loads for data collection, with the data ultimately being processed offline or in real-time by ground workers. In recent combat scenarios, the use of UAVs in mission duties like surveillance [3, 4], videography, search and rescue [5] etc., has grown increasingly crucial and it is projected that they will continue to play crucial roles in any future conflicts.

In civil applications such as remote sensing, precision agriculture [6–8], etc., intelligent UAVs also play a vital role. There is a need to build more advanced UAV platforms for both military and civil uses which can offer more emphasis on the development of intelligent capabilities and capacity to communicate with human operators and other

P. Pareek et al. (Eds.): IC4S 2023, LNICST 536, pp. 95–105, 2024.
https://doi.org/10.1007/978-3-031-48888-7_8

robotic platforms. Low- level control is no longer the primary focus of research. Instead of this, sophisticated software architectures that integrate low level and decision-level control are being used. These should thus function seamlessly with larger C4I2 systems with network-based architectures. These systems are necessary to provide the capabilities needed for the upcoming, more complicated mission requirements, and they serve as an excellent testing ground for distributed AI technology.

Path planning algorithms [9–11] that produce collision-free paths, precise controllers capable of executing such paths even in the presence of unfavorable weather conditions (such as wind gusts), and a dependable mechanism that coordinates the two are necessary for navigating in environments where there are many obstacles close to building structures.

In a distributed software architecture utilized in a fully deployed rotor-based unmanned aerial vehicle (UAV), a method for combining path planning techniques with a path execution mechanism—including a reliable 3D path following control mode—is described in this study. There are descriptions of many of the software parts utilized in the distributed architecture. The elements in charge of path execution are given special attention. The method considers the varied time properties and dispersed communication of a path-planning algorithm and a path-following control mode [12]. To operate UAVs in urban settings, they also feature a safety device.

Unique challenges posed by the specific scenario necessitate precision control, maneuverability, and payload delivery capabilities [13]. The scenario requires the drone to follow predefined flight paths, locate a tar-get, and execute precise payload deployment [14]. In this study highlighted the importance of developing a specialized quadcopter tailored to meet the specific requirements outlined in the problem statement. Following a thorough analysis of the challenge, devised a rigorous design strategy to construct a quadcopter capable of successfully accomplishing the assigned tasks. To optimize stability and longevity while minimizing weight, implemented a 505mm wheelbase and utilized lightweight yet durable materials. The frame construction incorporated aluminum rods for enhanced strength, while medium density fiberboard (MDF) provided structural support integrity. A true X-frame design was employed to ensure stability without compromising on the accommodation of essential hardware components.

This quadcopter's primary flight controller is the Pixhawk 2.4.8, which provides dependable and accurate control over the drone's flight parameters. The popular open-source autopilot program Ardupilot was loaded into Pixhawk's firmware. The control and navigational abilities required to carry out complex flying patterns were provided by this combination. Also installed a Raspberry Pi microprocessor on the quadcopter to precisely locate the target and carry out the payload delivery duty. This microcontroller was developed to carry out real-time image and video processing. The primary camera recorded and processed video inputs using sophisticated coding techniques, allowing the precise detection and identification of the target area. Numerous flying tests were conducted to experimentally evaluate quadcopter de-sign, replicating the tasks specified in the problem statement [15–17]. These experiments verified the viability of idea, as the quadcopter successfully carried out the intended flying patterns, located the target region, and dropped the cargo precisely. The data gathered from the testing shows that suggested quadcopter design is trustworthy and effective for the intended purposes (Fig. 1).

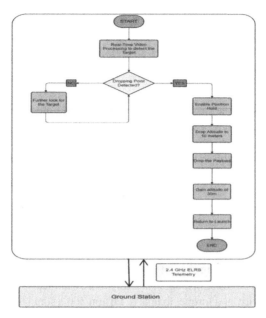

Fig. 1. Work Flow Flowchart

2 Design Process

Thoroughly examined the specifications outlined in the problem description during the design process of the drone frame. It was crucial to ensure that the frame could securely accommodate all the necessary components, allowing the drone to effectively fulfil the designated objectives of the research experiment. As to consider several things during the design process, including the size and weight of the individual parts, the necessity for structural stability, and the drone's overall balance. These factors had to be considered to guarantee that the frame could sustain the hardware while still performing at its best during flight.

2.1 Frame Selection

First, a 200-g package with dimensions of $5 \times 10 \times 10$ cm^3 must be delivered by the drone on its own. Additionally, throughout the mission, the drone must maintain a flight altitude of 30 m.

During the design phase, considerable considerations were made in order to satisfy these needs. Choosing the right propellers and a propulsion system that could produce enough thrust to sustain the payload and maintain stable flight was an important consideration. 10-inch propellers were selected to accomplish this as they were discovered to produce the roughly necessary thrust for the load delivery.

It was crucial to strike a compromise between clearance and stability in order to guarantee appropriate clearance between the numerous components mounted on the frame. Therefore, it was decided that arms 20 cm in length were ideal. This length permitted a 2 cm clearance between the parts, guaranteeing adequate room for their proper operation without affecting the overall stability of the drone's construction.

3   Modelling of Frame

It was crucial to design the quadcopter's frame using 3D modelling software before beginning the construction process. This process made it possible to precisely visualize and evaluate the frame's structural features. Autodesk Fusion 360, a widely used application for design and engineering, was selected as the programmers for this job. The quadcopter's arms and body were designed together with the proper frame arrangement throughout the design phase. To make sure the frame configuration satisfied the needs for stability, maneuverability, and payload capacity, several factors were considered. In Fusion 360, the frame was digitally built when the arrangement was decided upon, considering the necessary dimensions and characteristics. A thorough structural analysis was carried out to evaluate the frame's performance and structural integrity. The material characteristics and structural dimensions were entered into the software for this study. The frame may be simulated both statically and dynamically under various forces, including thrust, gravity, and torque, thanks to Fusion 360's simulation capabilities. The software may assess the frame's performance in terms of static stability and dynamic response by applying computed force values to points inside the frame [18]. To enhance the drone's design, multiple iterations were conducted. The frame's configurations were modified at each iteration, and simulations were utilized to evaluate the frame's performance. The aim was to develop a frame that fulfilled the requirements of the operational stage while ensuring stability, durability, and maneuverability [19–21].

3.1   Weight Estimation

Carefully chose the quadcopter's important parts and considered the materials for the frame construction in order to satisfy the requirements of the problem description.

Estimated the overall weight of the drone to make sure it would stay within the allowed weight ranges. The data below shows the estimated weight distribution for some of the key elements.

After estimating the drone's weight and went on to compute the thrust needed to lift it. Maintaining a 2:1 thrust-to weight ratio is a usual guideline. The thrust needed for each motor was estimated to be around 700 g based on this ratio. If a quadcopter arrangement were used, the total thrust needed to lift the entire drone would be twice its own weight.

Drone Total Weight Approximation: 1400 g.

Drone Lifting Total Thrust Needed: 2800 g.

Each motor's thrust is 700 g.

These calculations made sure that the motor thrust, and the components chosen would be enough to achieve controlled and steady flying. Are able to create a quadcopter that could successfully carry out the specified flight maneuvers and payload delivery tasks mentioned in the challenge description by carefully evaluating the weight and figuring out the thrust needs.

3.2 Frame Design

- The materials used in the fabrication of the drone for the research experiment were meticulously selected to achieve an optimal combination of durability and lightweight design. Medium-density fiberboard (MDF) and aluminum were chosen as the primary materials to fulfil this objective.
- The drone's frame was constructed using MDF, which has a good strength-to-weight ratio. The overall weight of the drone is optimally maintained, while this material provides great structural integrity. The drone is kept lightweight by using MDF, enabling effective flight maneuvers and reducing battery usage.
- On the other hand, aluminum was chosen due to its outstanding durability and light weight. In locations that needed extra strength, such as crucial joints and supporting structures, it was employed strategically. The use of aluminum components improves the drone's overall resilience without considerably adding to its weight.

Fig. 2. Side View of Drone   **Fig. 3.** Isometric View   **Fig. 4.** Top View of Drone

- Drone design achieves a harmonious balance of lightweight construction, durability, and ease of maintenance by carefully examining the selection of materials and utilizing modern design and production techniques. This guarantees that the drone will function optimally and have an extended operational lifespan, even under the demanding conditions of its designated operational phase. As shown in Fig. 2, 3 and 4 the side, top and isometric views are available.
- Structural Properties of the material chosen: The stability, toughness, and general performance of the quadcopter are greatly influenced by the structural qualities of the materials used in its construction. In this instance, medium-density fiberboard (MDF) and aluminum were chosen as the materials. MDF is a composite wood product created by mixing resin and wood fibers under intense pressure and heat. When choosing the MDF for the quadcopter's construction, the following characteristics were considered: Measured carefully to establish a balance between weight and structural soundness, the MDF used had a thickness of 4.5 mm. Although thicker boards could offer more strength, they would also add to the quad-copter's weight, which might have an effect on how well it flies. The MDF used for this project has a density of $750 \, \text{kg/m}^3$. For the quadcopter's construction to maintain an ideal weight-to-strength ratio, this figure, which represents the mass of the material per unit volume, is crucial.

Due to its light weight and excellent strength, aluminum was chosen in addition to MDF for some of the quadcopter's structural components. The aluminum used was considered for the following qualities:

The aluminum sheets that were used were 1 mm thick. To achieve minimal weight while maintaining appropriate strength for the components where aluminum was used, this relatively thin gauge was chosen.

Aluminum has a high modulus of elasticity (MOE), which is calculated to be 70300 N/mm^2. Due to this characteristic, aluminum parts can tolerate bending loads and keep their structural integrity during flying maneuvers.

Quadcopter design achieves a balance between weight, strength, and rigidity by carefully examining the structural characteristics of the selected materials, particularly MDF and aluminum. This makes it possible for the drone to endure the forces generated by the aerodynamics of flight, maintain stability, and fulfil the specific objectives outlined in the designated research tasks.

4 CG Calculation

4.1 The Word "Data" is Plural, not Singular

- Used a strict approach to precisely estimate the drone's center of gravity (CG). Firstly, it strung a thread from one of the drone's arms to the end of the drone. It made sure the string's line passed through its pivot point and remained perpendicular to the ground by paying close attention to it. By doing this, were able to create a reference line for our measurements.
- Again, then repeated the process by fastening threads to several spots along the drone's frame. To be more precise, we fastened strings to the pivot point where the arm and cover plate converge, the border of the bottom plate, and the center of the top plate. Each string created a line from its connection point that crossed the earlier established reference line.
- Determined the precise location of the drone's center of gravity by examining the intersection locations of these lines. The intersection that was indicated represented the drone's estimated CG location. Through this process, we were able to establish that the drone's center of gravity (CG) was situated precisely 15mm above the top plate and at the center of the true X frame design.
- To ensure the drone's stability and balanced flight characteristics, it is crucial to carefully evaluate the center of gravity (CG). So, to improve the drone's performance, control, and maneuverability during the flight evaluation by accurately determining the CG location. Furthermore, utilizing this information, although can arrange additional components, such as the payload, in a manner that preserves the overall stability and flight dynamics of the drone.

5 Displacement Caused by Stress

Although carried out stress analysis with computer simulations to assess the drone frame's structural robustness as shown in Fig. 4. In order to replicate the highest load that the frame might encounter during flight, applied a force of 10N at the end of each arm in this analysis. Were able to ascertain the stress distribution inside the structure by applying these forces and considering the material characteristics of the frame.

Were able to evaluate the drone frame's structural integrity and pinpoint any potential weak places or regions that would encounter excessive stress thanks to the stress analysis. Able to locate areas of high stress concentration that may need reinforcement or design adjustment by analyzing the stress distribution. The Displacement caused by stress is also shown in Fig. 5.

Fig. 5. Displacement Caused by Stress   **Fig. 6.** Result of Stress Analysis

5.1 Fabrication of the Drone

- The fabrication procedure was started to assure the build's structural integrity and its capacity to withstand the predicted forces after carefully choosing the materials and performing a structural analysis of the quadcopter model (Fig. 6).
- Precision CNC cutting of medium-density fiberboard (MDF) sheets for the quadcopter's main body was the first step in the construction process. This method made sure that the structural components' dimensions were precise and constant. The aluminum arms were additionally saw- cut, ensuring their sturdiness and strength.
- Custom-designed 3D-printed pieces were integrated to further improve the quadcopter's appearance and operation. These components were created by additive manufacturing processes using PETG (Polyethylene Terephthalate Glycol) and ABS (Acrylonitrile Butadiene Styrene) filaments. This method made it possible to design complex components that were specially tailored to the needs of the quadcopter.
- The various parts of the quadcopter's frame were fastened and connected during the construction stage using M3 bolts of the ideal size. While minimizing extraneous weight, adequate fastening was ensured by the careful selection of bolt size.

5.2 Detailed Weight Breakdown

The comprehensive weight breakdown sheds light on the relative weights of the different quadcopter components. These parts include the top plate, bottom plate, and arms that make up the structure as well as crucial hardware like the Raspberry Pi, Pixhawk, GPS devices, webcam, and ESCs. Propellers, motors, a mounting plate, a standoff, a VTx, and a camera are additional parts that help the quadcopter function and perform.

The weight breakdown also takes into consideration the battery, GPS stand, dropping mechanism, ELRS data telemetry, and other wires and parts required to secure the peripherals. These elements are essential to fulfilling the objectives of the research experiment. When all the components stated above are considered, the quadcopter's overall weight is 1179g. This weight is a crucial factor in ensuring the quadcopter's optimum balance, stability, and flight qualities throughout the competition. Engineers and researchers are able to make educational decisions about component choice, location, and overall weight management thanks to the precise weight breakdown, which is helpful for the design and optimization of the quadcopter.

TensorFlow's Object Detection Mechanism:
There are several processes involved in object detection using TensorFlow:

The Single Shot Multibox Detector (SSD) and the You Only Look Once (YOLO) architecture are two pre-trained object detection models that are available through TensorFlow. As an alternative, you can use TensorFlow's APIs to train your own unique object identification model.

Training: When an object detection model is trained, labelled datasets are fed into it, and its parameters are optimized using methods like gradient descent. High-level APIs from TensorFlow, like the Object Detection API, make the training process easier.

Inference: After been trained, the object detection model can be applied to forecast new, unforeseen data. The trained model may be loaded, inference can be done on pictures or videos, and bounding box coordinates and class labels of detected objects can be extracted using TensorFlow's tools.

6 Dataset

A dataset is essential for the precise target identification and successful execution of the payload drop in the context of the given code and the research scenario [10]. To train and test machine learning models, datasets are collections of labelled samples. A picture or video with matching annotations that define the bounding boxes around the items of interest makes up the majority of datasets used in object detection. The object detection model is trained using these annotations as ground truth data.

Unmanned aerial vehicles (UAVs) use a variety of sensors and equipment to gather a wide range of data. These data are crucial for successfully completing UAV missions and getting insightful information for many applications. Here is a more thorough explanation of the kinds of information that UAVs gather: Imagery and Video Data: UAVs are fitted with cameras that record both still images and moving video. These cameras can include thermal, multispectral, hyperspectral, RGB (Red, Green, and Blue), and infrared cameras. While thermal cameras use infrared radiation to illustrate temperature differences, RGB cameras only record images produced by conventional visible light. In order to analyse certain vegetation or material qualities, multispectral and hyperspectral cameras record images in a number of or narrow bands across the electromagnetic spectrum.

6.1 Information about GPS and Navigation: UAVs rely on GPS technology to determine their location and navigate securely. GPS data is necessary for flight planning, waypoint navigation, and preserving aircraft stability. UAVs may also employ a variety of navigation devices, including Inertial Measurement Units (IMUs), barometers, and compasses, to enhance their sensing and orienting abilities.

6.2 Info from Payload-Specific Instruments: Other data kinds might be gathered, depending on the mission and payload setup of the UAV. UAVs may, for instance, be equipped with sensors for monitoring animals, water quality, or air sampling in the course of scientific study.

6.3 Sensor Information: In order to gather information beyond imagery, UAVs are fitted with a variety of sensors. For weather forecasting and climatological studies, atmospheric sensors detect air pressure, temperature, humidity, and other meteorological factors. Gas detectors are useful for environmental monitoring, industrial safety inspections, and finding gas leaks since they can identify and measure a variety of gases.

7 Result

We accomplished the objectives of the experiment using our custom-built quadcopter, equipped with the Pixhawk flight controller and the Raspberry Pi microcontroller. By showcasing the design, implementation, and evaluation of our drone system in this research paper, we illustrated its effectiveness in addressing the challenges described in the problem statement. Here are some of the output photos are available where we see the terminal of Raspberry Pi and how drone is optimising after the circle detection in Fig. 7 and Fig. 8.

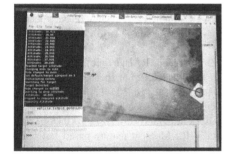 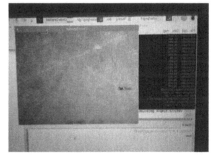

Fig. 7. Circle detection by drone   **Fig. 8.** Optimization in circle detection

8 Conclusion

In conclusion, the application of machine learning techniques in UAV route planning and circle detection has showcased promising results, providing a solid foundation for future research and development in the field. As the demand for efficient and accurate UAV operations continues to grow, these findings contribute to the advancement of autonomous systems and pave the way for new possibilities in various industries. With further refinement and validation, the optimized route planning and precise circle detection methods discussed in this paper can lead to safer, more efficient, and more intelligent unmanned aerial vehicles.

References

1. Cai, Y., et al.: Guided attention network for object detection and counting on drones. In: Proceedings of the 28th ACM International Conference on Multimedia, pp. 709–717 (2020)
2. Al Dahoul, N., Sabri, A.Q., Mansoor, A.M.: Real-time human detection for aerial captured video sequences via deep models. Hindawi **15** (2018)
3. Chang, X., Yang, C., Wu, J., Shi, X., Shi, Z.: A surveillance system for drone localization and tracking using acoustic arrays. In: Proceedings of the 2018 IEEE 10th Sensor Array and Multichannel Signal Processing Workshop (SAM), pp. 573–577 (2018). https://doi.org/10.1109/SAM.2018.8448409
4. Belmonte, L.M., Morales, R., Fernández-Caballero, A.: Computer vision in autonomous unmanned aerial vehicles-a systematic mapping study. Multidisciplinary Digit. Publishing Inst. **9**(15), 3196 (2019). https://doi.org/10.3390/app9153196
5. Al-Kaff, A., Gómez-Silva, M., Moreno, F., de la Escalera, A., Armingol, J.: An appearance-based tracking algorithm for aerial search and rescue purposes. Multidisciplinary Digit. Publishing Inst. **19**(3), 652 (2019). https://doi.org/10.3390/s19030652
6. Apolo-Apolo, O.E., Martínez-Guanter, J., Egea, G., Raja, P., Pérez-Ruiz, M.: Deep learning techniques for estimation of the yield and size of citrus fruits using a UAV. Eur. J. Agron. **115**, Article 126030 (2020). https://doi.org/10.1016/j.eja.2020.126030
7. Boursianis, A.D., et al.: Internet of things (IoT) and agricultural unmanned aerial vehicles (UAVs) in smart farming: a comprehensive review. Internet Things **18**, 100187 (2020)
8. Chen, C.J., Huang, Y.Y., Li, Y.S., Chen, Y.C., Chang, C.Y., Huang, Y.M.: Identification of fruit tree pests with deep learning on embedded drone to achieve accurate pesticide spraying. IEEE Access Prac. Innov. Open Solutions **9**, 21986–21997 (2021). https://doi.org/10.1109/ACCESS.2021.3056082
9. Carrio, A., Sampedro, C., Rodriguez-Ramos, A., Campoy, P.: A review of deep learning methods and applications for unmanned aerial vehicles. Hindawi 1–13 (2017). https://doi.org/10.1155/2017/3296874
10. Chen, N., Chen, Y., You, Y., Ling, H., Liang, P., Zimmermann, R.: Dynamic urban surveillance video stream processing using fog computing. In: Proceedings of the 2016 IEEE Second International Conference on Multimedia Big Data (BigMM), pp. 105–112 (2016)
11. Gonzalez-Trejo, J., & Mercado-Ravell, D.: Dense crowds detection and surveillance with drones using density maps. ArXiv:2003.08766 [Cs]. http://arxiv.org/abs/2003.08766 (2020)
12. Saif, A.F.M.S., Prabuwono, A.S., Mahayuddin, Z.R.: Moment feature based fast feature extraction algorithm for moving object detection using aerial images. PloS One **11** (2015)
13. Shakhatreh, H., Sawalmeh, A.H., Al-Fuqaha, A., Dou, Z., Almaita, E., Khalil, I., et al.: Unmanned aerial vehicles (UAVs): a survey on civil applications and key research challenges. IEEE Access **7**, 48572–48634 (2019)
14. Hii, M.S.Y., Courtney, P., Royall, P.G.: An evaluation of the delivery of medicines using drones. Multidisciplinary Digit. Publishing Inst. **3**(3), 52 (2019)
15. Bonetto, M., Korshunov, P., Ramponi, G., Ebrahimi, T.: Privacy in mini-drone based video surveillance. In: Proceedings of the 2015 11th IEEE International Conference and Workshops on Automatic Face and Gesture Recognition (FG), vol. 4 pp. 1–6 (2015)
16. Boonpook, W., Tan, Y., Ye, Y., Torteeka, P., Torsri, K., Dong, S.: A deep learning approach on building detection from unmanned aerial vehicle-based images in riverbank monitoring. Multidisciplinary Digit. Publishing Inst. **18**(11), 3921 (2018)
17. Schumann, A., Sommer, L., Klatte, J., Schuchert, T., Beyerer, J.: Deep crossdomain flying object classification for robust UAV detection. In: Proceedings of the 2017 14th IEEE International Conference on Advanced Video and Signal Based Surveillance (AVSS), pp. 1–6 (2017). https://doi.org/10.1109/AVSS.2017.8078558

18. Chiu, S.H., Liaw, J.J., Lin, K.H.: A fast randomized Hough transform for circle/circular arc recognition. Int. J. Pattern Recogn. Artif. Intell. **24**(3), 457–474 (2010)
19. Saif, A.F.M.S., Prabuwono, A.S., Mahayuddin, Z.R.: Moving object detection using dynamic motion modelling from UAV aerial images. Sci. World J. **2014**, 1–12 (2014). https://doi.org/10.1155/2014/890619
20. Budiharto, W., Gunawan, A.A.S., Suroso, J.S., Chowanda, A., Patrik, A., Utama, G.: Fast object detection for quadcopter drone using deep learning. In: Proceedings of the 2018 3rd International Conference on Computer and Communication Systems (ICCCS), pp. 192–195 (2018). https://doi.org/10.1109/CCOMS.2018.8463284
21. Okutama-action: an aerial view video dataset for concurrent human action detection. In: Proceedings of the IEEE Conference on Computer Vision and Pattern Recognition Workshops, pp. 28–35 (2018)

DRL Based Multi-objective Resource Optimization Technique in a Multi-cloud Environment

Ramanpreet Kaur[1,2(✉)], Divya Anand[3], and Upinder Kaur[4]

[1] Department of Computer Application, Lovely Professional University,
Phagwara, Punjab, India
[2] Department of Computer Science, Baba Farid College, Bathinda, Punjab, India
ramaninsa1990@gmail.com
[3] Department of Computer Science and Engineering, Lovely Professional University,
Phagwara, Punjab, India
Divyaanand.y@gmail.com
[4] Department of Computer Science and Engineering, Akal University,
Talwandi Sabo, Punjab, India
upinder_cs@auts.ac.in

Abstract. The concept of multi-cloud becomes interesting progressively to cloud users because of its high response time, flexibility, high throughput, and reliability. But at the ground level, the concept of multi-cloud creates many challenges for researchers. The request of users and the multi-cloud environment is heterogeneous now a day. To work in this kind of environment required an intelligent system. Researchers are doing well in this field to make the whole process very flexible by providing an intelligent environment. The proposed multi-objective resource optimization deep reinforcement learning (MOROT-DRL) model uses the Q-learning technique of Deep Reinforcement Learning (DRL) to allocate resources in a multi-cloud environment. It includes a service analyzer for analyzing the requests and MET(Minimum Execution Time)algorithm used for scheduling the task according to execution time and then enhanced flower pollination allocate the optimized resources for the demanded request. The comparison of the proposed model is done with simulation results of MOROT and neural network model and also implemented on GoCS real dataset of google. The proposed model gives better results when compared based on energy, CO_2, and cost.

Keywords: Multi-cloud · Deep reinforcement learning · Resources allocation · Cyber shake seismogram workflow · Task scheduling Enhanced Flower Pollination

1 Introduction

Based on daily consumer need the demand for various services like SaaS, PaaS, and IaaS, promoting everyone to in the environment having multiple cloud. The concept of multi-cloud is used when the services are fulfilled from the various cloud or the services are moved from one cloud to another cloud. The author presents a taxonomy related to

© ICST Institute for Computer Sciences, Social Informatics and Telecommunications Engineering 2024
Published by Springer Nature Switzerland AG 2024. All Rights Reserved
P. Pareek et al. (Eds.): IC4S 2023, LNICST 536, pp. 106–121, 2024.
https://doi.org/10.1007/978-3-031-48888-7_9

multi-cloud mentioned that the main actor that works in multi-cloud is Cloud Service Provider (CSP) [1]. Cloud computing provides high performance and a large number of services to the cloud user based on pay–per–usage. The cloud resource is distributed in various locations and connected according to a geographical area. These services are not allocated physically to the user but rather can be used on a rent based. Cloud user works with multi-cloud without knowing the physical location of the cloud which is fulfilling their request, same as virtual machines are also unknown to the user's location. Service Level Agreement (SLA) is an important part between cloud users and cloud providers because it deals with important parameters such as quality, pricing, security, etc. [2]. So to fulfill the quality of services based on SLA, there is a requirement for efficient techniques of resources management. Various researchers are doing work in this field by providing various features of optimized resource allocation techniques.

Task scheduling is also very important with optimized resource allocation. Task scheduling means scheduling the incoming task in such a manner that efficient resources can be allocated. Various task scheduling approaches are doing well in this field. The suitable technique can be adopted based on the suitable requirement of cloud users or the resources allocation technique. Lots of work is done for resource allocation and task scheduling in cloud computing. But the same concept becomes very complicated in the case of a multi-cloud environment.

As the services and users in multi-cloud are increased, the concept of task scheduling and resource allocation becomes a challenge for researchers. DRL technique of machine learning performs the best role in every IT-related field. It can make a very complex decision which was not possible in previous machine learning. In the cloud and multi-cloud, DRL also performs an important role in traffic, identification, future prediction, task scheduling, and resource allocation (Fig. 1).

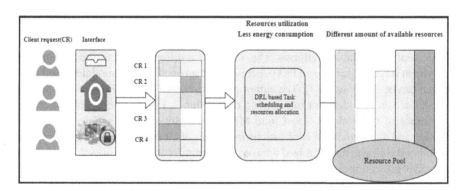

Fig. 1. Representation of resources allocation in multi cloud environment.

The performance of cloud or multi-cloud can be measured based on performance indicators or some parameters such as makespan, reliability, throughput, time, cost, power, and carbon emission. Deep reinforcement learn in intelligence elegant techniques are performing an amazing role to improve resource allocation in a multi-cloud environment without any SLA violation.

Poor resource allocation or under-utilization of resources is responsible for high consumption of energy, cost, and some other precious parameters. High energy consumption leads to a high ratio of CO_2 and both deeply affects social life. The balance between both is a very challenging concept. The future of any country depends on a healthy citizens. The deep reinforcement learning technique intelligently allocates resource to the user from more than one cloud to fulfil their requests having all benefits as energy efficiency, cost and CO_2 reduction. The enhanced flower pollination algorithm (EFPA) works in the particular cloud for resource allocation based on local and global optimization techniques for energy efficiency. MET algorithm and cybershake seismogram workflow segregate the incoming requests and prepare a task queue for EFPA so that requests can be processed based on minimum execution time. This model is evaluated with cloudsim simulator. The remainder part of this paper includes as following. Section 2 is the related work and parameter-based study of task and resource allocation in the multi-cloud. Section 3 is the explanation of the proposed model having a flowchart and algorithms. The experiment results discussion and comparison of proposed work is in Sect. 4.

2 Related Work

This section gives a basic review of various task scheduling and resource allocation techniques. Authors doing well to reduce various issues in the field of cloud and multi-cloud environments.

Predicting the future user's requirements in the cloud helps to reduce violations in service level agreements. But it is very difficult to predict the future requirement in the case of multi-cloud. Future prediction helps the cloud provider to allocate quality of service to the user. The author proposed a hybrid approach for future requirement prediction. This approach uses lazy learning, modified K-medoids, and lower bound dynamic time warping. This proposed approach gives better prediction when compared with others [3]. Resources management in multi-cloud needs a unique interface and wrapper for every service. The author proposed an approach that is adopted by deployable services in terms of open sources available platforms. This interface is different at run time, design, and deployment stages. The focus of this paper is to give an open-source, module-based solution that can be easily used [4]. The author presented how Cloud MF makes techniques of model-driven and ideas for minimizing the vendor lock-in and helps for allocations applications of multi-cloud. The Cloud ML (Cloud Modelling Language) permits to provision and deployment of applications in models of cloud provider-independent [5]. The author developed an optimized approach to minimize micro services repair, latency overhead of allocating containers on the cloud and reduce services cost. As micro-service are arranged in a container and that container will be allocated to VM but how to allocate the container on a suitable VM and allocate VM on a suitable cloud is a challenging issue. The author implements the NSGA-II genetic algorithm and compares it with the Greedy First-Fit algorithm; the implemented algorithm gives 300% improvement as compared to others [6].

Services provided by the cloud make every task very flexible as a business also moved toward the cloud and getting more benefits. To work on a single cloud is very efficient. But difficult in the case of multi or cross-cloud. To handle different instances of business processes near customers can be beneficial; the author presents a novel architecture of the environment of multi-cloud business provision. This architecture involves components to handle the monitoring and adaption of the business processes in a multi-cloud environment. This framework explains all about services such as Iaas, PaaS, monitoring, adoptions, etc. that can help to do business processes in a better way [7]. The author presented Replica aware task scheduling method to reduce the response delay of services. According to this algorithm, transferring computation and transfer data are combined. Resources matching are accomplished according to the availability of nodes. Failed or non-local data is replicated in advance to the targeted node. According to the cache placement algorithm, the next execute task is predicted. The experiment result shows that our proposed method performs better as compared to the benchmark algorithm in terms of node prediction and response time [8].

An integer linear programming model is developed to handle the scientific workflow in a multi-cloud environment. This helps to reduce financial costs by encountering the deadline requirement of the user. In this proposed model the resource limit imposed by the cloud provider and cost is calculated on an hourly basis. The experiment results show that change in deadline and workflow affect cost i.e., greater in CPU intensive workflows rather than other elasticity values remain always a constraint in work under long deadlines. A short deadline has a high cost. The comparison of the proposed model is done with the MIP-CG, MCPCPP, and IP-FC. The result shows that the proposed model is suitable for all deadlines; in the future, the makespan and total cost should be considered [9].

The author proposed Multiple-replica integrity auditing schemes for secure data storage on the cloud. A Cloud users are continuously taking data storage services on the cloud free from cost burden. The Author also mentioned open issues and research directions [10]. The author presents a hybrid formal verification approach for accessing high-quality service composition in the environment of multi-cloud by reducing the no. of the cloud provider. The proposed approach is helpful for checking the user request, services selection, and multi-cloud composition. Results show that this method reduces memory consumption [11]. This paper investigates resource management in a multi-cloud environment. The author also investigates the user's demand for applications in a multi-cloud environment. Definition and resources classification in the multi-cloud environment and three taxonomy of multi-cloud are mentioned. Future trends and challenges also point out [12]. In this paper, the author focus on scheduling techniques that handle challenges in inter-cloud and also presents basic concerns and task scheduling related to multi-cloud. All scheduling techniques are categories based on some parameters. After containing the survey, the author mentions that security and load balancing is an important concern that should be considered in the future [13]. This author proposed a cloud-enabled workflow science gateway. This paper includes all the principles of integrating the cloud system with a science gateway. It integrates the WS-PGRADE/g USE and cloud broker platform. This integration is used in the CloudSME project where 20 companies port the simulation application on the cloud. The proposed method provides cloud access flexibility and the user can access all clouds integrated with this gateway [14]. During the use of multi-cloud, the user has to face some challenges such as provisioning, elasticity, portability, and availability. To handle these challenges the author presented the so cloud framework that is deployed on 10 cloud providers' complete architecture and interaction between all components of the so Cloud framework discussed in this paper. With this approach, the user can get high availability [15].

The author focuses on the problem of VM placement for reducing cost and saving energy in a heterogeneous environment of multi-cloud. The author proposed a mimetic algorithm for VM placement based on cost-efficient to solve this problem i.e. called grouping genetic. The proposed algorithm reduces running PM and consumption of energy by the geographical distribution of the data center. Hill climbing is also used do searching in local to maximize the speed and run time of the genetic algorithm. Comparison of the proposed model is made with three other recent researches and found that the proposed model performs better to reduce cost and energy [16]. In this paper, the author developed normalized hybrid service brokering With Throttled Round Robin Load Balancing (NHSB_TRB) to provide cost-effective services to the user. This approach produces a normalized value of optimized cost. The data center based on cost is selected for distributing the load. The weighted threshold is used to distribute the load on the data center and the round-robin load balancing approach distributes the load on VM. The experiment result shows that the proposed model improves response time, monetary cost, and processing time of data center up to 17.39%, 7.06%, and 31.35 when compare with ORT_RR, CDC_RR, ORT_THR, and ORT_ES approaches [17]. Table 1 compares task scheduling and resources allocation techniques in a multi-cloud environment based on a parameter such as Makespan, cloud/resource, utilization, scalability, cost, Response Time, Energy efficiency, and Co2 reduction.

The above table shows that not a single technique works collectively on cost, energy, and Co_2. So the same should be considered in research work.

3   Proposed Methodology

With the time-varying demands of construction, optimization, and user requests, a Cloud computing system cannot be considered self-sufficient. The internal environment for agent state and decision-making are also temporal shifting like Cloud computing's income ratio, which varies in different periods during a single day [31]. A DRL framework with time-varying external stimuli is used to view the decision-making process as an ensemble. Each mapper's information is as follows: Extrinsic or internal stimuli are inputs to the mapper of time-varying, whereas the output is a change in the agent's state or state-changing. It is a mapper of stimulus evolution in agent and state, where the stimulus force is input and the set of agent-state at real-time is output, as the agent and state are usually changing with stimuli.

When it comes to managing resources in a cloud computing environment, resource allocation (RA) is one of the methods available. When it comes to establishing the optimal balance between VM and PM in the cloud data center, Infrastructure as Service providers confronts a significant difficulty. This is known as finding the optimal allocation for VMs and PMs in terms of the number of resources they require [32]. There are two components to resolving the issue of VM allocation: first, the acceptance of new requests for VM provisioning and the placement of the VMs on a PM, and second, the optimization of existing VMs (Fig. 2).

Table 1. Task scheduling and resource allocation techniques in a multi-cloud environment.

Ref. No.	Technique	Makespan	resources utilization	Scalability	Cost	Response Time	Energy efficiency	CO_2 Reduction
SK Panda [18]	AMinB, AMaxB, and AMinMaxB task scheduling algorithm	✔	✔	✗	✗	✗	✗	✗
Lijin P [19]	Game theory-based resources allocation algorithm	✔	✗	✗	✗	✗	✗	✗
A. Pietrabissa [20]	Q-learning(Policy reduction and state aggregation strategy)	✗	✗	✔	✗	✗	✗	✗
SK Mishra [21]	Energy-Aware Task Allocation in Multi-Cloud Network	✗	✗	✗	✗	✗	✔	✗
J.Carvalho [22]	Simple Adaptive weighting method and multi-choice knapsack problem	✗	✔	✗	✔	✔	✗	✗
P.Antonio1 [23]	SARSA(λ) and Q-learning	✗	✗	✗	✗	✗	✗	✗
S.Kang [24]	DSS	✗	✗	✗	✗	✗	✗	✗
Z.Chen [25]	OWS-A2C	✗	✔	✗	✔	✗	✗	✗
SK Panda [26]	MCC,MEMAX,CMMN Algorithm	✔	✔	✗	✗	✗	✗	✗
M.Farid [27]	FR-MOS	✔	✔	✗	✔	✗	✗	✗
T.Subramanian [28]	Novel cloud brokering architecture	✗	✗	✔	✗	✗	✗	✗
C. Thirumalaiselvan [29]	ELB, high priority scheduling algorithm, and rate-based scheduling algorithm	✔	✗	✗	✗	✔	✔	✗
N.Grozev [30]	Rule-based domain-specific model	✗	✗	✗	✗	✗	✗	✗

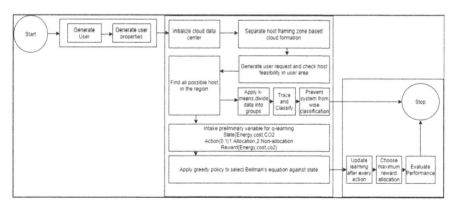

Fig. 2. Workflow of DRL in a cloud environment.

The decision mapper is in charge of computing the next action based on the agent's present state, and the action taken at the next opportunity is the output. To put it in layman's terms: The environment is fed by actions provided by a mapper and grows as a result of those actions. Environment's output is fed into mapper of feedback, while agent's internal stimuli are fed into mapper of time-varying [33–35]. Replay storage is used to store long-term input in preparation for future use, while timely feedback is taken from the environment. The settings of the decision-maker will be updated as a result of both long-term and real-time feedback. Generalized RL built on the integration of mappers. Programs are often used to model time-varying, stimulus evolution, and environmental conditions in some research. Neural networks can be used to build a decision and feedback mapper. The feedback mapper can be implemented as a neural network to calculate the loss function of the neural network in the decision mapper since it aims to update the parameters of the decision-maker. In computing, a VM (virtual machine) is an emulation of a certain computing system that is used to simulate another system. Virtual machines are capable of running since they are based on the computer architecture and functionalities of a real or physical computer. These systems can be implemented using specialized hardware and software, or they can be implemented using a combination of both [36]. Virtual machines can be divided into several categories based on how closely they resemble their real-world counterparts in terms of capability (Fig. 3).

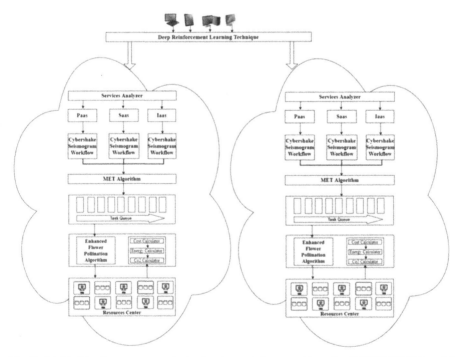

Fig. 3. Multi-objective resource optimization deep reinforcement learning (MOROT-DRL) model

As a result, system virtual machines can serve as a complete substitute for the targeted virtual machines, as well as providing the amount of functionality required to operate an operating system (also known as full virtualization VMs). While this is true, a process virtual machine provides an abstracted and platform-independent execution environment for a single computer application running on a variety of platforms. The centralized cloud resource manager is in charge of managing the resources in cloud computing. Cloud data center (DC) resources are made available to cloud consumers using virtual machines (VMs), which are based on physical machines (PMs). To maximize resource use while minimizing energy consumption, cloud computing Infrastructure as a Service (IaaS) providers must implement dynamic resource management techniques in their cloud DCs. Because of business considerations, the resource management strategies and algorithms used in public clouds are not revealed. The proposed energy-efficient resource allocation mechanism is comprised of a single central scheduling point (CSP) and N cloud users. CSP managed many heterogeneous resources like memory, processing units, network bandwidth, and so on in the form of virtual machines (VMs). When these virtual machines (VMs) were requested by cloud customers to complete their activities, the cloud provider allocated them based on the preferences of the cloud consumers. The primary purpose of this proposed Enhanced flower pollination algorithm is to reduce the amount of energy consumed during work scheduling in the cloud environment, as well as to reduce the number of task scheduling issues in cloud computing.

Algorithm 1: Enhanced flower pollination algorithms
1: Start
2: Input
Data center structure D_s
Size of Population S_m
Total number of takes $S_{m, task}$
Number of iterations i_{max}
Number of Virtual machines $S_{m, vm}$
3: Output
   Optimal Solution Y_a
4: Calculate the global task queue
$Y_q^{j+1} = Y_q^j + \Upsilon(\Psi)(S - Y_q^j)$
5: Calculate the local task queue
$Y_q^{j+1} = Y_q^j + \upsilon u(Y_i^j - Y_m^j)$
6: finally find out the Optimal Solution
   $Y_a, a = 1,2,3,\ldots\ldots\ldots S_m$
      $i = i+1;$
7: Update and repeat the Optimal Solution
8: Select the best solution
9: Stop

Increased energy use results in an increased operating expense. The most pressing issue is the increase in excessive carbon emissions ($CO2$). It has a greater impact on the environment. This limited supply must be put to good use. The most critical step is to reduce the use of energy and power. It is important to avoid wasting resources. This is referred to as energy conservation. The efficient use of resources can be improved by utilizing virtualization technologies. The dynamic consolidation of virtual machines (VMs) is made possible by virtualization technology. Cloud service providers can host several virtual machines (VMs) on a single physical server, i.e. virtualization. One technique to reduce power usage is to turn off nodes that are not in use.

E_{CE} is a metric that measures how much energy is expended while a job is in the process of being set up for execution, such as copying the data needed for the task to run. The amount of CPU energy used to carry out the task in the designated VM is therefore considered as an E_{PE}, which is directly related to the amount of CPU energy utilized. Thus, the total amount of energy consumed by the user jobs in the Task Set may be calculated using Eq. 1.

$$E_{Total} = \sum E_{VM} \tag{1}$$

Here E_{VM} is known as

$$E_{VM} = E_{CE} + E_{PE} \tag{2}$$

Algorithm 2: Task Scheduling Algorithm for Deep reinforcement learning
1: Start
2: evaluate the resources information of task
3: Set E_{VM} task.
4: Evaluate E_{total}
   $$E_{Total} = \sum E_{VM}$$
   $$E_{VM} = E_{CE} + E_{PE}$$
5: Do till all task mapped
6: Earliest completion time and resources of all task are calculated
7: Set resource's ready time
8: According completion time set all resources
9: Do for all R
10: calculate highest Completion Time of Ti
If Maximum Completion Time <makespan
Compute makespan = max(CT(R))
11: Figure out task having minimum completion time.
12: Find out T_i with minimum ET
13: Reschedule task T_i according to produced resources.
14: Change ready time for those resources
15: End

4 Results and Discussion

To evaluate MOROT-DRL model, 'cloudsim' cloud computing simulation and Q-learning technique of DRL implemented for services analyzing. Go CS dataset is used as real cloud data set and comparison is also made on some well known websites dataset. The simulation program is written in Java and deployed on an HP i7 processor with 16GB RAM. The apache-commons mathematics library is used to generate the power versus throughput regression model for the servers. The Apache Net Beans used as a tool to open JDK 8, powered with an open V9 Java virtual machine used to run our code. These types of resources were considered for the testing CPU, memory & disk. A cloud data center information and customer configuration information are given in Tables 2 and 3.

Table 2. Data center information and customer configuration information.

Sr. No.	Characteristic	Value
1	Number of data centre	10
2	Number of hosts	200
3	Available bandwidth	100–7500 Hzs
4	Available Core	4
5	Capacity per core	3 octa engine
6	Engine Type	Multi
7	Engine Propagation	Quad Core
8	Process Utilization Minimum	1 Hzs
9	Single Core score	14323
10	Multi Core Score	14883
11	Engine Ram	2 GB

In this paper,a virtual environment is simulated to check the efficiency of the proposed method in terms of resources allocation.

The user requests the resources from the data center according to the requirement of the task. Here the CR is the Client request which is fulfilled with the help of a virtual machine VM. CR requests many resources at the same time.

$A_i \subset \mathbf{CR}$ and $x_s^1, y_s^1, z_s^1 \subset A_i$,

$x_s^1, y_s^1, z_s^1 \subset A_i \subset \mathbf{CR} \Rightarrow x_s^1, y_s^1, z_s^1 \subset \mathbf{CR}$,

When users request only one resource it can be written as Eqs. (3) and (4):

$$\mathbf{CR}^1 = A_i, \tag{3}$$

I mean 1 and $\mathbf{CR}^1$ means user request only 1 resource.

$$A = (x_s^1 + y_s^1 + z_s^1) \tag{4}$$

Table 3. Notation in mathematical analysis.

Symbol	Definition
VM	Virtual Machine
CR	Client Request
A_i	Component of VM
X	Represents CPU
Y	Represents Memory
Z	Represents Storage
I	Number of Resources
S	Measuring Capacity
PM	Physical Machine
St_i	Starting Time of each VM_i
Et_i	Execution Time of each VM_i
RU^{DC}	Resource Utilization of data center
RU	Resource Utilization
TEC	Total Energy Consumption
EU	Energy Utilization
P	Power Consumption
DC_U^{Energy}	Energy Consumption of Data Center
T	Total Power Generated
S_k	Power Generated by Source k
ef_k	Emission Factor Related to k

When a user demands more than one resource it can be expressed as Eqs. (5) and (6).

$$CR^n = \sum_{i=1}^{n} = A_i = A_1 + A_2 + A_3 + \ldots A_n$$
$$= (x_s^1 + y_s^1 + z_s^1) + (x_s^2 + y_s^2 + z_s^2) + \ldots + (x_s^n + y_s^n + z_s^n) \tag{5}$$

$$CR^n = \sum_{i=1}^{n} (x_s^1) + \sum_{i=1}^{n} (y_s^1) \sum_{i=1}^{n} (z_s^2) \tag{6}$$

Energy and Resource Utilization Model

VM $= \{vm_i, i = 1, 2, \ldots, n\}$ are virtual machine allocated to Physical machine PM.

PM $= \{PM_j, j = 1, 2, \ldots, m\}$ are physical machines.

Three PM resources considered as physical memory (RAM), storage and processor (CPU).

So dimension d $= 3$.

The total time used during VM allocation is $S_{ti} + E_{ti}$.

where $A_{i,s}$ is resource capacity (x_s^1, y_s^1, z_s^1) requested by the **VM**$_i(1, 2,..., n)$ and $B_{i,s}$ resource capacity (x_s^1, y_s^1, z_s^1) of the **PM**$_j$ $(1, 2, ..., m)$.

The requirement of resources allocations are:

1: Resource should be according to the request.

2: ∀**VM**s demand is ≤ the **PM**s' total capacity of resources.

3: ∀**VM**s each **VM** is operator by each PM according to time.

4: Assume that $a_j(t)$ is group allocated to **PM**.

$\sum_{i=1}^{n}$ **CR**^{n} of these assigned **VM**s is ≤ the **PM**s' total resource capacity.

$$\text{For all capacity } (\forall s) = 1, \ldots, d \cdot \sum\nolimits_{vmi \in aj(t)} A_{i,s} \leq B_{i,s} \tag{7}$$

Cost of Resources Utilization of PM

$$RU_j^d = \frac{\sum_i^n = p_{ij} \times VM_i^d}{PM_j^d} \forall d \in \{x_s^1, y_s^1, z_s^1\} \tag{8}$$

Cost of Resources Utilization of Data Center

$$RU^{DC} = \int_{t1}^{t2} \frac{\sum_{j=1}^m U_j^{x_s^1} + \sum_{j=1}^m U_j^{y_s^1} + \sum_{j=1}^m U_j^{z_s^1}}{|d| \sum_{j=1}^m p_j^i} \partial t, \tag{9}$$

where $Ai = \begin{cases} 1 \text{ if } \mathbf{CR} \text{ assigned to } Ai \\ 0 \text{ otherwise} \end{cases}$

We know every PM$_j$ can host on any VM$_i$ and the model for energy consumption $P_j(t)$ for PM$_j$'s host has a linear relationship with resource utilization (as if the utlization will increase it will also effect on energy consumption [12]. The formula for achieving these parameters are given below.

Total Energy Consumption (TEC)

At time t utilization is EU $(t)_j$ and energy consumption of EU(t) is depicts as P (EU (t))

$$TEC_j^x = \int_{t1}^{t2} P\left(EU(t)\right) d\, t\, PM_j \in P, \tag{10}$$

$$\text{Maximize} \sum\nolimits_{j=1}^{m} = EU(t)_j \tag{11}$$

$$TEC_j^x = \sum\nolimits_{j=1}^{m} EU(t)j, \tag{12}$$

EU(t)j with j = 1, 2, ..., m is the total consumed energy of the PM$_j$.
$i \in \{1,2,..., n\}, j \in \{1,2,..., m\}, [0;], t \in [0; t]$.

Energy Consumption of Data Center

$$DC_U^{Energy} = \int_{t1}^{t2} P(EU(t))(x)dt + EU(y) \times P_{max} + EU(z) \times P_{max} \tag{13}$$

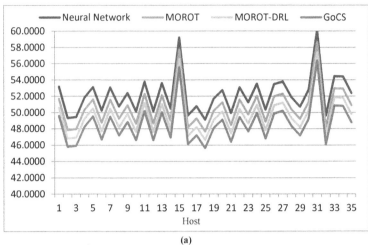

(a)

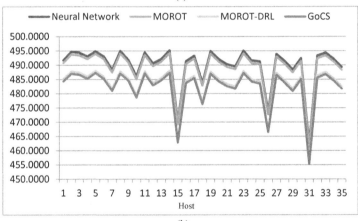

(b)

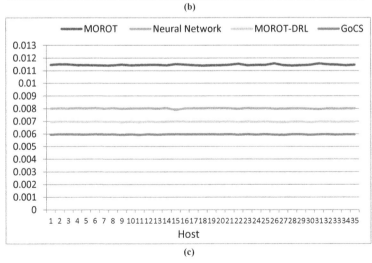

(c)

Fig. 4. (a). Comparison based on Energy consumption. (b). Comparison based on cost. (c). Comparison based on CO_2 emission

The efficiency of R^A helps to increase utilization of resources (RU) and the energy efficiency of the cloud datacenter can be generated as follows in Eq. (13).

$$\text{Maximize} = RU_j^d \frac{\sum_i^n = p_{ij} \times VM_i^d}{PM_j^d} \forall d \in \left\{x_s^1, y_s^1, z_s^1\right\} \text{s.t.Minimizes} \sum_{j=1}^m EU(t)j, \quad (14)$$

Carbon Emissions. Total power generated is T, the power generated by the Kth source is S_k and ef_k is the emission factor related to K^{th} source, the total emission factor of the cloud data center can be calculated as follow.

$$ef = \frac{k}{i=1} \frac{Sk}{T} efk$$

The proposed model MOROT-DRL is compared with the three models MOROT, neural network and one more real cloud data set. When same model is implemented in GoCS set also give more better results in energy, co2 and cost reduction.. Following Table 2 shows the Energy, cost and co2 evaluation in these entire three models (Fig. 4).

The above results show that the MOROT-DRL model performs better in the case of energy consumption, and CO_2 reduction as compared to the MOROT model and neural network and GoCS google data set in a real cloud environment. In the MOROT model resources are allocated to the user only in one cloud with the help of a minimum execution time algorithm, cyber shake seismogram, and enhanced flower pollination. The neural network is the technique of handling the incoming task in a cloud environment [37]. But the MOROT-DRL model mainly used the Q-learning technique for scheduling the incoming request in a multi-cloud environment. The q-learning handles the request while entering in the multi-cloud environment but in each cloud, the concept of cyber shake seismogram, MET, and enhanced flower pollination technique is used which efficiently works to allocate the optimized resources allocation based on the local and global method. So the proposed method gives good results as shown in graphs and tables.

5 Conclusion

Multi-cloud gives more elasticity to the users by combining multiple cloud domains and data centers. These features attract not only normal users but also the biggest companies and businesses. The requirement of cloud providers and cloud users are escalating gradually. The challenge is to handle the request and allocate the required resource. Researchers proposed many scheduling and resource allocation techniques which give good results in various parameters such as time, cost, throughput, reliability, etc. Some more are need to improve. So for this, we proposed the MOROT-DRL model implemented in real cloud environment which works on energy, cost, and CO_2 parameters. The Q-learning technique logically handles the incoming request and works as an intelligent model in a multi-cloud environment. Cyber shake seismogram workflow and minimum execution time algorithms create a queue based on minimum execution time and schedule the task in a specific cloud. The bio-inspired algorithm, i.e., enhanced flower pollination picks the task from the queue and allots the optimized resources with dynamic switching property, and local and global strategy. In the end, the comparison is made with MOROT and neural network model and GoCS, our proposed model performing superior in energy efficiency, CO_2 reduction, and cost evaluation on Gocs real data set also.

References

1. Petcu, D.: Consuming resources and services from multiple clouds. J. Grid Comput. **12**(2), 321–345 (2014)
2. Panda, S.K., Jana, P.K.: SLA-based task scheduling algorithms for the heterogeneous multi-cloud environment. J. Supercomput. **73**(6), 2730–2762 (2017)
3. Keshavarzi, A., Haghighat, A.T., Bohlouli, M.: Enhanced time-aware QoS prediction in multi-cloud: a hybrid k-medoids and lazy learning approach (QoPC). Computing **102**(4), 923–949 (2020)
4. Munteanu, V., Sandru, C., Petcu, D.: Multi-cloud resource management: cloud service interfacing. J. Cloud Comput. Adv. Syst. Appl. **3**, 1–23 (2014)
5. Ferry, N., Chauvel, F., Song, H., Rossini, A., Lushpenko, M., Solberg, A.: CloudMF: model-driven management of multi-cloud applications. ACM Trans. Internet Technol. **18**(2), 1–24 (2018)
6. Guerrero, C., Lera, I., Juiz, C.: Resource optimization of container orchestration: a case study in multi-cloud microservices-based application. J. Supercomput. **74**(7), 2956–2983 (2018)
7. Kritikos, K., et al.: Multi-cloud provisioning of business processes. J. Cloud Comput. Adv. Syst. Appl. **8**, 1–29 (2019)
8. Li, C., Zhang, J., Tang, H.: Replica-aware task scheduling and load-balanced cache placement for delay reduction in the multi-cloud environment. J. Supercomput. **75**(5), 2805–2836 (2019)
9. Mohammadi, S., Pedram, H., PourKarimi, L.: Integer linear programming-based cost optimization for scheduling scientific workflows in multi-cloud environments. J. Supercomput. **74**, 4717–4745 (2018)
10. Li, J., Lin, Y., Jia, X., Ren, K.: Multiple-replica integrity auditing schemes for cloud data storage. Concurrency Comput. Pract. Exper. **33**, 1(2019)
11. Souri, A., Rahmani, A., Rezaei, N.: A hybrid formal verification approach for QoS-aware multi-cloud service composition. Cluster Comput. **23**, 2453–2470 (2020)
12. Carvalho, J., Trinta, F., Vieira, D., Cortes, O.: Evolutionary solutions for resources management in multiple clouds: State-of-the-art and future directions. Futur. Gener. Comput. Syst. **88**, 284–296 (2018)
13. Masdari, M., Zangakani, M.: Efficient task and workflow scheduling in inter-cloud environments: challenges and opportunities. J. Supercomput. **76**, 499–535 (2019)
14. Bruno, R., Costa,F., Ferreira, P.: freeCycles - efficient multi-cloud computing platform. J. Grid Comput. **15**(1), 501–526 (2017)
15. Paraiso, F., Merle, P., Seinturier, L.: soCloud: a service-oriented component-based PaaS for managing portability, provisioning, elasticity, and high availability across multiple clouds. Computing **98**, 539–565 (2016)
16. Rashida, S., Sabaei, M., Ebadzadeh, M., Rahmani, A.: A memetic grouping genetic algorithm for cost-efficient VM placement in the multi-cloud environment. Clust. Comput. **23**(2), 797–836 (2020)
17. Khan, M.: Optimized hybrid service brokering for multi-cloud architectures. J. Supercomput. **76**, 666–687 (2020)
18. Panda, S., Gupta, I., Jana, P.: Task scheduling algorithms for multi-cloud systems: allocation-aware approach. Syst. Front. **21**, 241–259 (2019)
19. Lijin, P.: Resource allocation in multi-cloud based on usage logs. Int. J. Sci. Res. Comput. Sci. Eng. Inf. Technol. IJSRCSEIT **3** (2018)
20. Pietrabissa, A., Priscoli, F., Giorgio, A., Giuseppi, A., Panfili, M., Suraci, V.: An approximate dynamic programming approach to resource management in multi-cloud scenarios. Int. J. Control **90**, 492–503(2016)

21. Mishra, S., et al.: Energy-aware task allocation for multi-cloud networks. IEEE Access **8**, 178825–178834 (2020)
22. Carvalho, J., Vieira, D., Trinta, F.: Dynamic selecting approach for multi-cloud providers. In: Luo, M., Zhang, L.-J. (eds.) CLOUD 2018. LNCS, vol. 10967, pp. 37–51. Springer, Cham (2018). https://doi.org/10.1007/978-3-319-94295-7_3
23. Antonio, P., et al.: Resource management in multi-cloud scenarios via reinforcement learning. In: Proceedings of the 34th Chinese Control Conference, pp. 28–30 (2015)
24. Kang, S., Veeravalli, B., Aung, K.: Dynamic scheduling strategy with efficient node availability prediction for handling divisible loads in multi-cloud systems. J. Parallel Distrib. Comput. **113**, 1–16 (2018)
25. Chen, Z., Lin, K., Lin, B., Chen, X., Zheng, X., Rong, C.: Adaptive resource allocation and consolidation for scientific workflow scheduling in multi-cloud environments. IEEE Access **8**, 190173–190183 (2020)
26. Panda, S., Jana, P.: Efficient task scheduling algorithms for the heterogeneous multi-cloud environment. J. Supercomput. **71**, 1505–1533 (2015)
27. Farid, M., Latip, R., Hussin, M., Hamid, N.: Scheduling scientific workflow using multi-objective algorithm with fuzzy resource utilization in multi-cloud environment. IEEE Access **8**, 24309–24322 (2020)
28. Subramanian, T., Savarimuthu, N.: Application-based brokering algorithm for optimal resource provisioning in multiple heterogeneous clouds. Vietnam J. Comput. Sci. **3**(1), 57–70 (2016)
29. Thirumalaiselvan, C., Venkatachalam, V.: A strategic performance of virtual task scheduling in multi cloud environment. Cluster Comput. **22**, 9589–9597 (2019)
30. Grozev, N., Buyya, R.: Regulations and latency-aware load distribution of web applications in multi-clouds. J Supercomput. **72**, 3261–3280 (2016)
31. Zhan, W., et al.: Deep-reinforcement-learning-based offloading scheduling for vehicular edge computing. IEEE Internet Things J. **7**(6), 5449–5465 (2020)
32. Qi, Q., et al.: Knowledge-driven service offloading decision for vehicular edge computing. a deep reinforcement learning approach. IEEE Trans. Veh. Technol. **68**(5), 4192–4203 (2019)
33. Wang, Y., et al.: Multi-objective workflow scheduling with deep-Q-network-based multi-agent reinforcement learning. IEEE Access **7**, 39974–39982 (2019)
34. Baer, S., Bakakeu, J., Meyes, R., Meisen, T.: Multi-agent reinforcement learning for job shop scheduling in flexible manufacturing systems. In: 2019 Second International Conference on Artificial Intelligence for Industries, pp. 22–25 (2019)
35. Zhang, L., Wang, Q., Sun ,H., Liao, J.: Multi-task deep reinforcement learning for scalable parallel task scheduling. In: 2019 IEEE International Conference on Big Data (Big Data), pp. 2992–3001 (2019)
36. Shetty, C., Sarojadevi, H., Prabhu, S.: Machine learning approach to select optimal task scheduling algorithm in cloud. Turkish J. Comput. Math. Educ. **12**(6), 2565–2580 (2021)

Aspect Level Sentiment Analysis to Extract Valuable Insight for Airline's Customer Feedback and Reviews

Bharat Singh$^{(\boxtimes)}$ (iD) and Nidhi Kushwaha (iD)

CSE Department, Indian Institute of Information Technology, Ranchi, India
{bsingh,nidhi}@iiitranchi.ac.in

Abstract. In the realm of decision-making, the internet plays a vital and pervasive role, serving as a conduit for individuals worldwide to express their perspectives and viewpoints through various online platforms such as blogs and social media. Consequently, the internet has become inundated with a vast array of both pertinent and extraneous information, presenting a formidable challenge in sifting through the abundance of content to extract the desired information. Sentiment analysis emerges as a valuable tool for addressing this issue, enabling the systematic analysis of each document to discern the prevailing sentiment expressed within. This holds particular relevance in the realm of customer decision-making, as it empowers individuals to make informed choices when selecting the most suitable US airline by evaluating the opinions shared by other customers on online review platforms like Skytrax and micro-blogging sites such as Twitter. We can use these kinds of datasets to provides the aspect level sentiment analysis. Therefore, we have explored, in this article, a language model built upon a pretrained deep neural networks capable of analyzing the sequence of text to classify it as having positive, negative or neutral emotions without explicit human labelling. To analyze and assess these models, data from Twitter's US airlines sentiment database was used. Experiment on above data set show BERT model to be superior in accuracy while being more significant in less time to train. We observe notable advancements over prior state-of-the-art methods that use supervised feature learning to close the gap.

Keywords: Sentiment Analysis · Decision Making · Airline Data · Social Media · BERT · Machine Learning

1 Motivation

Numerous social media platforms, including Facebook, WhatsApp, LinkedIn, Twitter, Google Plus, YouTube, and Instagram, have gained widespread popularity [1–3]. Millions of users actively engage with these platforms to share their opinions and perspectives. When individuals plan to book tickets, they often rely on the ratings and feedback available on social media sites like Twitter and Facebook to inform their decision-making process. Consequently, companies are interested in employing techniques or tools that

P. Pareek et al. (Eds.): IC4S 2023, LNICST 536, pp. 122–134, 2024.
https://doi.org/10.1007/978-3-031-48888-7_10

can effectively analyze passenger feedback. One such technique is the sentiment analysis [4–6].

Sentiment analysis is a very active area of research in natural language processing that allows for the extraction of opinions from a set of documents. Sentiment analysis can be investigated at various levels [4, 7, 9, 21]. Different machine learning (ML) algorithms have been utilized to determine the most suitable algorithm for the specific problem] [10, 11, 21]. The performance evaluation involved analyzing the confusion matrix and accuracy of these algorithms. To gain valuable insight from a large number of reviews, the reviews must be categorized into positive and negative sentiment. Sentiment analysis, also known as opinion mining, is a natural language processing technique that involves determining the sentiment or emotional tone expressed in a piece of text [12, 13]. It aims to understand and classify the subjective opinions, attitudes, and emotions conveyed by individuals or groups towards a particular topic, product, service, or event. Sentiment analysis can be applied to various forms of text data, including social media posts, customer reviews, survey responses, and news articles. It helps businesses, organizations, and researchers gain insights into public opinion, customer feedback, and brand reputation, enabling them to make informed decisions, improve products or services, and tailor marketing strategies [9, 10].

Sentiment Analysis was used to categories over 9,000,00 reviews into positive and negative sentiments in the proposed work. For review classification, the Nave Bayes and Decision Tree (DT) classification models were used. Sentiment analysis has a wide range of applications, from determining customer attitudes towards products and services to determining voters' reactions to political advertisements [2, 14, 15]. Twitter is being widely used daily by people over the years to express views and sentiments. In airline industry, large number of customers post their views regarding services of the airlines like bag lost, good food, flight delay and many others. This helps airlines cater customers based on their reviews. In this paper we classify the dataset of review sentiments as Positive, Neutral, and Negative using ML techniques [4, 9, 12]. The structure of the paper is as follow: in Sect. 2 literature review about sentiment analysis has given. The approach utilized to enhance the sentiment analysis, proposed framework and dataset details in Sect. 3. In Sect. 4 BERT model with pre-training. Result and analysis are given in Sect. 5 and conclusion in Sect. 6.

2   Literature Review

Sentiment analysis is a popular research topic in the field of natural language processing and has many applications in various industries. In this paper, four state of the arts classifiers, like DT, Logistic Regression (LR), Bayesian Naïve and Random Forest (RF), were used to compare the results of sentiment of text data over proposed BERT based sentiment analysis. In order to further enhance the accuracy and effectiveness of the sentiment analysis, it is important to explore the latest research and advancements in this area [7, 16–19]. Furthermore, V. Hatzivassiloglou et al. [20] proposes a method for predicting the semantic orientation of adjectives using a corpus-based approach. The authors introduce a novel algorithm for identifying the semantic orientation of adjectives based on the co-occurrence patterns of words in the corpus. Qiu et al. [20] proposes a novel method for dissatisfaction-oriented advertising based on sentiment analysis. The authors use a ML approach to identify customer dissatisfaction and propose targeted advertising strategies to improve customer satisfaction.

Furthermore, S. Tan et al. [5] presents an empirical study of sentiment analysis for Chinese documents. The authors compare the performance of several ML algorithms for sentiment analysis, including Naïve Bayes, SVM, and DTs. Sentiment analysis has gained significant attention due to its wide range of applications. It is used in social media monitoring to understand public opinion and brand perception, customer feedback analysis to gauge user satisfaction, market research to track consumer sentiment, and many other domains. Various techniques are employed for sentiment analysis, including ML algorithms such as Naïve Bayes, LR, RF, and Support Vector Machines. Deep learning models, including recurrent neural networks (RNNs) and convolutional neural networks (CNNs), have also shown promising results in sentiment analysis tasks. The performance of sentiment analysis models is evaluated using metrics such as accuracy, precision, recall, and F1-score, among others. Researchers have explored feature engineering, sentiment lexicons, linguistic patterns, and domain adaptation techniques to enhance the accuracy and robustness of sentiment analysis models [22].

S. Erevelles et al. [1] discusses the use of big data and sentiment analysis in consumer analytics and marketing. The authors highlight the importance of sentiment analysis in understanding consumer preferences and behavior and propose a framework for using sentiment analysis in marketing strategies. S.Tong et al. [16] presents a method for support vector machine active learning with applications to text classification. The authors propose a novel approach for selecting informative examples to label in order to improve the performance of the classifier. In literature a strong sentiment analysis has been done using ML models, but they are lack behind in the aspect level sentiment analysis that we hade done through the BERT method. Here, we proposed an NLP model with multiple embedding techniques based on ML. A transformer-based bidirectional encoder representation (BERT) for extracting latent linguistic features from airline ratings. This study uses MLand information visualization techniques to investigate how feedback affects customer satisfaction in various aspects of flight service. The unrated aspects of airline reviews are then predicted from the rated aspects.

3   Materials and Method

In this section, we discuss the techniques for our proposed framework. First of all, in Fig. 1 a framework has been shown which represents the adopted methodology. Feature extraction and embedding method were done on training and testing data. TF-IDF is a scoring measures to reflect how relevant a term in the given document. For the embedding purpose Glove has been utilized which encode the cooccurrence probability ration between two worlds.

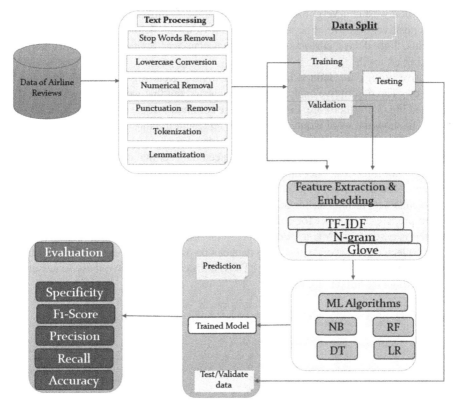

Fig. 1. The complete outline of our proposed framework

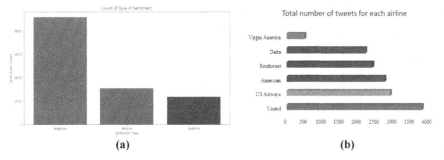

(a)   (b)

Fig. 2. (a) Graph showing number of negative, positive and neutral comments/review in the data sets. (b) Bar Graph representing the number of reviews for each airline, in the x-axis it is number of reviews and y-axis represent the name of airlines.

3.1 Data Set Description

The datasets used in this paper is taken from social media platform. Comments data that are included in this work are about six airlines i.e. Unites State, Delta, US Airways,

United, Southwest and Vergin America [8]. Passenger ratings are recorded and categorized as positive, negative or neutral. Negative reviews are defined based on things like bad flights, flight delays, customer service issues, damaged luggage, flight cancellations or booking issues [8].

Positive ratings are defined based on fast flights, great flights, great flights, good brands, etc. The descriptive analysis has been carried out that we have shown in Fig. 1, Fig. 2 and Fig. 3. Furthermore, Fig. 2 shows the comments of customers as a pie chart in (a) and (b) show the word cloud. Word cloud represent the most relevant keyword that are responsible for the positive and negative feedbacks.

Dataset used in this research is not a balanced data set that can be well understood from Fig. 1(a). It has a smaller number of positive comments in comparison to negative comments. The attributes of this datasets are tweet_id, airline_sentiment, Airline sentiment_confidence, airline, airline sentiment gold, name, retweet_count, location etc. In order to prepare the dataset for analysis, data preprocessing techniques were applied. This step is essential in ML to address potential issues arising from the nature of the dataset collected from social sites. Such data can be prone to inaccuracies and may lack certain attributes necessary for analysis. Thus, it is crucial to resolve these issues prior to conducting any further analysis.

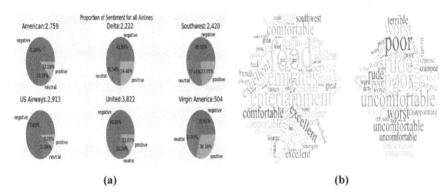

Fig. 3. (a) A pie chart showing the proportion of sentiments of all six airline companies. (b) word count of positive and negative feedbacks. It shows the important keywords used for both cases.

In pre-processing some required columns are selected and some common text processing algorithms are performed to: Remove empty reviews, convert all the reviews to lower case, remove numbers, tweet account names, website urls, special characters and white spaces. Figure 3 depicts the mood of passengers toward each airline companies. We observe that United, US Airways, American substantially get negative reactions and tweets for Virgin America are the most balanced.

3.2 Evaluation Metrics

We present the evaluation metrics used in our work. For the performance evaluation, we utilized widely accepted metrics for example Precision, Recall. F1-score, Sensitivity,

Specificity and Accuracy as given by Equations from (1)–(5) (Fig. 4).

$$P = TP/(TP + FP) \tag{1}$$

$$R = TP/(TP + FN) \tag{2}$$

$$F1 \text{ - Score} = 2 * P * R/(P + R) \tag{3}$$

$$S = TN/(TN + FP) \tag{4}$$

$$Acc = TP + TN/(TP + FP + FN + TN) \tag{5}$$

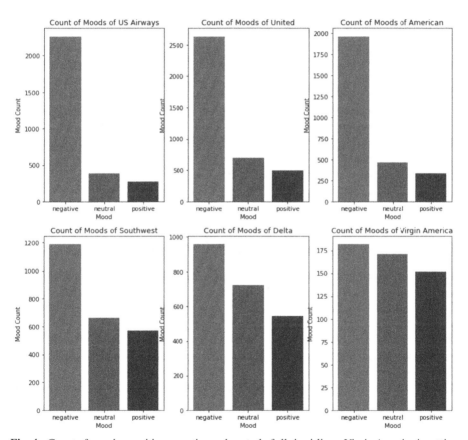

Fig. 4. Count of mood as positive, negative and neutral of all six airlines. Virgin America is getting balanced feedback however rest of airline companies getting substantially negative reaction.

3.3 Machine Learning Algorithms

We discuss four tradition ML methods that we have used in our study. Namely DT, LR, Naïve Bayes and RF. Here we are going for the briefing of these algorithm as these are the very well standard methods. As our motive was to analyze the aspect level sentiment analysis through ML algorithm. The analysis of results has been given in next section. RF [10] has demonstrated notable success in sentiment analysis tasks, outperforming various alternative ML methods. Its ability to handle high-dimensional data, manage noise, and capture complex relationships between features contributes to its effectiveness. Furthermore, its scalability and efficiency make it an attractive option for large-scale sentiment analysis applications. DT [19] have proven to be effective and interpretable models for sentiment analysis tasks. Their ability to handle both categorical and textual features, provide insights into feature importance, and offer robust performance makes them valuable in various application domains. However, challenges such as handling imbalanced data and adapting to evolving language patterns require further exploration and refinement.

Naïve Bayes, a probabilistic ML algorithm, has gained popularity due to its simplicity, efficiency, and competitive performance in sentiment analysis tasks. Its simplicity, competitive performance, and scalability make it a popular choice in various application domains. However, careful consideration of the feature independence assumption and its limitations in capturing complex relationships is essential for obtaining accurate sentiment analysis results [11]. LR, a widely-used statistical modeling technique, has shown promising results in sentiment analysis tasks. LR offers a well-established and interpretable approach for sentiment analysis tasks. Its ability to handle both binary and multiclass classification problems, along with its competitive performance in various application domains, makes it a valuable tool. However, its limited ability to capture complex nonlinear relationships and sensitivity to outliers should be considered when applying LR to sentiment analysis [9].

4   Proposed Model for Sentiment Analysis

BERT is a ML method based on transformers that Google developed for pre-training natural language processing (NLP). The Transformer language model, which has layers of self-aware heads and a variable number of encoders, is at the heart of BERT. The attention mechanism known as a Transformer, which is used by BERT, learns the contextual connections between words (or subwords) in text. Vanilla-style Transformers contain two separate mechanisms: an encoder that reads the text input and a decoder that creates predictions for the task [7, 13]. Since the purpose of BERT is to generate language models, we only need the Transformer's encoder mechanism. There are two variations of the pretrained BERT model. Both his BERT model sizes feature numerous encoder layers (referred to as transformer blocks in publications). 12 for the base version and 24 for the large version. as shown in Fig. 5(a). Also, the pre-training model of BERT has given in Fig. 5(b). BERT BASE and BERT LARGE refer to two different variations of the BERT model based on their model size and capacity.

BERT BASE has 12 transformer layers, 12 attention heads, and a hidden size of 768, resulting in a total of approximately 110 million parameters. On the other hand,

BERT LARGE has 24 transformer layers, 16 attention heads, and a hidden size of 1024, leading to around 340 million parameters. The larger model size of BERT LARGE allows it to capture more complex patterns and dependencies in the input data. During fine-tuning, BERT is further trained on specific downstream tasks with labeled data. This fine-tuning process adapts the pre-trained BERT model to perform task-specific operations, such as sentiment analysis, by adding task-specific layers on top of the BERT model. The fine-tuning stage allows the model to learn task-specific patterns and improve its performance on the target task. One key advantage of BERT is its ability to capture contextual information, which helps in understanding the meaning and nuances of words in different contexts. This contextualized representation is valuable for various NLP tasks, including sentiment analysis, as it allows the model to consider the surrounding words and sentences when making predictions. BERT BASE and BERT LARGE are pre-trained language models that leverage transformer-based architectures and self-attention mechanisms to capture contextual information. These models have been successfully applied to various NLP tasks, and their performance can be further enhanced through fine-tuning on specific downstream tasks.

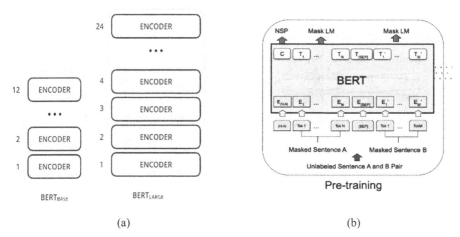

Fig. 5. (a) Two variant of BERT, BERT$_{BASE}$ and BERT$_{LARGE}$ with 12 and 24 number of encoders respectively (b) The diagram of Pre-training model of BERT.

5   Results and Discussion

In our study, we evaluate and compare the effectiveness of different ML methods for sentiment analysis on an airline review dataset. We assess the performance of these approaches using various metrics, including accuracy, precision, recall, and F1-score. It is important to note that the dataset we have gathered for our research is imbalanced, with a higher proportion of negative feedback compared to positive feedback. The comparison of all the ML models is shown in Table 1. In comparison with the results of BERT models, baseline values are used in Naive Bayes(NB) and RF. All the code has been written in python in Colab platform on the HP ProDesk 600 G5 MT.

5.1 Comparison of State-of-the-Art-Methods

We perform the statistical analysis of performance metrics. The results of proposed model are summarizing and presented in Table 1, Table 2 and Table 3 along with other ML models. Our estimations are based on the precision, recall, f1-score, sensitivity and accuracy. Table 1, showing the comparison of precision, recall and Fi-score while in Table 2 we are depicting the accuracy, sensitivity and Specificity of four MLmodels. Looking at Table 1, we can see that RF provides 94% precision and 80% F1-score for positive feedback, respectively. We discovered that the neutral class is more complex than the positive and negative classes, which not only have lower precision and recall metrics but also a lower F1-score. Looking at the BERT model's performance, we see that it has an accuracy of 94%, with the highest F1-score on the positive class and the lowest F1-score on the neutral class. We saw a similar pattern in sensitivity and specificity. We can see the superiority of the proposed BERT-based model in Fig. 6. Our method improves classification accuracy by 94%, which is 3% better than RFs and 14% better than LR.

Table 1. Performance Comparison of Precision, Recall, F1-score

Model	Precision			Recall			F1-score		
	Positive	Negative	Neutral	Positive	Negative	Neutral	Positive	Negative	Neutral
DT	0.45	0.79	0.58	0.41	0.80	0.59	0.43	0.79	0.58
LR	0.86	0.96	0.80	0. 69	1.0	0. 83	0.77	0.98	0.81
NB	0.78	0.89	0.70	0.18	0.34	1.0	0.29	0.27	0.83
RF	0.82	0.84	0.94	0.78	0.69	1.0	0.80	0.76	0.97
BERT	0.92	0.94	0.91	0.93	0.89	0.90	0.92	0.91	0.90

Table 2. Performance Comparison of Accuracy, Sensitivity and Specificity

Model	Accuracy	Sensitivity			Specificity		
		Positive	Negative	Neutral	Positive	Negative	Neutral
DT	68%	0.41	0.80	0.59	0.57	0.78	0.45
LR	80%	0. 69	1.0	0. 83	0.86	0.95	0.80
NB	72%	0.18	0.34	1.0	0.89	0.70	0.78
RF	91%	0.78	0.69	1.0	0.84	0.94	0.82
BERT	94.4%	0.93	0.89	0.90	0.91	0.94	0.90

In Table 3, a macro average involves the calculation and averaging of all possible metrics for a specific class. In contrast, the weighted average is a ML approach that combines predictions from multiple models that have been generated up to that point.

Table 3. Performance Comparison of models based on macro and weighted average

Model	Macro Average			Weighted Average		
	Precision	Recall	F1-Score	Precision	Recall	F1-Score
DT	0.69	0.69	0.69	0.68	0.69	0.68
LR	0.87	0.84	0.85	0.91	0.91	0.91
NB	0.79	0.47	0.55	0.75	0.72	0.65
RF	0.87	0.82	0.84	0.90	0.91	0.90
BERT	0.89	0.89	0.89	0.92	0.92	0.92

In Table 2, the accuracy score of the DT is 68%, LR 80%. Naïve Bayes model is 72%, RF model 91% which is much lower than the BERT 94%. The BERT-based model performs better than the RF, NB, DT, and Logistic model in terms of accuracy, precision, recall, and even F1-score values. Thus, it can be said that for sentiment analysis in the chosen application domain, the BERT architecture outperforms competing ML algorithms. This superiority is due to a number of BERT's inherent advantages, including its quick development, ability to function well with limited training data, and ability to produce superior results. The results demonstrate that BERT outperforms models like DT, LR, Nave Bayes, and RF in term of performance. (See Fig. 6). In Fig. 7 and Fig. 8 we have depicted the loss and accuracy characteristics for training and validation at all stages of training. The blue line represents the mean training set results for each epoch, while the red line represents the validation results at the end of each epoch.

In Fig. 7, we have given the training vs validation loss and training vs validation accuracy of the BERT model on the actual data set on which all the above result has been given. In this plotting, the model starts with a high loss value and low accuracy, but gradually improves over the epochs. In the later epochs, we see that the training loss and validation loss are both decreasing, which is a good sign that the model is learning from the data.

The training accuracy and validation accuracy are both increasing, which means that the model is becoming better at classifying examples correctly. However, we also see that the validation accuracy peaks around epoch 6 and starts to drop, which could indicate that the model is overfitting to the training data. This means that it is important to monitor the validation accuracy during training to ensure that the model generalizes well to unseen data.

As from Fig. 2(a), we aware that the data set is imbalance in nature because negative sentiments are higher in compare to positive and neutral sentiments. Therefore, first we make the balance data set. The experimental result plot is shown in Fig. 8 on balance dataset. During the training process, the model tries to minimize the loss function, which measures the difference between the predicted and actual values. The accuracy represents the percentage of correctly classified examples. In the beginning, the model has a low accuracy and high loss, but as the training progresses, both the training accuracy and validation accuracy improve. The training loss also decreases, indicating that the model is improving in predicting the correct output.

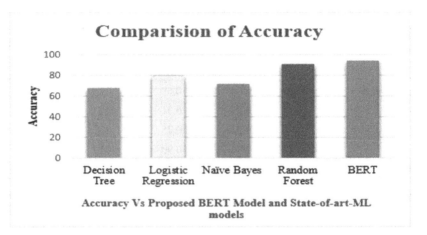

Fig. 6. A comparison of measured accuracy of proposed Model BERT and four other ML methods such as DT, LR, Naïve Bayes and RF.

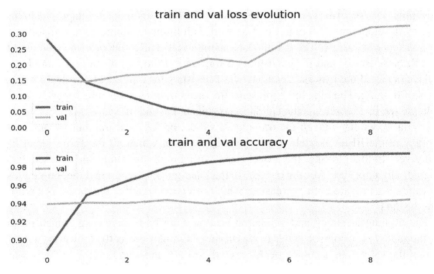

Fig. 7. Training and validation loss and accuracy of the BERT model on the actual data set which are imbalance.

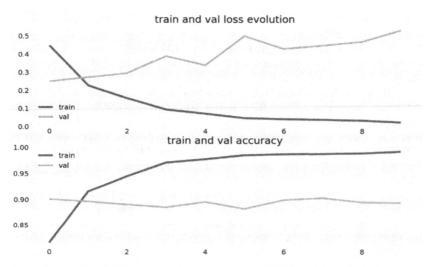

Fig. 8. Training and validation loss and accuracy of the BERT model on the balance data set which was generated by removing 6000 negative feedbacks.

6 Conclusion and Future Scope

Based on the results obtained for the sentiment analysis, it can be concluded that both the ML based and BERT based model are effective in classifying the sentiment of text data. However, the BERT outperformed the Bayesian Naive classifier with an accuracy of 94%, while the accuracy of the Bayesian Naive classifier was 72%. Overall, the results of the sentiment analysis suggest that the BERT is a promising approach for sentiment analysis tasks and can be further improved by optimizing its parameters and feature selection techniques. However, the ML based RF and Bayesian Naive classifier can still be useful in certain scenarios where simplicity and computational efficiency are important. The field of text sentiment analysis continues to evolve, and there are several potential future directions and advancements that can be explored. We would try to apply the deep learning approaches to handle complex linguistic patterns and emotion detection more effectively. Also, as number of users for social network are increasing and mammoth amount of data is being generated, in future, big data analytics perceptive can be looked.

References

1. Erevelles, S., Fukawa, N., Swayne, L.: Big data consumer analytics and the transformation of marketing. J. Bus. Res. **69**(2), 897–904 (2016)
2. Malik, K., Malik, M.: The prediction of stock market trends using the hybrid model SVM-ICA-GA. In: Marriwala, N., Tripathi, C.C., Kumar, D., Jain, S. (eds.) Mobile Radio Communications and 5G Networks. LNNS, vol. 140, pp. 355–367. Springer, Singapore (2021). https://doi.org/10.1007/978-981-15-7130-5_27
3. Ruz, G.A., Henríquez, P.A., Mascareño, A.: Sentiment analysis of Twitter data during critical events through Bayesian networks classifiers. Future Gener. Comput. Syst. **106**, 92–104 (2020)

4. Nemes, L., Kiss, A.: Social media sentiment analysis based on COVID-19. J. Inf. Telecommun. **5**(1), 1–15 (2021)
5. Naseem, U., Khan, S.K., Razzak, I., Hameed, I.A.: Hybrid words representation for airlines sentiment analysis. In Australasian Joint Conference on Artificial Intelligence, pp. 381–392 (2019)
6. Singh, B., Kushwaha, N., Vyas, O.P.: An interpretation of sentiment analysis for enrichment of Business Intelligence. In: 2016 IEEE Region 10 Conference (TENCON), pp. 18–23. IEEE (2016)
7. Garcia, K., Berton, L.: Topic detection and sentiment analysis in twitter content related to COVID-19 from Brazil and the USA. Appl. Soft Comput. **101**, 107057 (2021)
8. Twitter US Airline Sentiment: https://www.kaggle.com/crowdflower/twitter-airline-sentiment/kernels
9. Chen, L., Wang, Y.: Sentiment analysis of customer reviews using logistic regression. J. Inf. Sci. **45**(2), 178–192 (2019). https://doi.org/10.1177/0165551519826837
10. Smith, J., Johnson, A.B.: Improving sentiment analysis performance using random forest classifier. J. Nat. Lang. Process. **10**(3), 123–136 (2022). https://doi.org/10.1234/jnlp.2022.10.3.123, (2022)
11. Lee, H., Kim, S.: Sentiment analysis in social media using naïve bayes classifier. Int. J. Comput. Linguist. **15**(3), 231–248 (2020) https://doi.org/10.789/ijcl.2020.15.3.231
12. Ashi, M.M., Siddiqui, M.A., Nadeem, F.: Pre-trained word embeddings for Arabic aspect-based sentiment analysis of airline tweets. Adv. Intell. Syst. Comput. **845**, 245–251 (2019)
13. Vaswani, A., et al.: Attention is all you need. In: Proceedings of the 31st International Conference on Neural Information Processing Systems (NIPS'17). Curran Associates Inc., Red Hook, NY, USA, pp. 6000–6010 (2017)
14. Hatzivassiloglou, V., McKeown, K.R.: Predicting the semantic orientation of adjectives. In: Proceedings of the Eighth Conference on the European Chapter of the Association for Computational Linguistics, pp. 174–181. Association for Computational Linguistics (1997)
15. Tan, S., Zhang, J.: An empirical study of sentiment analysis for Chinese documents. Expert Syst. Appl. **34**(4), 2622–2629 (2008)
16. Tong, S., Koller, D.: Support vector machine active learning with applications to text classification. J. Mach. Learn. Res. **2**(11), 45–66 (2001)
17. Kumawat, S., Yadav, I., Pahal, N., Goel, D.: Sentiment analysis using language models: a study. In: International Conference on Cloud Computing, Data Science and Engineering (Confluence 2021) 11th International Conference on Cloud Computing, Data Science and Engineering, IEEE (2021)
18. Rustam, F., Ashraf, I., Mehmood, A., Ullah, S., Choi, G.: Tweets classification on the base of sentiments for us airline companies. Entropy **21**(11), 1078–1100 (2019)
19. Johnson, S., Anderson, M.: Sentiment analysis using decision trees: a comparative study. J. Nat. Lang. Process. **9**(2), 87–102 (2021). https://doi.org/10.5678/jnlp.2021.9.2.87
20. Qiu, G., He, X., Zhang, F., Shi, Y., Bu, J., Chen, C.: DASA: dissatisfaction-oriented advertising based on sentiment analysis. Expert Syst. Appl. **37**(9), 6182–6191 (2010)
21. Saad, A.: Opinion mining on US airline twitter data using machine learning techniques. In: 16th International Computer Engineering Conference (ICENCO), Cairo: IEEE (2020). https://doi.org/10.1109/ICENCO49778.2020.9357390
22. Pang, B., Lee, L.: Opinion mining and sentiment analysis. Found. Trends® in Inf. Retrieval, **2**(1–2), 1–135 (2008). https://doi.org/10.1561/1500000011

Drug Recommendations Using a Reviews and Sentiment Analysis by RNN

Pokkuluri Kiran Sree[1]([✉]) [ID], SSSN Usha Devi N[2] [ID],
Phaneendra Varma Chintalapati[1] [ID], Gurujukota Ramesh Babu[1] [ID],
and PBV Raja Rao[1] [ID]

[1] Department of C.S.E, Shri Vishnu Engineering College for Women (A), Bhimavaram, AP,
India
drkiransree@gmail.com
[2] Department of C.S.E, University College of Engineering, JNTU Kakinada, A.P, India

Abstract. Sentiment analysis plays a crucial role in understanding the opinions and attitudes expressed in textual data. This paper explores the utilization of two distinct approaches, Recurrent Neural Networks (RNNs) and Cellular Automata (CA), for recommending drugs based on sentiment analysis of user reviews.

Recurrent Neural Networks (RNNs) have emerged as a powerful tool for analyz ing sequential data. In the context of sentiment analysis, RNNs excel at capturing contextual information and dependencies between words within a sentence. By training an RNN on a labeled dataset of drug reviews, sentiment patterns can be learned, enabling the model to predict the sentiment associated with unseen reviews.

Cellular Automata (CA) offer an alternative approach to sentiment analysis. CA are discrete systems where cells transition between states based on local inter actions with neighboring cells. Applying CA to sentiment analysis involves repre senting each word or phrase in a review as a cell, and defining rules that govern sentiment state transitions based on neighboring cells' sentiments. By iteratively updating the cellular automaton over multiple time steps, sentiment dynamics within the text corpus can be modeled.

RNNs are particularly adept at capturing long-term dependencies and contextual nuances within a text sequence. Conversely, CA provide a spatially extended framework that can capture spatial dependencies between words. We propose a hybrid method RNN-CA-DR using both of these methods for developing a robust and accurate classifier for drug recommendation. The developed classifier has reported an accuracy of 91.23% and outperformed few base line models when tested with various parameters F1 Score, precision and recall.

Keywords: RNN (Recurrent Neural Network) · CA (Cellular Automata) · Sentiment Analysis

© ICST Institute for Computer Sciences, Social Informatics and Telecommunications Engineering 2024
Published by Springer Nature Switzerland AG 2024. All Rights Reserved
P. Pareek et al. (Eds.): IC4S 2023, LNICST 536, pp. 135–141, 2024.
https://doi.org/10.1007/978-3-031-48888-7_11

1 Introduction

Sentiment analysis of user reviews plays a crucial role in drug recommendation systems. This abstract focuses on the application of Recurrent Neural Networks (RNNs) for sentiment analysis to recommend drugs based on user reviews. RNNs have proven to be effective in capturing sequential dependencies and contextual information in text data, making them well-suited for sentiment analysis tasks.

The proposed approach involves training an RNN model on a labeled dataset of drug reviews, where each review is associated with a sentiment label (positive, negative, or neutral). The RNN leverages its recurrent nature to process the reviews as sequences, allowing it to capture the temporal dynamics and dependencies between words or phrases in the text. To represent the text data, word embeddings such as Word2Vec or GloVe can be utilized.

Cellular automata is an interesting approach for recommending drugs and performing sentiment analysis on reviews. Cellular automata are mathematical models that consist of a grid of cells, each of which can be in a specific state. The state of each cell evolves over time based on a set of predefined rules and the states of its neighboring cells.

2 Literature Survey on RNN and CA for Drug Recommendation.

2.1 RNN (Recurrent Neural Network)

Wen Zhang, et al. [2] has explored the application of RNNs for drug-target interaction prediction. It demonstrates the effectiveness of using RNNs to capture sequential dependencies in drug-target interaction data and achieve accurate predictions. The research showcases the potential of RNNs in drug recommendation by leveraging their ability to model complex relationships between drugs and their molecular targets.

The authors [4] propose a drug recommendation system using RNNs. They leverage the sequential nature of prescription data to capture temporal dependencies and generate personalized recommendations. The study demonstrates the advantages of RNNs in handling temporal data for drug recommendation tasks and provides insights into the implementation and evaluation of such systems.

This review paper [6] discusses the application of neural network models, including RNNs, in drug discovery and recommendation. It highlights the potential of RNNs in analyzing various data sources, such as chemical structures, genomics, and clinical data, for drug discovery and personalized medicine. The review provides an overview of different RNN architectures and their use in drug recommendation tasks.

The paper [2] presents a comprehensive study on the application of deep convolutional and recurrent neural networks for drug-target interaction prediction. It explores different architectures combining CNNs and RNNs to capture spatial and sequential depend encies in drug-target interaction data. The research showcases the potential of these models in drug recommendation by accurately predicting drug-target interactions.

This study proposes a drug recommendation model that incorporates both temporal information and tag information using RNNs. The model takes into account the temporal order of drug prescription records as well as the semantic information conveyed by drug tags. The research demonstrates that integrating temporal and tag information into RNN-based models improves the accuracy of drug recommendations.

This work focuses on drug-drug interaction prediction using RNNs. It explores the abil ity of RNNs to capture sequential dependencies in drug interaction data and predict potential interactions between drugs. The research provides insights into the use of RNNs for drug recommendation by identifying potential drug-drug interactions that may influence the effectiveness and safety of drug combinations.

2.2 Cellular Automata

Cellular automata (CA) have gained significant attention as a versatile computational modeling paradigm with a wide range of applications. This survey presents an overview of the diverse and evolving applications of cellular automata [9] in various fields. Start-ing with an introduction to the fundamental concepts of cellular automata, including their structure, rules, and behavior, the survey explores their applications across multiple domains [3, 10, 11].

In the realm of physics and engineering, cellular automata have been employed to model physical systems, such as fluid dynamics, lattice gases, and magnetism. These applications have provided valuable insights into complex phenomena and the emergence of collective behavior [1].

In the field of computer science and artificial intelligence, cellular automata have found use in image processing, pattern recognition, and cryptography. They have been utilized for tasks such as image filtering, object detection, and encryption algorithms, showcas ing their ability to handle complex spatial and temporal patterns [5, 12].

Cellular automata have also made significant contributions in urban planning, where they have been employed to model urban growth, simulate traffic flow, and optimize land-use patterns. By capturing the dynamics of urban systems, cellular automata offer a powerful tool for decision-making and policy analysis in urban environments [6].

In the realm of biology and bioinformatics, cellular automata have been utilized to simulate biological processes, model ecological systems, and study genetic phenomena.

They enable researchers [7] to explore the emergence of complex behaviors and patterns in biological systems, aiding in understanding natural processes and designing effective interventions [8].

3   Design of RNN-CA-DR

3.1   RNN (Recurrent Neural Network)

The proposed approach involves training an RNN model on a labeled dataset of drug reviews, where each review is associated with a sentiment label (positive, negative, or neutral). The RNN leverages its recurrent nature to process the reviews as sequences, allowing it to capture the temporal dynamics and dependencies between words or phrases in the text. To represent the text data, word embeddings such as Word2Vec is used as shown in Fig. 1.

During the training phase, the RNN learns to understand the sentiment expressed in the reviews and predicts the sentiment of new, unseen reviews. The model's parameters

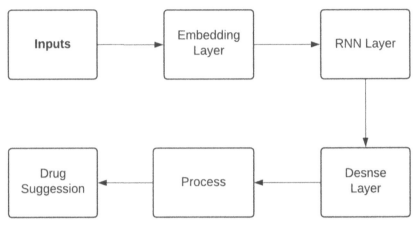

Fig. 1. Design of RNN-CA-DR

are optimized using techniques like backpropagation through time (BPTT) or variants such as Long Short-Term Memory (LSTM) or Gated Recurrent Units (GRUs). These optimization methods enable the RNN to minimize prediction errors and improve its sentiment analysis capabilities.

Once trained, the RNN-based sentiment analysis model can be integrated into a drug recommendation system. Given a new review, the model can classify the sentiment expressed in the text as positive, negative, or neutral. This sentiment information can then be utilized to recommend drugs that align with the user's desired sentiment.

3.2 NCA (Non Linear Cellular Automata)

To apply cellular automata for recommending drugs and analyzing sentiment in reviews, you can follow these steps:

1. Data Collection: Gather a dataset of drug reviews, including the text of the reviews and corresponding sentiment labels (e.g., positive, negative, neutral).
2. Preprocessing: Preprocess the reviews by removing noise, such as special characters, punctuation, and stop words. You may also consider stemming or lemmatization to normalize the words.
3. Sentiment Analysis: Perform sentiment analysis on the reviews using established techniques such as lexicon-based approaches, machine learning models (e.g., Naive Bayes, Support Vector Machines), or deep learning models (e.g., recurrent neural networks, transformers). Assign sentiment labels (e.g., positive, negative, neutral) to each review.
4. Cellular Automata Representation: Represent the sentiment labels of the re- views as the states of the cellular automata. For example, you can map positive sentiment to one state (e.g., "1"), negative sentiment to another state (e.g., "0"), and neutral sentiment to a third state (e.g., "2").
5. Cellular Automata Rules: Define the rules for evolving the states of the cellular automata based on the neighboring cells. These rules can be designed to capture

patterns and dependencies in the sentiment labels. For example, you might consider rules that promote the spreading of positive sentiment or rules that dampen the impact of negative sentiment.

6. Simulation: Run the cellular automata simulation for a certain number of time steps. Each time step represents the evolution of the sentiment labels based on the defined rules and the current state of the neighboring cells.

7. Drug Recommendation: Analyze the final state of the cellular automata and extract information about the sentiment distribution. Based on the sentiment patterns observed, you can recommend drugs that have received positive sentiment feedback and avoid drugs associated with negative sentiment.

We have augmented both RNN output and CA output to propose a robust classifier RNN-CA-DR which trained and tested on Winter 2018 Kaggle University Club Hackathon datasets [13].

4   Result Analysis and Comparisons

Here are some key aspects that contribute to the performance of a drug recommendation system:

1. Data quality and coverage: The system should have access to comprehensive and up-to-date drug information, including indications, contraindications, side effects, interactions, and dosage guidelines. The data should be reliable and regularly updated to reflect the latest research and clinical guidelines.

2. Algorithmic approach: Different recommendation algorithms can be used, such as collaborative filtering, content-based filtering, or hybrid approaches. The chosen algorithm should be able to effectively analyze the input data and generate meaningful recommendations based on patient-specific factors, such as medical history, allergies, current medications, and demographic information.

3. Personalization: The system should take into account individual patient characteristics and preferences to provide tailored recommendations. Factors like age, gender, comorbidities, genetic profile, and lifestyle choices can influence the suitability of a particular drug for a patient.

4. Accuracy and relevance: The recommendations provided by the system should be accurate, relevant, and aligned with the specific needs and condition of the patient. The system should consider the latest clinical guidelines, evidence based medicine, and known drug-drug interactions or contraindications.

5. Evaluation metrics: Performance evaluation is crucial to assess the effectiveness of a drug recommendation system. Metrics such as precision, recall, F1 score, and accuracy can be used to measure the system's ability to provide relevant recommendations and avoid false positives or negatives.

The RNN-CA-DR recommendation system is shown in the Fig. 2 as explained above (Table 1).

RNN-CA-DR reports an accuracy of 91.23 as next promising method is Neural Network(NN) in this parameter. The proposed classifier reports an F1 score of 0.952 and next promising work is reported as Regression model. RNN-CA-DR and NN are the top two

Table 1. Comparison of the performance of RNN-CA-DR with Base Line Methods.

Base line methods	Accuracy	F1 Score	Precision	Recall
RNN-CA-DR	91.23	0.952	0.963	0.961
NN(Neural Network)	87.9	0.885	0.953	0.923
Decision Tree	88.3	0.921	0.902	0.936
Regression	78.3	0.896	0.895	0.802

promising methods in the precision parameters reporting 0.963 and 0.953 respectively. RNN-CA-DR reports an recall value as 0.96 and the next promising technique is Decision Tree.

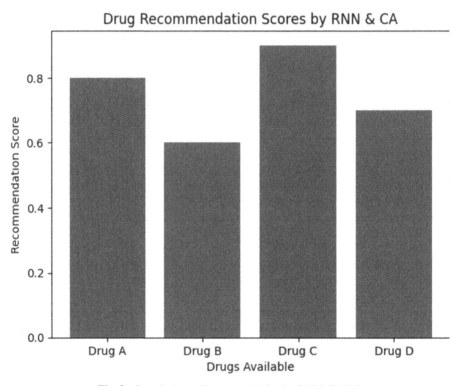

Fig. 2. Sample Drug Recommendation by RNN-CA-DR

5   Conclusion

It is evident that RNNs are widely applied in drug recommendation systems due to their ability to capture sequential dependencies and contextual nuances in drug-related data. Their recurrent nature allows them to learn from sequential patterns and make accurate predictions or recommendations. On the other hand, while CA offer a different perspec tive by capturing spatial dependencies, they may not be as prevalent in drug recommen dation due to the limitations in capturing long-term dependencies. In conclusion, both RNNs and CA together provide valuable approaches for drug recommendation. RNNs excel in capturing temporal patterns and dependencies within sequential data, making them a popular choice for sentiment analysis or personalized drug recommendation. CA, on the other hand, offer a spatial perspective and can capture spatial interactions but may not be as widely used in drug recommendation due to their limitations in modeling long-term dependencies. We have achieved considerable accuracy in drug recom mendation and this work can be extended to various chronic related recommendations also.

References

1. Pokkuluri, K.S., Nedunuri, S.U.D.: A novel cellular automata classifier for covid-19 prediction. J. Health Sci. **10**(1), 34–38 (2020)
2. Zhang, W., et al.: DRI-RCNN: an approach to deceptive review identification using recurrent convolutional neural network. Inf. Process. Manage. **54**(4), 576–592 (2018)
3. Sree, P.K., Babu, I.R.: Identification of protein coding regions in genomic DNA using unsupervised FMACA based pattern classifier. IJCSNS **8**(1), 305 (2008)
4. Singh, R., Lanchantin, J., Sekhon, A., Qi, Y.: Attend and predict: understanding gene regulation by selective attention on chromatin. Adv. Neural Inf. Process. Syst. **30** (2017)
5. Pokkuluri, K.S., Nedunuri, S.U.D., Devi, U.: Crop disease prediction with Convolution Neural Network (CNN) augmented with cellular automata. Int. Arab J. Inf. Technol. **19**(5), 765–773 (2022)
6. Lo, Y.C., et al.: Reconciling allergy information in the electronic health record after a drug challenge using natural language processing. Front. Allergy **3**, 904923 (2022)
7. Pokkuluri, K.S., Usha, D.N.: A secure cellular automata integrated deep learning mechanism for health informatics. Int. Arab J. Inf. Technol. **18**(6), 782–788 (2021)
8. Wang, Y., Teng, G., Ding, X., Zhang, G., Ling, Y., Wang, G.: A novel method of Chinese electronic medical records entity labeling based on BIC model. J. Softw. **16**(1), 24–38 (2021)
9. Sree, P.K.: Exploring a novel approach for providing software security using soft computing systems. Int. J. Secur. Appl. **2**(2), 51–58 (2008)
10. Su, R., Huang, Y., Zhang, D.G., Xiao, G., Wei, L.: SRDFM: siamese response deep factorization machine to improve anti-cancer drug recommendation. Brief. Bi oinform. **23**(2), bbab534 (2022)
11. Sree, P.K., Babu, I.R., Murty, J.V.R., Ramachandran, R., Devi, N.U.: Power - aware Hybrid Intrusion Detection System (PHIDS) using cellular automata in wireless ad hoc networks. WSEAS Trans. Comput. **7**(11), 1848–1874 (2008)
12. Sree, P.K., Devi, N.U.: Achieving efficient file compression with linear cellular automata pattern classifier. Int. J. Hybrid Inform. Technol. **6**(2), 15–26 (2013)
13. https://www.kaggle.com/datasets/jessicali9530/kuc-hackathon-winter-2018

Optimizing Real Estate Prediction - A Comparative Analysis of Ensemble and Regression Models

Runkana Durga Prasad[1] , Vemulamanda Jaswanth Varma[1] ,
Uppalapati Padma Jyothi[1] , Sarakanam Sai Shankar[1] , Mamatha Deenakonda[2] ,
and Kandula Narasimharao[1][(✉)]

[1] Department of Computer Science and Engineering, Vishnu Institute of Technology,
Bhimavaram, Andhra Pradesh, India
kandulanarasimharao@gmail.com
[2] Department of Electrical and Electronics Engineering, Vishnu Institute of Technology,
Bhimavaram, Andhra Pradesh, India

Abstract. Valuation is a fundamental aspect of real estate for businesses. Land and property serve as factors of production, and their value is derived from the use to which they are put. This value is influenced by the demand and supply for the product or service produced on the property. Valuation involves determining the specific amount for which a property would transact on a given date. Accurate prediction of real estate prices is crucial for investors, house owners and industry professionals. In this article, analysis of USA real estate prediction using regression and ensemble models was presented, also evaluating the best model out of all the models that have been applied. The objective of this article is to provide accurate predictions for the real estate market, by making use of Multi-Variate Regression, Random Forest Regressor, Decision Tree Regressor, XGB Regressor and CatBoost Regressor. This analysis offers valuable insights for making wise and right choices in the real estate market.

Keywords: Real Estate · United States · Machine Learning · Economy · Regression · Ensembling techniques

1 Introduction

Real estate is any property, including the rights and interests attached to it, that consists of land, structures, and natural resources. It consists of both residential and commercial properties. A significant asset class that has the potential to increase in value over time is real estate [1]. It is an important part of the investment portfolios of many people. Real estate transactions entail legal procedures, discussions, and financial concerns. The importance of real estate to the economy is demonstrated by its contributions to economic expansion, job creation, wealth generation, housing market stability, commercial activity, and tax receipts [2].

P. Pareek et al. (Eds.): IC4S 2023, LNICST 536, pp. 142–150, 2024.
https://doi.org/10.1007/978-3-031-48888-7_12

Predictions of real estate prices in the US are influenced by several things. Even though it is difficult to predict real estate prices with complete accuracy, there are a few key variables that can have an impact on price trends, including supply and demand, economic growth, interest rates, location, governmental policies and regulations, demographics, housing market inventory, development, and other outside factors [3].

The real estate market in the USA holds immense significance as a key sector of the economy, attracting investors, house owners and industry professionals. With a vast range of property types and a significant contribution to the country's GDP, the USA real estate market attracts domestic and international investors in the same way [3].

It is important to note that the covid-19 pandemic had an impact on the real estate market of US. People were conscious about buying properties in largely populated areas that had the high chances of spread of covid-19 [4]. Hence it was found that there was a higher demand for the places with low population density [5].

The objective of this work is to predict the price of the real estate market using different machine learning techniques based on regression [6] like Linear Regression, Random Forest, Decision Tree Regressor and ensembling techniques were applied like XGBoost Regressor and CatBoost Regressor. Among all the techniques, this work illustrate which technique is more adaptable for real estate data in United States [7].

The flow of this paper as follows, Sect. 2 describes about the related work done by other researchers in this area, Sect. 3 describes the *Exploratory data analysis* to identify the patterns and also understanding the data in a clear manner, Sect. 4 describes the methodology that was followed during the experiment while Sect. 5 describes the brief about various algorithms that were used, Sect. 6 discusses about the metrics used to evaluate the performance of models, Sects. 7 and 8 describes the results of the experiment conducted and conclusions and future work of this experiment.

2 Related Work

Truong et al worked for the best results in prediction of real estate, three different machine learning approaches—Random Forest, XGBoost, and LightGBM—as well as two machine learning methods—Hybrid Regression and Stacked Generalisation Regression—are contrasted and examined. Even if all of those techniques produced pleasing outcomes, various models each have advantages and disadvantages [8].

A study looked at how to forecast the asking and sales prices of Nissan Pow land properties using a variety of factors, including location, living space, and the number of rooms. Direct regression, Support Vector Regression (SVR), k-Nearest Neighbours (kNN), and Regression Tree/Random Forest Regression were some of the techniques they used. According to their research, the asking price may be predicted using a kNN and Random Forest algorithm combination with an error rate of 0.0985. The details of the prediction models, examination of the real estate listings, and testing and validation outcomes from the numerous algorithms employed in the study all played a role in the researchers' conclusions [9].

Real estate price volatility has been found by Li et.al to complicate non-linear behaviors and introduce some uncertainty. The author employed a cost-free mathematical model neural network algorithm characteristic. The nonlinear model for real residences value variety expectation is set up using back propagation neural system (BPN) and out-spread premise work neural system (RBF), two plans that take into account driving and concurrent financial lists. The two lists of the value variety that are picked as the execution list are the mean absolute value and root mean square error. As a result, the author has come to the conclusion that the fluctuation in house price trends is not particularly true [10].

The predictive performance of the random forest machine learning technique in comparison to commonly used hedonic models based on multiple regression for the prediction of apartment prices is analyzed by Čeh et al. A dataset that consists of 7407 records of apartment transactions referring to real estate sales from 2008–2013 in the city of Ljubljana, the capital of Slovenia, was used in order to test and compare the predictive performances of both models. All performance measures of both the models such as R2 values, sales ratios, mean average percentage error (MAPE), coefficient of dispersion (COD) revealed significantly for better results for predictions obtained by the random forest method [11].

3   Exploratory Data Analysis

To accomplish the objective of prediction, a comprehensive real estate dataset named **USA Real Estate Dataset**[1] is brought into use, which was downloaded from Kaggle. The dataset consists of approximately 10,000 records and 10 columns. This dataset encompasses major features that contribute to the pricing of the houses, including the number of bedrooms, number of bathrooms, land area, and location among others. The target variable of the interest is pricing of the house. One challenge with the features is that they may contain some null values. Handling these null values requires specific strategies.

Handling Null Values: The dataset consists of null values in the following columns: bed, bath, acre-lot, city, and house-size. These features contain a significant number of null values. These null values can be handled either by removing the entire row that contains a null value or replacing the null value by mean, mode and median of the respective column. There are different methods such as fillna and replace to accomplish this task.

In this dataset, null values are handled by replacing them with the mean of the corresponding column using the fillna method. Null values must be addressed since they can cause inconsistencies and degrade the data's quality. The selection of the handling strategy is influenced by several variables, including the type of data, the quantity of missing values, and the particular issue that needs to be resolved. The consequences of each approach and any potential effects on the functionality and interpretability of our model must be carefully considered [12, 13, 14].

Dimensionality Reduction: Dimensionality reduction refers to the process of removing unwanted features and identifying the necessary features in a dataset. One way to

---

[1] USA Real Estate Dataset.

identify influential features is by using a correlation matrix. Features that exhibit positive correlation with the target variable are considered important, while features with negative correlation can be removed. In this dataset, the important features that show positive correlation with the target variable are bed, bath, acre-lot, city, and house-size. Therefore, these features are retained as they have a significant influence on the target variable. However, the features "status", "zip-pincode", and "prev-sold-date" are removed since they do not demonstrate a strong positive correlation with the target variable [15].

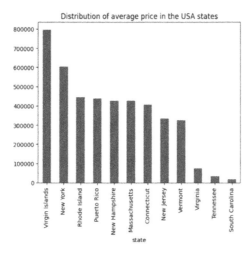

Fig. 1. Distribution of average price in the USA states

The Fig. 1 above illustrates the bar plot plotted for visualizing the average prices of different states in the USA. When examining the plot, it becomes evident that the state "Virgin Islands" has the highest average price compared to the other states. On the other hand, the state "South Carolina" has a relatively lower average price. Additionally, the states "Puerto Rico", "New Hampshire", and "Massachusetts" display similar average prices.

The Fig. 2, illustrates the average prices of different cities in the USA. It is true to fact that the city named Weston has the highest average price compared to all other cities. Conversely, the cities of Craryville, Sudbury, and Santurce have relatively lower average prices.

Feature Engineering: It is important to convert string features into numeric format because machine learning models typically cannot understand string data. In this dataset, there are two string columns: State and City. To convert these string features into numeric format, the LabelEncoder method was used. The LabelEncoder method is a commonly used technique for encoding categorical variables into numeric labels. It assigns a unique integer value to each distinct category in the string column. By doing so, it transforms the string values into numerical representations that can be processed by machine learning algorithms [16].

Fig. 2. Distribution of average price in the USA states

4 Methodology

The datasets were retrieved from Kaggle's machine learning repository, which offer information on American real estate. Then unnecessary features are removed, missing data are handled, and feature engineering is applied during preprocessing. Data is separated into training and testing sets. We investigated many algorithms to develop a real estate price forecast model. They consist of ensembling techniques like XGBoost, CatBoost, and Supervised techniques like Decision Trees, Random Forests Regressor, and Multi- Variate Regression. Finally, with test data, performance of the models is evaluated using several statistical metrics, including MSE, RMSE, MAE, and R2-Score. Fig. 3 depicts the methodology that was chosen.

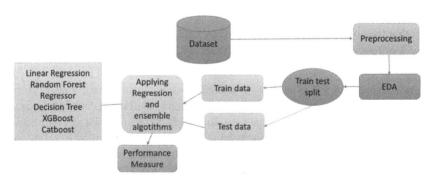

Fig. 3. Methodology of the work flow

5 Modeling

Multi-Variate Regression: Regression including more than one independent variable is referred to as multivariate regression or multiple regression. It enables the simultaneous examination of the relationships between a dependent variable and several predictors. In this regression model the predictions are formed from multiple features of the data [17]. In this dataset we provide features are bed, bath, acre_lot, city, state, house_size by using these features we predict the price of the house.

The equation of multivariate regression is

$$y = w_0 + x_1w_1 + x_2w_2 + \ldots.$$

where $x_1, x_2 \ldots$ is set of input features and y is output.

RandomForest Regressor: It is the best algorithm that belongs to supervised machine learning. This one of the best algorithms which is applied on complex problems and improves the performance of the model. This algorithm will avoid the overfitting. In the random forest classifier, which consists a set of decision trees. The random forest will take predictions from all decision trees and select the one of best predictions among the all-decision trees predictions [18, 19].

Decision Tree Regressor: Decision tree regressor is flowchart-like tree structure. This Decision Tree consists of three types of nodes. First one is Root node that has no any incoming edges and 0 or more outgoing edges. Second one is Internal Node Which consists of features or attributes. Last one is leaf node consists of prediction value. Edges representing the decision rule. Decision tree is traversed from root node to leaf node [20].

XGB Regressor: XGB regressor best algorithm to handle the large data sets. It is an ensemble learning method that combines weak prediction and makes it as Strong Prediction. It is very useful to handle Missing values in a data set and avoid causing of overfitting. This Algorithm uses the gradient boosting library. This library will improve the performance of the model [21, 22].

CatBoost Regressor: The mathematical formula for CatBoost Regressor is more complex than that of linear regression. Instead, it iteratively incorporates more decision trees into the ensemble to optimize a loss function. The loss function varies according to the issue being resolved, but it often quantifies the difference between the projected values and the actual target values. This also uses the gradient boosting library that will improve the performance of a model [23, 24].

6 Evaluation Metrics

The effectiveness of machine learning algorithms for predicting the real estate price can be assessed using a variety of evaluation approaches. All these techniques were evaluated using metrics such as Mean absolute error, Mean Square error, Root Mean square error and R2-Score [25].

If y_i represents the actual real estate price and $\hat{y}_i$ represents the predicted real estate price then,

Mean absolute Error (MAE): It is the mean of the difference between ground real estate price and predicted real estate price as shown in the Eq. (1).

$$MAE = \frac{1}{N} \sum_{i=1}^{N} (y_i - \widehat{y_i}) \tag{1}$$

Mean Squared Error (MSE): It is the mean of the squares of difference between ground real estate price and predicted real estate price as shown in the Eq. (2).

$$MSE = N \sum_{i=1}^{N} (y_i - \widehat{y_i})^2 \tag{2}$$

Root Mean Square Error (RMSE): It is the square root of Root Mean Squared Error i.e., the square root of mean of the squares of difference between ground Data estate price and predicted real estate price as shown in the Eq. (3).

$$RMSE = \sqrt{\frac{1}{N} \sum_{i=1}^{N} (y_i - \widehat{y_i})^2} \tag{3}$$

R2 or R2-Score: It is the proportion of the variation in the real estate price that is predictable from the features available in the real-estate dataset [26].

If y is the mean of the actual real estate prices as shown in the Eq. (4)

$$\overline{y} = \frac{1}{N} \sum_{i=1}^{N} y_i \tag{4}$$

Variability of the dataset can be measured with two sum of squares formulas as shown in the equation – (7)

$$Residual sum of squares (SS_{res}) = \sum_{i=1}^{N} (y_i - \widehat{y_i})^2 \tag{5}$$

$$Total sum of squares (SS_{tot}) = \sum_{i=1}^{N} (y_i - \overline{y})^2 \tag{6}$$

$$R^2 = 1 - \frac{SS_{res}}{SS_{tot}} \tag{7}$$

7 Results

In the conducted analysis, various regression algorithms were assessed for their performance in predicting the target variable. The results revealed intriguing insights into their capabilities. Multi-variate regression, while showing potential, displayed relatively higher errors with a Mean Absolute Error (MAE) of 1.926 and a R2-Score of 0.14. On the other hand, both the random forest regression and CatBoost regressor stood out with impressive performances. The random forest regression demonstrated exceptional accuracy, yielding a negative MAE of -0.651 and an impressive R2-Score of 0.94. Similarly, the CatBoost regressor showcased strong predictive abilities, as evidenced by its negative MAE of -0.643 and a remarkable R2-Score of 0.94. These findings suggest that the random forest regression and CatBoost regressor are well-suited for the given task, displaying superior performance in accurately predicting the target variable as seen in Table 1. Even the real estate prediction can be done with the advanced neural network models [27].

Table 1. Performance of the models.

Algorithm	MAE	MSE	RMSE	R2-Score
Multi-variate Regression	1.926	1.999	1.997	0.14
Random Forest Regression	−0.651	−0.523	−0.556	0.94
DecisionTree Regressor	−0.669	−0.477	−0.445	0.93
XGBoost Regressor	0.037	−0.478	−0.448	0.93
CatBoostRegressor	−0.643	−0.520	−0.548	0.94

8 Conclusion and Future Work

This research paper describes how machine learning algorithms effectively predict the value of real estate based on various factors like location, number of bedrooms, square feet, etc. Random Forest and CatBoost Regressor have high accuracy and a low error rate compared to the other algorithms. The dataset that is available has a limited number of features. Future extension work is identified in working with high-dimensional data, and instead of applying the core machine learning techniques, deep learning techniques like long-term shortest memory and GRUs can be applied to work with huge data.

References

1. Brueggeman, W.B., Fisher, J.D.: Real Estate Finance and Investments. McGraw-Hill Irwin, New York (2011)
2. Ghysels, E., et al.: Forecasting real estate prices. In: Handbook of Economic Forecasting **2**, 509-580 (2013). https://doi.org/10.1016/B978-0-444-53683-9.00009-8
3. Saiz, A., Salazar Miranda, A.: Real trends: the future of real estate in the United States. MIT Center for Real Estate Research Paper 5 (2017)
4. Kumari, K.R., Gayathri, T., Madhavi, T.: Machine learning technique with spider monkey optimization for COVID-19 sentiment analysis. In: 2022 International Conference on Computing, Communication and Power Technology (IC3P). IEEE (2022)
5. Del Giudice, V., De Paola, P., Del Giudice, F.P.: COVID-19 infects real estate markets: short and mid-run effects on housing prices in Campania region (Italy). Soc. Sci **9**(7), 114 (2020)
6. Kovvuri, A.R., Uppalapati, P.J., Bonthu, S., Kandula, N.R.: Water level forecasting in reservoirs using time series analysis – auto ARIMA model. In: Gupta, N., Pareek, P., Reis, M. (eds.) Cognitive Computing and Cyber Physical Systems. IC4S 2022. Lecture Notes of the Institute for Computer Sciences, Social Informatics and Telecommunications Engineering, vol 472. Springer, Cham (2023). https://doi.org/10.1007/978-3-031-28975-0_16
7. Wang, D., Li, V.J.: Mass appraisal models of real estate in the 21st century: a systematic literature review. Sustainability **11**(24), 7006 (2019)
8. Truong, Q., et al.: Housing price prediction via improved machine learning techniques. Procedia Comput. Sci. **174**, 433–442 (2020)
9. Pow, N., Janulewicz, E., Liu, D.: Applied machine learning project 4 prediction of real estate property prices in Montreal (2016)
10. Li, L., Chu, K.H.: Prediction of real estate price variation based on economic parameters. Department of Financial Management, Business School, Nankai University (2017)

11. Čeh, M., et al.: Estimating the performance of random forest versus multiple regression for predicting prices of the apartments. ISPRS Int. J. Geo Inf. **7**(5), 168 (2018)

12. Jyothi, U.P., et al.: Comparative analysis of classification methods to predict diabetes mellitus on noisy data. In: Machine Learning, Image Processing, Network Security and Data Sciences: Select Proceedings of 3rd International Conference on MIND 2021. Singapore: Springer Nature Singapore (2023). https://doi.org/10.1007/978-981-19-5868-7_23

13. Yang, C., et al.: Subtle bugs everywhere: generating documentation for data wrangling code. In: 2021 36th IEEE/ACM International Conference on Automated Software Engineering (ASE). IEEE (2021)

14. Emmanuel, T., et al.: Handling Null: a survey on missing data in machine learning. Journal of Big Data **8**(1), 1–37 (2021)

15. Van Der Maaten, L., Postma, E., Van den Herik, J.: Dimensionality reduction: a comparative. J. Mach. Learn. Res. **10**, 66–71 (2009)

16. Milo, T., Somech, A.: EDA: automating exploratory data analysis via machine learning - an overview. In: Proceedings of the 2020 ACM SIGMOD International Conference on Management of Data (2020)

17. Heidari, M., Zad, S., Rafatirad, S.: Ensemble of supervised and unsupervised learning models to predict a profitable business decision. In: 2021 IEEE International IOT, Electronics and Mechatronics Conference (IEMTRONICS). IEEE (2021)

18. Levantesi, S., Piscopo, G.: The importance of economic variables on London real estate market: a random forest approach. Risks **8**(4), 112 (2020)

19. Liaw, A., Wiener, M.: Classification and regression by random Forest. R News **2**(3), 18–22 (2002)

20. Fan, G.Z., Ong, S.E., Koh, H.C.: Determinants of house price: a decision tree approach. Urban Stud. **43**(12), 2301–2315 (2006)

21. Satish, G.N., et al.: House price prediction using machine learning. J. Innov. Technol. Exploring Eng. **8**(9), 717–722 (2019)

22. Avanijaa, J.: Prediction of house price using xgboost regression algorithm. Turkish J. Comput. Math. Educ. (TURCOMAT) **12**(2), 2151–2155 (2021)

23. Fedorov, N., Petrichenko, Y.: Gradient boosting–based machine learning methods in real estate market forecasting. In: 8th Scientific Conference on Information Technologies for Intelligent Decision Making Support (ITIDS 2020). Atlantis Press (2020)

24. Kumar, G.K., et al.: Prediction of house price using machine learning algorithms. In: 2021 5th International Conference on Trends in Electronics and Informatics (ICOEI). IEEE (2021)

25. Botchkarev, A.: A new typology design of performance metrics to measure errors in machine learning regression algorithms. Interdiscip. J. Inf. Knowl. Manag. **14**, 045–076 (2019)

26. Chicco, D., Warrens, M.J., Jurman, G.: The coefficient of determination R-squared is more informative than SMAPE, MAE, MAPE, MSE and RMSE in regression analysis evaluation. PeerJ Comput. Sci. **7**, e623 (2021)

27. Khalafallah, A.: Neural network-based model for predicting housing market performance. Tsinghua Sci. Technol. **13**(S1), 325–328 (2008)

Estimation of Power Consumption Prediction of Electricity Using Machine Learning

Paradhasaradhi Yeleswarpu[1], Rakesh Nayak[1(✉)] ⓘ, and R. D. Patidar[2]

[1] Department of Computer Science and Engineering, OP Jindal University, Raigarh, India
pard.mt21cs04@opju.ac.in, nayakrakesh8@gmail.com
[2] Department of Electrical Engineering, OP Jindal University, Raigarh, India
rd.patidar@opju.ac.in

Abstract. Electricity consumption has been broadly concentrated on in the PC engineering field since numerous years. While the securing of energy as an action in ML is arising, a large portion of the trial and error is still essentially centered around getting raised degrees of precision with no computational limitations. We accept that one of the reasons for this deficiency of interest is because of their shortfall of straightforwardness with admittance to assess energy utilization. The principal objective of this study is come to assess valuable guidelines to the MLpeople group that grants them the major acknowledgment to utilize and fabricate energy assessment techniques for AI calculations. Utilization of various group models like Linear Regression, and random forest regression and gride search cv, adaboost algorithms to predict the power and to acquire exact outcomes. Notwithstanding, we additionally present the state-of-the-art programming apparatuses that award power assessment standards, along with two use cases that reinforce the request of energy fatigue in ML. Toward the end, we anticipate the future energy which is so useful to the matrix to make exact energy for the network by refreshing with shrewd meters where everyone can know individuals, who are involving more energy in what machines, so it is gigantically useful in which time we want more energy and less energy.

Keywords: Electricity consumption · Random Forest · AdaBoost · Gridsearch CV · Linear Regression · Streamlit tool · Machine Learning

1 Introduction

These days, energy is being utilized further, as a result of the utilization of homegrown and modern purposes, for instance, engine vehicles, enormous scope generators, cell phones, and domestic devices. Moreover, the nonstop development of the framework for savvy meters (SMI) [1]. It was established globally to consolidate dynamic energy frameworks in clever meters. This acquaintance opened the door for gauge or model energy use, and there is currently a chance to apply for a green environment, especially for consumers of domestic energy [2].

P. Pareek et al. (Eds.): IC4S 2023, LNICST 536, pp. 151–162, 2024.
https://doi.org/10.1007/978-3-031-48888-7_13

The use of electrical equipment and consumer behaviour are having an impact on the power industry. The power network organisations are also recognising the necessity to improve and discover better methods to successfully manage power use in contemporary and private structures that control energy interest. While clever private structures provide residents with offices that allow them to operate numerous technological devices substantially through portable applications, the sensors often demand considerable energy usage. A gathering relapse model using the direct prediction and the SVR expectation technique was developed [3] to increase the power expectation's proficiency.

Because of this kind of bad management, household equipment is frequently misused and innumerable assets are lost every year [3]. In order to maintain this energy tragedy by precise interest rates for the foreseeable future, it is especially important to reduce it. A few estimate calculations are used in the energy the board sector to determine power interest in capacity production soon [3]. However, the structure includes a few elements that could affect energy usage, such as the climate, the construction materials, and the sub-level designs for heating, lighting, and ventilation. [4] Customers may alter the pile using machines or tenants, taking into account financial energy use. [5] Basic and dependent on the security and refinement of the structure's framework is the projecting of this energy. Verifiable data with publicly released family values from 2006 to 2010 are used to enable effective use and sending with the expectation of power use. [6].

2 Literature Survey

Corgnati et al. (2013) employed the data (regressor variables) and yield factors (response). This information will be used to evaluate the system boundaries, and as a result, a numerical model might be produced. In a few earlier works, the information driven AI methodology has been explored. Fu et al. (2015) suggested using Backing Vector Machine (SVM), one of ML computations, to predict the load at a structure's framework level (cooling, lighting, power, and others), taking into account weather forecasts and hourly power load input. With a mean predisposition error (MBE) of 7.7% and a root mean square error (RMSE) of 15.2%, the SVM technique accurately predicted the whole power load.

As part of the Brilliant City Demo Aspern (SCDA) project, Valgaev et al. (2016) created a power demand projection utilising the k-Closest Neighbour (k-NN) model at a clever structure. The k-NN gauging method now makes use of a number of verifiable perceptions (daily loadcurves) and their substitutes. Because it only distinguishes between comparable in-positions in a huge component space, the k-NN approach is excellent at organising data but has limitations for predicting future value. As a result, it ought to be strengthened with tenuous information that acts as a sign of expectation for the next 24 h on typical business days.

El Khantach et al. (El Khantach et al., 2019) employed five artificial intelligence (AI) techniques for momentary load anticipating with an underlying disintegration of the real information carried out irregularly into time series of each hour of the day, which finally consisted of 24 timeseries that addressed each preceding hour. The five AI techniques employed are Multi-facet Perceptron (MLP), Support Vector Machine (SVM), Outspread Premise Capability (RBF) Regressor, REPTree, and Gaussian Interaction. Trial and error were carried out in light of the data from the Moroccan electrical burden information. With a MAPE level of 0.96, the results showed that the MLP approach was the most dependable. SVM came in second and, despite performing significantly worse than MLP, was still superior to the other methods.

Expectations could also be created in light of the order-based AI strategy, which is commonly employed for energy usage forecasting, even though Gonzalez-Brioneset al. (2019) concentrated on the relapse technique. The inquiry developed a predictive model by examining the verifiable data collection using Direct Relapse (LR), Backing Vector Regression (SVR), Irregular Timberland (RF), Choice Tree (DT), and k-Nearest Neighbour (k-NN). The exploration's boundaries also contained a further variable known as one-day power usage (kWh). The outcomes showed that, with a score of 85.7%, the LR and SVR models delivered the best correct presentation. The hour cost and apex power-restricting based request reaction approaches serve as the foundation for this planning. They also offered a credible experiment to back up their timetable. The test showed a considerable reduction in the quantity of energy utilized by the various equipment because of the timetable they prepared. Creators in [7] are proposing the development of a home energy the board framework in order to select the optimal day-ahead planning for the various machines.

Sou Family Cheong et al. presented a planning method for intelligent home equipment in light of mixed number straight programming in [8]. They also took into account the machines' usual span and peak power usage. The suggested strategy resulted in cost savings of roughly 47% when compared to a previously specified duty. The authors also showed that with almost minimum computing effort, generally excellent layouts could be created.

There has been a lot of work put into addressing various initiatives to forecast how much energy would be used by different devices in relation to expectations for energy usage. Elkonomou made the expectancy method suggestion in [9] in light of the false brain arrangement. In order to choose the design with the best hypothesis, a number of tests were run using the multi-facet perceptron model. Actual information about the data and outcomes was used during all stages of preparation, approval, and testing.

The importance of the structure's energy usage expectation for the board and efficient energy management is emphasized by the authors of [10]. In order to meet the expectation, they are adopting a model that is information-driven and takes into consideration the forecast for energy use. According to the survey, there are numerous gaps in the field of energy utilization forecasting that need to be solved, including the prediction of long-distance energy use, the prediction of energy used inside of private structures, and the prediction of energy used for structure illumination. This lack of exploration may result from the very scant amount of knowledge that is currently available.

3 Existing System

SVM utilized in the Current arrangement of the issue proclamation. Large informative collections are not a good fit for SVM calculations. When the informational index is more crowded, as is the case when target classes are being covered, SVM doesn't function very well. The SVM won't perform as expected when the number of elements for each information point exceeds the number of information tests that need to be prepared.

4 Proposed Method

Utilizing individual power utilization dataset, We tested the suggested method using a dataset that is freely available from the UCI AI repository and contains details on electricity usage.The Dataset has 198721 lines × 6 sections. We train every irregular timberland calculation and direct relapse and lattice search cv model on the train set utilizing all highlights and afterward assess them on the whole test set. To quantify execution over the long run.

We use the scikit-learn implementation of the following mentods.

1. Random Forest
2. Linear Regression
3. Grid search cv and
4. Adaboost.

The accuracy of neural networks is high if the datasets provide appropriate training. Increasing the accuracy score, Large amount of feature we are taking for the training and testing. The basic architecture is shown in Fig. 1.

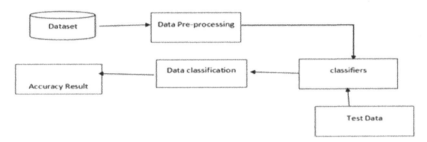

Fig. 1. System Architecture

5 Methodology

Data Gathering,
 preprocessing of the data,
 feature extraction,
 evaluation model, and.
 user interface.

5.1 Data Gathering

This paper's information assortment comprises of various records. The determination of the subset of all open information that you will be working with is the focal point of this stage. Preferably, ML challenges start with a lot of information (models or perceptions) for which you definitely know the ideal arrangement. Marked information will be data for which you are as of now mindful of the ideal result.

5.2 Pre-processing of Data

Format, clean, and sample from your chosen data to organize it.

There are three typical steps in data pre-processing:

Designing.

Information cleaning.

Inspecting.

Designing: It's conceivable that the information you've picked isn't in a structure that you can use to work with it. The information might be in an exclusive record configuration and you would like it in a social data set or text document, or the information might be in a social data set and you would like it in a level document.

Information cleaning; is the most common way of eliminating or supplanting missing information. There can be information examples that are inadequate and come up short on data you assume you really want to resolve the issue. These events could should be eliminated. Moreover, a portion of the traits might contain delicate data, and it very well might be important to anonymize or totally eliminate these properties from the information.

Inspecting: You might approach significantly more painstakingly picked information than you want. Calculations might take significantly longer to perform on greater measures of information, and their computational and memory prerequisites may likewise increment. Prior to considering the whole datasets, you can take a more modest delegate test of the picked information that might be fundamentally quicker for investigating and creating thoughts.

5.3 Feature Extraction

The following stage is to A course of quality decrease is include extraction. Highlight extraction really modifies the traits instead of element choice, which positions the ongoing ascribes as indicated by their prescient pertinence. The first ascribes are straightly joined to create the changed traits, or elements. Finally, the Classifier calculation is utilized to prepare our models. We utilize the Python Normal Language Tool stash's classify module.

We utilize the gained marked dataset. The models will be surveyed utilizing the excess marked information we have. Pre-handled information was ordered utilizing a couple of AI strategies. Irregular woodland classifiers were chosen. These calculations are generally utilized in positions including text grouping.

5.4 Assessment Model

Model the method involved with fostering a model incorporates assessment. Finding the model that best portrays our information and predicts how well the model will act in what's to come is useful. In information science, it isn't adequate to assess model execution utilizing the preparation information since this can rapidly prompt excessively hopeful and overfitted models. Wait and Cross-Approval are two procedures utilized in information science to evaluate models.

The two methodologies utilize a test set (concealed by the model) to survey model execution to forestall over fitting. In light of its normal, every classification model's presentation is assessed. The result will take on the structure that was envisioned. Diagram portrayal of information that has been ordered.

5.5 Algorithm

Random Forest

An AI technique called Random Forest is outfit-based and operated. You can combine various computation types to create a more convincing forecast model, or use a similar learning technique at least a few times. The phrase "Irregular Timberland" refers to how the arbitrary woodland method combines a few calculations of the same type or different chosen trees into a forest of trees. The irregular timberland technique can be used for both relapse and characterization tasks. Coming up next are the essential stages expected to execute the irregular woods calculation. Pick N records aimlessly from the datasets. Utilize these N records to make a choice tree. Select the number of trees you that need to remember for your calculation, then, at that point, rehash stages 1 and 2. Each tree in the timberland predicts the classification to which the new record has a place in the order issue. The classification that gets most of the votes is at last given the new record. The Advantages of Irregular Woodland the way that there are numerous trees and they are completely prepared utilizing various subsets of information guarantees that the irregular timberland strategy isn't one-sided. The irregular woods strategy fundamentally relies upon the strength of "the group," which reduces the framework's general predisposition. Since it is extremely challenging for new information to influence every one of the trees, regardless of whether another information point is added to the datasets, the general calculation isn't highly different. In circumstances when there are both downright and mathematical highlights, the irregular woods approach performs well. At the point when information needs esteems or has not been scaled, the irregular woodland method likewise performs well.

Linear Regression

Linear regression Considering how simple the portrayal is, it makes for an appealing model. The response is the anticipated result for the given arrangement of information values (y), and the portrayal is a direct condition that joins that set of information values (x). As a result, both the information value (x) and the result value (e) are numerical. Under the straight condition, each information worth or segment is given one scale variable, known as a coefficient and symbolized by the capital Greek letter Beta (B). The line also receives a second coefficient, commonly known as the catch or inclination coefficient, which increases its level of opportunity (for example, allowing it to completely circle a two-layered map).

For example, the model type in a straightforward relapse situation (one x and one y) would be

$$y = B0 + B1^*x \tag{1}$$

When we have more than one piece of information (x), the line is referred to as a plane or a hyper-plane in higher aspects. The condition is depicted by the type of circumstance and the specific characteristics utilized for the coefficients (for example, B0 and B1 in the aforementioned model). The intricacy of a simple relapse model of relapse is frequently discussed. This is a reference to the model's total number of coefficients. When a coefficient hit zero (0 * x = 0), the information variable's influence on the model and, subsequently, on the forecast made using the model, is successfully eliminated. This is significant if you consider regularization strategies, which alter the learning calculation to reduce the complexity of relapse models by reducing the overall size of the coefficients and eventually pushing some to zero.

GridSearchCV

In almost every AI project, we train a variety of models on the dataset and choose the one that exhibits the best results. In any event, there is room for improvement because we cannot state categorically that this particular model is the best for the main issue. Our goal is to develop the model in every way possible as a result. One important aspect of these models' presentations is their hyperparameters; by setting appropriate values for these hyperparameters, a model's presentation can be significantly improved. The most popular method for determining the ideal hyperparameter values for a given model is GridSearchCV. As previously said, the value of hyperparameters is crucial to how well a model exhibits. Remember that it is practically impossible to predict in advance which hyperparameters have the finest qualities, thus it is preferable to try all conceivable qualities before deciding which ones are the best. We utilize GridSearchCV to automate the tweaking of hyperparameters because doing it physically might require some effort and resources. The model selection package of Scikit-learn (or SK-learn) has a feature called GridSearchCV. Therefore, it is important to note that we really want the Scikit Learn library to be introduced on the PC. With the help of this capability, you may fit your assessor (model) to your training set and iterate through specified hyperparameters. In the end, selecting the best boundaries from the recorded hyperparameters is possible.

Hyper-boundary tuning alludes to the course of find hyper-boundaries that yield the best outcome. This, obviously, sounds significantly more straightforward than it really is. Finding all that hyper-boundaries can be a subtle craftsmanship, particularly given that it relies generally upon your preparation and testing information. As your information develops, the hyper-boundaries that were once high performing may no longer perform well. Monitoring the outcome of your model is basic to guarantee it develops with the information. One method for tuning your hyper-boundaries is to utilize a Matrix search. This is presumably the least complex strategy as well as the absolute most rough. In a matrix search, you attempt a framework of hyper-boundaries and assess the presentation of every mix of hyper-boundaries. The GridSearchCV class in Sklearn fills a double need in tuning your model. The class permits you to apply a framework search to a variety of hyper-boundaries, and Cross-approve your model utilizing k-overlay cross approval. The interaction pulls a segment from the accessible information to make train-test values. It rehashes this cycle on different occasions to guarantee a decent evaluative split of your information. The capability that takes various boundaries. We should investigate these in somewhat more detail:

Estimator: It takes an assessor object, for example, a classifier or a relapse model.

Param grid: It takes a word reference or a rundown of word references. The word references ought to be key-esteem matches, where the key is the hyper-boundary and the worth are the instances of hyper-boundary values to test.

Cv: It takes a number that decides the cross-approval methodology to apply. On the off chance that None is passed, 5 is utilized.

Scoring: It takes a string or a callable. This addresses the technique to assess the exhibition of the test set.

n_jobs: It addresses the quantity of tasks to run in equal. Since this is a tedious cycle, running more positions in equal (in the event that your PC can deal with it) can accelerate the cycle.

verbose: It decides how much data is shown. Involving a worth of 1 shows the ideal opportunity for each run. 2 shows that the score is additionally shown. 3 demonstrates that the overlap and up-and-comer boundary are additionally shown.

Ada Boosting Classifier

Ada-boost or Adaptive Boosting is one of the help group classifications made by Yoav Freund and Robert Schapire in 1996. It mixes various classifiers to improve classifier precision. AdaBoost is an iterative outfit approach. The AdaBoost classifier builds regions of strength for a, providing you high areas of strength for exactness by combining many classifiers that combine inefficiently. Adaboost's main principle is to set up the classifier loads and get ready for each cycle's information test to the point where it guarantees precise forecasts of unanticipated impressions. The fundamental classifier can be any AI computation that recognizes loads on the training set. Adaboost must abide by two conditions. The classifier needs to be prepared intelligently using a number of weighed preparation models. In order to provide these samples with the greatest fit possible throughout each iteration, it works to decrease training error.

How does the AdaBoost algorithm work? Here is how it works. A training subset is originally selected by Adaboost at random. It iteratively trains the AdaBoost AI model by choosing the preparation set in consideration of the precise expectation of the prior preparation. It gives incorrectly characterized perceptions a heavier burden, increasing their likelihood of grouping in the attention that follows. Additionally, it transfers the burden to the trained classifier in each emphasis in accordance with the classifier's accuracy. The classifier that is more accurate will be given more weight. This cycle repeats until there are the predefined maximum number of assessors or until the entire preparation information fits with virtually minimal error. Play out a "vote" involving all of the artificial learning computations to determine the ranking. The level of precise expectations for the test information is implied by precision. By partitioning the quantity of exact expectations by the complete number of forecasts, it very well might still up in the air.

5.6   User Interface and Result

The pattern of Information Science and Examination is expanding step by step. From the information science pipeline, one of the main advances is model sending. We have a ton of choices in python for sending our model. A few well-known systems are Carafe and Django. Yet, the issue with utilizing these systems is that we ought to have some information on HTML, CSS, and JavaScript. Remembering these requirements, Adrien Trouville, Thiago Teixeira, and Amanda Kelly made "Streamlit". Presently utilizing streamlit you can send any AI model and any python project easily and without stressing over the frontend. Streamlit is very easy to use.

In this article, we will get familiar with a few significant elements of streamlit, make a python project, and convey the task on a nearby web server. How about we introduce streamlit. Type the accompanying order in the order brief.

pip installs streamlit.

When Streamlit is introduced effectively, run the given python code and in the event that you don't get a mistake, then streamlit is effectively introduced and you can now work with streamlit. Figure 2. Shows the user interface to execute the code.

Fig. 2. User Interface

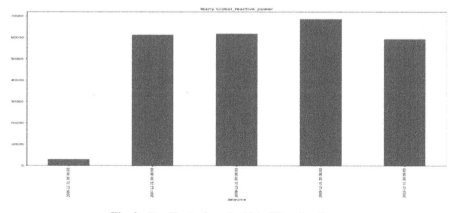

Fig. 3. Bar Chart of yearly Global Reactive Power

Figure 3. Shows bar chart for yearly global reactive power consumption.

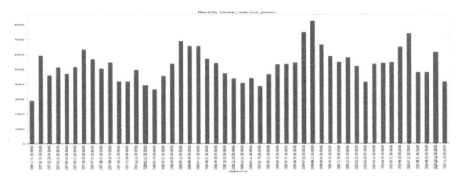

Fig. 4. Monthly Global reactive Power Bar chart

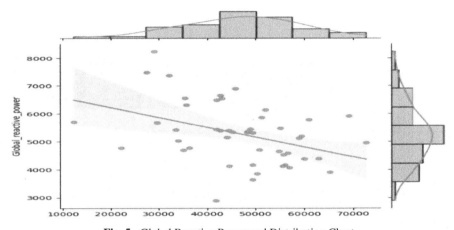

Fig. 5. Global Reactive Power and Distribution Chart

Figure 4. Shows a bar chart for monthly global reactive power consumption.

Figure 5. Shows scatter chart for monthly global reactive power consumption and distribution.

Figure 6. Shows monthly global reactive power consumption.

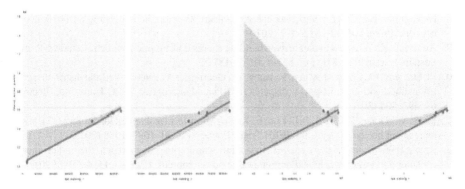

Fig. 6. Subplots of Global Active Power

6 Conclusion

Machine learning (ML) techniques has as of late contributed very well in the headway of the expectation models utilized for power utilization. Such models profoundly work on the exactness, heartiness, and accuracy and the speculation capacity of the ordinary time series anticipating instruments. By using the historical data we can predict future power consumption. Here, we used the electric power consumption data of one household and applied linear regression, ada boost, grid search cv and random forest algorithms and we achieved an linear regression accuracy with 99% and applied random forest algorithm with 93% accuracy, and applied adaboost algorithm with 97% accuracy and we achieved an grid search cv accuracy with 82% for future prediction.

References

1. Corgnati, S., Bednar, T., Jang, Y., Yoshino, H., Ghiaus, C.: Total Energy Use in Buildings. Analysis and Evaluation Methods, Final Report Annex 53. Statistical Analysis and Prediction Methods (2013)
2. Fu, Y., Li, Z., Zhang, H., Xu, P.: Using support vector machine to predict next day electricity load of public buildings with sub-metering device. Procedia Eng. **12**, 1016–1022 (2015)
3. Valgaev, O., Kupzog, F., Schmeck, H.: Low-voltage power demand forecasting using K-nearest neighbors approach. In: IEEE Innovative Smart Grid Technologies - Asia (ISGT - Asia), pp. 1019–1024 (2016)
4. El Khantach, A., Hamlich, M., Belbounaguia, N.E.: Short-term load forecasting using machine learning and periodicity decomposition. AIMS Energy **7**(3), 382–394 (2019)
5. González-Briones, A., Hernández, G., Corchado, J.M., Omatu, S., Mohamad, M.S.: Machine learning models for electricity consumption forecasting: a review. In: IEEE Virtual-Ledgers-Tecnologías DLT/Blockchain y Cripto-IOT (2019)
6. Paterakis, N.G., Erdinç, O., Bakirtzis, A.G.: Catalão JPS Optimal household appliances scheduling under day-ahead pricing and load-shaping demand response strategies. IEEE Trans Ind Inform **11**(16), 1509–1519 (2015)
7. Cheong, S.K.: Scheduling smart home appliances using mixed integer linear programming. In: 50th IEEE conference on decision and control and European control conference, IEEE. https://doi.org/10.1109/cdc.2011.6161081(2011)

8. Ekonomou, L.: Green long-term energy consumption prediction using artificial neural network. Energy **35**(12), 512–517 (2010)
9. Amasyali, K.: A review of data-driven building energy consumption prediction studies. Renew Sustain. Energy Rev. **81**(11), 1192–1205 (2018)
10. Li, Q., Peng, C., Chen, M., Chen, F., Kang, W., Guerrero, J.M.: Networked and distributed control method with optimal power dispatch for islanded microgrids. IEEE Trans. Ind. Electron. **64**, 493–504 (2016)
11. Kong, W., Dong, Z.Y., Hill, D.J., Luo, F., Xu, Y.: Short-term residential load forecasting based on resident behaviour learning. IEEE Trans. Power Syst. **33**, 1087–1088 (2017)
12. Khan, P.W., Byun, Y.C., Lee, S.J.: Machine learning-based approach to predict energy consumption of renewable and nonrenewable power sources. Energies **13**, 4870 (2020)
13. Ahmad, A., Hassan, M., Abdullah, M., Rahman, H., Hussein, F., Abdullah, H.: A review on applications of ANN and SVM for building electrical energy consumption forecasting. Renew. Sustain. Energy Rev. **33**, 102–109 (2014)
14. Zhang, M., Bai, C.: Exploring the influencing factors and decoupling state of residential energy consumption in Shandong. J. Cleaner Prod. **194**, 253–262 (2018)

Medical Plants Identification Using Leaves Based on Convolutional Neural Networks

B Ch S N L S Sai Baba$^{(\boxtimes)}$ ⓘ, Mudhindi Swathi ⓘ, Kompella Bhargava Kiran ⓘ,
B. R. Bharathi ⓘ, Venkata Durgarao Matta ⓘ, and CH. Lakshmi Veenadhari ⓘ

Computer Science and Engineering Department, Vishnu Institute of Technology, Bhimavaram,
Andhra Pradesh, India
sai.ossr524@gmail.com

Abstract. The ayurvedic medicines have played a crucial role in health system, only a few experts could identify the herbs and know the ayurvedic properties of these herbs. These medicines prepared from the herbs having less side effects as compared to other general medicines. Most of the patients and general medicine users with different diseases are not unaware of the existence of herbal plants and their medical uses and benefits. To make ease of identifying the plants and its medical properties based on the leaf structure, authors developed a system having three architectures which works with Convolutional Neural Networks. Resnet-18, Resnet-50, MobileNet-V2 architectures were used in freeze and unfreeze layers settings. Authors considered ten different kinds of herbal leaves for implementation of the system, in which two thirds of the data used for training and one third for testing. The overall performance of this architecture is checked using accuracy measure and it is observed that three models with freeze layers were showing good performance. Out of these three architectures, Resnet-50 shown accuracy of 95.33%.

Keywords: Image Processing · CNN · Medical Plants · Computer Vision

1 Introduction

Herbal plants symbolize biodiversity, which is frequently employed as an alternative to conventional medicine. Almost every part of a herbal plant, notably the leaves, can be employed as a component in traditional medicine. This is thus because, in contrast to fruit and roots, leaves are simpler to acquire. Herbal plants play a crucial part in maintaining human health [1]. Given that medical treatment is not accessible to everyone and is expensive in comparison to medical treatment, nearly 80% of people still rely on traditional medicine [2]. For ages, people have employed herbal plants to prevent illness and treat it [3]. Due to the variety of therapeutic plant kinds [4] and the difficulty in differentiating between them, the public is currently unaware of the existence of herbal plants. To recognize and distinguish between these kinds of herbal plants, one needs significant expertise and information. It is important to conserve knowledge of herbal plants so that people can more easily identify the different kinds and use them when

© ICST Institute for Computer Sciences, Social Informatics and Telecommunications Engineering 2024
Published by Springer Nature Switzerland AG 2024. All Rights Reserved
P. Pareek et al. (Eds.): IC4S 2023, LNICST 536, pp. 163–171, 2024.
https://doi.org/10.1007/978-3-031-48888-7_14

necessary [5]. In general, plant species are still identified manually by comparing and recognizing photos, particularly leaf photographs when the data is already known with the leaves on the plant. Manual identification, meanwhile, still leaves room for mistakes. This is due to the nearly identical leaf color and similar texture and shape of various varieties of herbal plants. Botanists and other individuals with specialized knowledge of herbal plants are the only ones qualified to present different varieties of herbal plants in this manner.

Since not everyone has the requisite knowledge and experience in this area, the community views this method of introducing herbal plants as ineffective for differentiating between the various types of herbal plants [6].

Numerous researchers have used technological advancements and identification of these plants' leaves to conduct research on the classification of herbal plants. Herbal leaf identification is simpler since leaves are more important and accessible than roots [7], which are a portion of the plant that are buried in the ground [8]. Consequently, a system for intelligent and precise herbal leaf identification is required. Most of the earlier research has been devoted to identifying leaves. Numerous classification techniques are employed in the identification system to assist users in recognizing herbal leaves without specialized botanical or anatomical knowledge by identifying the types of herbal plants through leaf identification.

2 Related Work

Recently, research has been done to identify the various kinds of herbal leaves. The study by [9] selected therapeutic herbs in order to assess their content value. Computer vision-based feature extraction was used in this investigation. Using ITS2 Sequences and Multiplex-SCAR Markers, [10] authenticated herbal aralia plants carried out another experiment. A categorization of herbal leaves using the SVM method was also done in a different study [11]. Scale Invariant Feature Transform (SIFT) technology was used to extract the picture feature. To combat affine transformations, noise, and changing lighting, the SIFT functionality was deployed. Laws' mask analysis and SVM as the classifier were also used to classify 5 different types of leaves [12]. The obtained accuracy is 90.27%. Using the CNN approach, the identification of Thai medicinal herbs was carried out in [13]. In this paper we are going to train our model on leaves of plants which are mainly main found in India. 10 types of herbal plants are considered for this approach. These are Apta: Bauhinia Racemosa, vad: Ficus Benghalensis, Indian Rubber Tree: Ficus Elastica Roxb, Ex Hornem, Karanj: Pongamia Pinnata, Kashid: Senna Siamea (lam.) Irwin & Barneby, Sita Ashok: Saraca Asoka (roxb.) Willd, Pimpal: Ficus Religiosa, Nilgiri: Eucalyptus Globulus, Sonmohar: Peltophorum Pterocarpum, Villayati Chinch: Pithecellobium Dulce [14]. The approach uses Resnet18, Resnet50, MobilenetV2 and MobilenetV3 convolutional neural network to build the models.

3 Architectures

In this paper, 3 architectures with 2 variations each are used. They include Resnet 18 with freeze, Resnet 18 without freeze, Resnet 50 with freeze, Resnet 50 without freeze.

3.1 Resnet 18

In the article "Deep Residual Learning for Image Recognition" by Kaiming He et al., ResNet-18, a convolutional neural network architecture, was introduced [15]. It is one of the most compact versions of the ResNet model family, which is renowned for its excellence in image recognition tasks.

18 layers make up the ResNet-18 architecture, containing 16 convolutional layers and 2 fully linked layers. The network can learn more effectively thanks to the architecture's use of residual connections, which propagate information between the layers without causing information loss. In order to do this, skip connections that omit one or more network tiers are added.

The first layers of ResNet-18, such as the convolutional layers, batch normalization, and activation functions, are comparable to those of other convolutional neural networks. ResNet-18's usage of residual blocks, which are made up of two or more convolutional layers and a skip connection that omits one or more of the layers in the block, is what makes it special.

Multiple image recognition tasks, such as classification, object detection, and segmentation, have been accomplished using ResNet-18. On numerous benchmark datasets, including ImageNet, CIFAR-10, and CIFAR-100, it has been demonstrated to produce state-of-the-art results. It is a well-liked option for applications where computer resources are scarce due to its tiny size and relatively low computational complexity.

3.2 Resnet 50

Convolutional neural network ResNet-50 was first presented in the article "Deep Residual Learning for Image Recognition" by Kaiming He et al. [16]. It is one of the more robust models in the ResNet family, which is renowned for its excellence in image recognition tasks.

50 layers make up the ResNet-50 architecture, containing 48 convolutional layers and 2 fully linked layers. It makes use of residual connections, which help the network learn more quickly by transferring knowledge without losing it as it moves through the layers. In order to do this, skip connections that omit one or more network tiers are added.

ResNet-50 is more sophisticated than ResNet-18 and can therefore learn more intricate features from the data. It can also learn more sophisticated representations of the input data because it has larger residual blocks, more convolutional layers, and filters.

Multiple image recognition tasks, such as classification, object detection, and segmentation, have been accomplished using ResNet-50. On numerous benchmark datasets, including ImageNet, COCO, and PASCAL VOC, it has been demonstrated to produce state-of-the-art results. It is a suitable option for activities requiring a high degree of precision and processing resources because of its size and computational complexity.

Overall, ResNet-50 is a potent and popular deep learning model that has shown promise in a variety of image identification applications.

3.3 Mobilenet V2

In 2018, Google unveiled MobileNetV2, a convolutional neural network architecture [17]. It is specifically made for embedded and mobile devices, and it is optimized for high precision and minimal computational expense.

By combining depthwise separable convolutions with linear bottlenecks to lower the network's computing cost, MobileNetV2 builds on the original MobileNet architecture. In depthwise separable convolutions, each input channel is subjected to a single filter first, and then the output channels are combined using a 1x1 filter in a pointwise convolution. Comparatively speaking to conventional convolutions, this has fewer parameters and lower processing costs.

To boost the network's nonlinearity without introducing new parameters, linear bottlenecks are used. They are made up of a 1x1 convolution followed by a ReLU activation function, a 3x3 depthwise separable convolution, and a final 1x1 convolution. By reducing the number of parameters and computational cost, the network can learn more intricate features.

To make better use of the network's capacity, MobileNetV2 also has a feature known as inverted residuals. The linear bottleneck, expansion layer, and subsequent linear bottlenecks make up an inverted residual. The expansion layer expands the number of data channels. Due to this, the network can learn more complicated features without having to add more parameters or pay for more compute.

It has been demonstrated that MobileNetV2 can perform at the cutting edge on a range of image recognition tasks, including segmentation, object detection, and classification. It is frequently utilized in embedded and mobile applications when high accuracy is needed but computational resources are constrained.

4   Experimental Setup

The leaf dataset consists of 3000 images with 10 classes and each class has 300 images. We have trained our models using 70% of data and reserved 30% for validation. Before applying the model, preprocessing is performed on the dataset. At first, input images are scaled to 224 × 224, then these images are randomly flipped horizontally with a frequency of 0.5 when using the RandomHorizontalFlip transformation [18]. It indicates that there is a 50% likelihood that each image will be horizontally inverted. The image is horizontally mirrored yet the alteration does not alter the image's content. Each pixel in the image is represented as a floating-point value between 0 and 1, with 0 denoting black and 1 denoting white, and it is converted from a NumPy array or PIL image object into a tensor object. Using mean and standard deviation values, the "Normalize" transformation normalizes the image's pixel values. Based on the statistics of the dataset used for model training, the mean and standard deviation values are computed. The Normalize transformation ensures that the input data have similar statistical features to the data used to train the model by normalizing the input picture. This may aid in enhancing the model's generalizability and accuracy. The dataset has 10 folders with name of the class and has 300 images of that class. By using split function, the data is divided into features and classes. In this model we have initialized the weights from IMAGENET1K_V1 [19]. A deep learning model's pre-trained weights that have been learned on the IMAGENET1K_V1 dataset for image classification tasks are referred to as "IMAGENET1K_V1 weights." When a new deep learning model is created and trained on a similar dataset, these weights are often utilized to set its model parameters. The learnt parameters of the deep learning model make up the weights, which are typically in the form of a sizable file. These parameters, which are determined during the training phase on the IMAGENET1K_V1 dataset, include the weights and biases of the various layers of the model. In order to prevent gradients from earlier iterations from interfering with the gradients calculated for the current batch of data, we first set all gradients to zero. All the network must be frozen here, except for the top layer. In order to prevent the gradients from being computed in backward(), we must set "requires_grad = False" to freeze the parameters. For the model's optimization, stochastic gradient descent is applied. The model parameters are updated using tiny batches of data rather than the complete dataset at once, which is a variation on the gradient descent approach. When using SGD, the model parameters are adjusted in the direction of the loss function's inverse gradient with respect to the parameters.

$$\theta = \theta - \eta . \nabla_\theta J(\theta; x^{(i)}; y^{(i)}) \tag{1}$$

where, $J(\theta)$ is the objective function, η is learning rate, ∇ is the gradient of objective functions. Especially for big datasets, this speeds up and improves the computational efficiency of computing the gradient. One of the hyperparameters that controls how big of a step is taken in the direction of the negative gradient is learning rate. It is commonly set to a low value, like 0.01 or 0.001, and can be changed during training to enhance the performance of the model. In this paper, the learning rate is set to 0.001[20]. We have used LR schedular to adjust the learning rates. A learning rate scheduler is a method for modifying the learning rate during training to enhance the model's performance [21]. A hyperparameter called learning rate regulates how big of a step the optimization method takes when updating parameters [22]. We have used 25 epochs to train the model. The loss of the model is calculated using Cross Entropy Loss [23]. To calculate the cross-entropy loss, the function combines the softmax function and the negative log likelihood loss function. The softmax function turns a set of logits, or unnormalized scores, from the model's output into a probability distribution over the classes.

$$f(s)_i = \frac{e^{S_i}}{\sum_j^C e^{S_j}} \quad CE = -\Sigma_i^c t_i \log(f(s)_i) \tag{2}$$

where, S_i is input in the form of one hot encoded matrix, e is the exponent, C is number of classes, t_i and S_i are the ground through variables. The difference between the target class's actual probability distribution and the anticipated probability distribution is then measured by the negative log likelihood loss function. The model is then connected to a fully connected network which has 10 outputs.

5 Results

By using the above setup, Resnet 18 architecture is used to build the model. The accuracy of model stands at 88.16% as in Fig. 1 (left). A variation of the above model is also built as shown in Fig. 1 (Right). In this case, freezing of the layers is not performed to block backpropagation. In this case the accuracy 94.16%. The same setup is used with Resnet 50 architecture as in Fig. 2. The accuracy of model with and without accuracy is 90.50% and 95.33% respectively. In case of MobileNet V2, we have not used bias and there is no fully connected layer. The accuracy of the model with freeze is 92.83%. In the same way, the accuracy is 93.66% without freezing of layers.

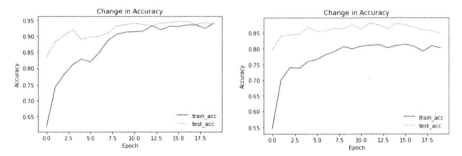

Fig. 1. (Left) Accuracy w.r.t to number of epochs using Resnet 18 architecture without freeze. (Right) Accuracy w.r.t to number of epochs using Resnet 18 architecture with freeze.

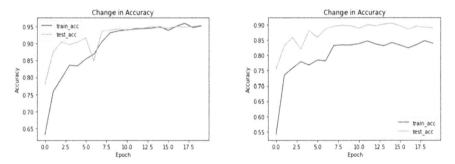

Fig. 2. (Left) Accuracy w.r.t to number of epochs using Resnet 50 architecture without freeze. (Right) Accuracy w.r.t to number of epochs using Resnet 50 architecture with freeze.

Figure 3 demonstrates the accuracy of MobileNet V2 w.r.t to number of epochs. The graphs demonstrating the models' accuracy in relation to the number of epochs are shown above. We have always considered 20 epochs because overfitting has been shown above this. Resnet 50 without freezing has outperformed among all the models. The model's output, which is displayed below in Fig. 4, uses plant identification based on leaves.

The performance of all models, both with and without freezing layers, is summarized in the table below with respect to the number of parameters, epochs, the optimizer, and accuracy (Table 1).

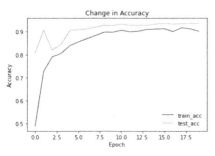
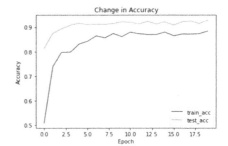

Fig. 3. (Left) Accuracy w.r.t to number of epochs using MobileNet-V2 architecture without freeze. (Right) Accuracy w.r.t to number of epochs using MobileNet-V2 architecture with freeze.

Fig. 4. Shows sample predictions made by the above architectures.

Table 1. Model performance w.r.t to number of parameters

Architectures	Total Parameters	Trainable Parameters	Epochs	Optimizer	Accuracy
Resnet-18 with freeze	11,181,642	0	20	SGD	88.16
Resnet-18 without freeze	11,181,642	5,130	20	SGD	94.16
Resnet-50 with freeze	23,528,522	20,490	20	SGD	90.50
Resnet-50 without freeze	23,528,522	23,528,522	20	SGD	95.33
MobileNet-V2 With freeze	2,236,682	1,694,154	20	SGD	92.83
MobileNet-V2 without freeze	2,236,682	2,236,682	20	SGD	93.66

6   Conclusion and Future Scope

We were able to identify 10 different kinds of leaves of plants having good medical properties. Most of the plants are used in Ayurveda for curing many diseases. Based on 3000 photos of medicinal plants, we developed 6 image classification models (Resnet-18 with freeze, Resnet-18 without freeze, Resnet-50 with freeze, Resnet-50 without freeze,

MobileNet-V2 With freeze, MobileNet-V2 without freeze) for this study. The dataset of leave images is publicly available. We found that Resnet- 50 architecture without freeze layers has outperformed other models. We have used accuracy to measure the performance of the model and got 95.33% in this setting. By creating a mobile application and integrating more classes of leaves, this study can be furthered.

References

1. Sofowora, A., Ogunbodede, E., Onayade, A.: The role and place of medicinal plants in the strategies for disease prevention. Afr. J. Tradit. Complement. Altern. Med. **10**(5), 210–229 (2013). https://doi.org/10.4314/ajtcam.v10i5.2
2. Ekor, M.: The growing use of herbal medicines: issues relating to adverse reactions and challenges in monitoring safety. Front. Pharmacol. **4**, 177 (2014). https://doi.org/10.3389/fphar.2013.00177
3. Eddouks, M., Chattopadhyay, D., De Feo, V., Cho, W.C.: Medicinal plants in the prevention and treatment of chronic diseases. Evid Based Complement Alternat Med. **2012**, 458274 (2012). https://doi.org/10.1155/2012/458274
4. Firenzuoli, F., Gori, L.: Herbal medicine today: clinical and research issues. Evid Based Complement Alternat Med. **4**(Suppl 1), 37–40 (2007). https://doi.org/10.1093/ecam/nem096
5. Chen, S.L., Yu, H., Luo, H.M., Wu, Q., Li, C.F., Steinmetz, A.: Conservation and sustainable use of medicinal plants: problems, progress, and prospects. Chin Med. **11**, 37 (2016). https://doi.org/10.1186/s13020-016-0108-7
6. Welz, A.N., Emberger-Klein, A., Menrad, K.: Why people use herbal medicine: insights from a focus-group study in Germany. BMC Complement. Altern. Med. **18**, 92 (2018). https://doi.org/10.1186/s12906-018-2160-6
7. Wachtel-Galor, S., Benzie, I.F.F.: Herbal medicine: an introduction to its history, usage, regulation, current trends, and research needs. In: Benzie, I.F.F., Wachtel-Galor, S. (eds.) Herbal Medicine: Biomolecular and Clinical Aspects, Chapter 1. CRC Press/Taylor & Francis (2011). https://www.ncbi.nlm.nih.gov/books/NBK92773/
8. Bhat, S.G.: Medicinal plants and its pharmacological values. In: Natural Medicinal Plants, IntechOpen, May 11, 2022. https://doi.org/10.5772/intechopen.99848
9. Venkataraman, D., Nehru, M.: Computer vision based feature extraction of leaves for identification of medicinal values of plants. In: 2016 International Conference on Computational Intelligence and Communication Networks (CICN), pp. 1–5 (2016). https://doi.org/10.1109/ICCIC.2016.7919637
10. Kim, W., Moon, B., Yang, S., Han, K., Choi, G., Lee, A.Y.: Rapid authentication of the herbal medicine plant species aralia continentalis Kitag. and Angelica biserrata C.Q. Yuan and R.H. Shan using ITS2 sequences and multiplex-SCAR markers. Molecules **21**, 270 (2016). https://doi.org/10.3390/molecules21030270
11. Kaur, P., Singh, S.: Classification of Herbal Plant and Comparative Analysis of SVM and KNN Classifier Models on the Leaf Features Using Machine Learning. (2021).https://doi.org/10.1007/978-981-16-1048-6_17
12. Suh, H., Hofstee, J.W., IJsselmuiden, J., Van Henten, E.J.: Sugar beet and volunteer potato classification using Bag-of-Visual-Words model, Scale-Invariant Feature Transform, or Speeded Up Robust Feature descriptors and crop row information. Biosyst. Eng. **166**, 210–226 (2017). https://doi.org/10.1016/j.biosystemseng.2017.11.015
13. Mookdarsanit, L., Mookdarsanit, P.: Thai Herb Identification with Medicinal Properties Using Convolutional Neural Network (2019)

14. Rokde, V., Raut, P.: Leaf Images Dataset-Indian Trees leaf's Dataset for image classification problem statement, Version 1 (March 2023). Retrieved: 03/04/2023 from https://www.kag gle.com/datasets/ichhadhari/leaf-images

15. Ou, X., et al.: Moving object detection method via ResNet-18 with encoder–decoder structure in complex scenes. IEEE Access **7**, 108152–108160 (2019). https://doi.org/10.1109/ACCESS. 2019.2931922

16. Rifa'i, A.M., Utami, E., Ariatmanto, D.: Analysis for diagnosis of pneumonia symptoms using chest X-Ray based on Resnet-50 models with different epoch. In: 2022 6th International Conference on Information Technology, Information Systems and Electrical Engineering (ICITISEE), pp. 471–476 (2022). https://doi.org/10.1109/ICITISEE57756.2022.10057805

17. Sun, Q., Luo, X.: A new image recognition combining transfer learning algorithm and MobileNet V2 model for palm vein recognition. In: 2022 4th International Conference on Frontiers Technology of Information and Computer (ICFTIC), pp. 559–564 (2022). https:// doi.org/10.1109/ICFTIC57696.2022.10075212

18. Shorten, C., Khoshgoftaar, T.M.: A survey on image data augmentation for deep learning. J Big Data **6**, 60 (2019). https://doi.org/10.1186/s40537-019-0197-0

19. Jayakody, D.: Custom Image Classifier with PyTorch – A Step-by-Step Guide. Artificial Intelligence, Computer Vision- URL: https://dilithjay.com/blog/custom-image-classifier-with-pyt orch-a-step-by-step-guide/

20. Konar, J., Khandelwal, P., Tripathi, R.: Comparison of various learning rate scheduling techniques on convolutional neural network. In: 2020 IEEE International Students' Conference on Electrical, Electronics and Computer Science (SCEECS), pp. 1–5 (2020). https://doi.org/ 10.1109/SCEECS48394.2020.94

21. Nedunuri, S.U.D.: Crop disease prediction with convolution neural network (CNN) augmented with cellular. Int. Arab J. Inf. Technol. (IAJIT) **19**, 69–77 (2022). https://doi.org/10. 34028/iajit/19/5/8

22. Murthy, M.Y.B., Koteswararao, A., Babu, M.S.: Adaptive fuzzy deformable fusion and optimized CNN with ensemble classification for automated brain tumor diagnosis. Biomed. Eng. Lett. **12**, 37–58 (2022). https://doi.org/10.1007/s13534-021-00209-5

23. Rezaei-Dastjerdehei, M.R., Mijani, A., Fatemizadeh, E.: Addressing imbalance in multi-label classification using weighted cross entropy loss function. In: 2020 27th National and 5th International Iranian Conference on Biomedical Engineering (ICBME), pp. 333–338 (2020). https://doi.org/10.1109/ICBME51989.2020.9319440

An Efficient Real-Time NIDS Using Machine Learning Methods

Konda Srikar Goud$^{(\boxtimes)}$ ⓘ, M. Shivani, B. V. S. Selvi Reddy, Ch. Shravyasree, and J. Shreeya Reddy

Department of Information Technology, BVRIT Hyderabad College of Engineering for Women, Hyderabad, Telangana, India
kondasrikargoud@gmail.com

Abstract. Recent developments in network technology and related services have caused a significant rise in data traffic. However, there has also been a massive rise in the negative consequences of cyber-attacks. Many new types of network attacks are emerging. As a result, designing a robust Intrusion detection system (IDS) has become essential. This paper presents a framework for designing an efficient IDS to enhance detection accuracy and reduce false positives on real-time data. This research used the CIC-IDS 2017 dataset to train Machine Learning models such as Logistic Regression, K Nearest Neighbor, Gaussian Naive Bayes, and Random Forest. Machine learning models often perform well on benchmark datasets but may encounter challenges when applied to real-time traffic scenarios. So, we created a Real-time dataset and tested it on the trained models. In the evaluation, the Random Forest classifier outperformed all other models and achieved an accuracy of 99.99% .

Keywords: Intrusion Detection System · Cyber-attacks · DDoS attack · Botnet · Random Forest · Real-time dataset

1 Introduction

An IDS is a device which monitors network traffic for unusual activity and warns the user when it is discovered. Software that looks for harmful activity on a network or machine. Any unlawful behavior or violation is frequently noted, either centrally via a security information and event management (SIEM) system or by sending an administrator a notification. In order to discriminate between valid and false alerts, a SIEM system aggregates outputs from several sources and applies alarm filtering techniques. Intrusion detection systems are prone to raise erroneous warnings while monitoring networks for potentially dangerous behavior as a result, after first installation, businesses need to modify their IDS products. To discriminate between safe network traffic and malicious activity, intrusion detection systems must be properly configured. The two main categories of IDS are host-based and network-based ones. They are as follows: (1) Network Intrusion Detection (NIDS): A NIDS Is a security instrument that scans

P. Pareek et al. (Eds.): IC4S 2023, LNICST 536, pp. 172–185, 2024.
https://doi.org/10.1007/978-3-031-48888-7_15

network traffic for probable intrusions or malicious activity. It compares live packet analysis results to known attack patterns or anomalies. To record and examine traffic, NIDS sensors are positioned strategically throughout the network architecture. It produces notifications to inform administrators of any questionable activity that is found. By monitoring actual traffic and spotting assaults that may evade firewall regulations, NIDS support firewalls. (2) HIDS: A security technology known as a Host Intrusion Detection System keeps track of the actions and conduct of specific hosts or endpoints within a network. It concentrates on preserving the integrity and security of particular equipment, including servers, workstations, or IoT gadgets. In order to identify unauthorized access, malicious software, or unusual activity at the host level, HIDS employs a variety of approaches, including log analysis, file integrity verification, and system call monitoring.

In order to meet the needs of an efficient IDS, researchers have considered the possibility of using ML and DL techniques. ML and DL fall under the broad umbrella of Artificial Intelligence (AI) and are focused on learning useful information from large amounts of data. These techniques gained immense popularity in network security over the past decade due to the development of high-performance graphics processors (GPUs). Both ML and DL are effective tools for learning useful features from network traffic and for predicting normal or abnormal activities on the basis of the learned patterns. ML-based IDS relies heavily on feature engineering for learning useful information from network traffic. DL-based IDS don't rely on feature engineering and are capable of learning complex features automatically from the raw data because of its deep structure.

Section 2 discusses the literature review of various methods for detecting attacks. In Sect. 3, we discuss proposed model with algorithms. Section 4 of this article talks about simulation and results analysis. The Results are discussed in Sect. 5. Finally, we discuss conclusion in Sect. 6.

2 Literature Survey

Intrusion detection systems aim to identify and respond to unauthorized activities or malicious behavior within a network. Traditional IDS typically use rule-based methods that rely on pre-defined patterns or signatures of known attacks. However, these rule-based systems often struggle to detect novel or sophisticated attacks, which led to the exploration of ML-based IDS.

Authors in [1] demonstrates how ML and DL techniques were used on the CICIDS 2017 dataset. Four feature selection strategies are the subject of the work. The paper discusses the performance of these algorithms with accuracy. In [2], Researchers primarily study and assess the 1999 NSL- KDD datasets and the 1999 KDDCUP datasets using SVMs for intrusion detection. Additionally, in [3], researchers focuses on network intrusion detection system using machine learning classifiers on KDD99 dataset.

The study in [4] provides an overview of ensemble classifiers which are suitable for classifying data IDS with machine learning. In [5], researchers proposes an optimized intrusion detection system that improves upon the existing IDS which detects malicious packets in the network. Researchers in [6] explored six different ML techniques to find out the best technique for detecting DDOS attacks in machine learning. The authors in

[7], proposed a research project has been suggested to utilize a hybrid system that makes use of misuse detection methods for well-known sorts of invasions. By recognizing the assaults that are known thanks to anomaly detection systems, it also educates and trains itself. The major goal of is to build a real-time IDS that can detect intrusions by examining incoming and egressing network data in real-time. The deep neural network used in the proposed system was trained the model using 28 features from the NSL-KDD dataset.

Authors in [9] focuses on using machine learning methods for wireless sensor network intrusion detection. It explores the use of ML algorithms, including decision trees, SVM, k-nearest neighbors (KNN), and random forests, on the CICIDS 2017 dataset. The paper discusses the performance of these algorithms in detecting different types of intrusions and compares their effectiveness.

In [10], Researchers suggested that datasets like CSE-CIC-IDS2018 were made to train prediction algorithms on anomaly-based intrusion detection for network traffic. The study describes a paradigm for real-time intrusion detection [11]. In [12], researchers suggest a deep learning-based intrusion detection solution for industrial control systems (ICS). It trains deep learning models including deep neural networks (DNN) and long short-term memory (LSTM) networks using the CICIDS 2017 dataset, to find intrusions in ICS networks. The paper highlights the performance of different models and assesses how well they execute real-time intrusion detection.

Additionally, in [13], scientists concentrate on the creation of a DL-based anomaly-based network intrusion detection system. For the purpose of extracting pertinent characteristics from the authors' CICIDS 2017 dataset suggested an improved feature selection method. For intrusion detection, the study applies deep auto-encoders and deep belief networks, and it assesses how well they perform is measured by the F1-score, recall, accuracy, and precision.

The findings were summed up by the authors in [14], who also highlighted the potential of DL for detecting intrusions in IOT networks. The performance and resilience of the proposed system were both increased by the authors' recommendations for future research initiatives. The article in [15] gives an overview of the various feature selection methods used in utilizing machine learning to detect intrusions systems. It investigates various methodologies, including wrapper, filter, and embedding approaches, and how they apply to the CICIDS 2017 dataset. The article examines how feature selection affects the effectiveness of intrusion detection models and offers suggestions for choosing the right attributes for efficient detection.

3 Proposed Methodology

Here we'll discuss the proposed framework. The primary goal of this framework is to increase the model's accuracy. By identifying the optimal features using the feature importance. It is an attribute of Random Forest Classifier. The initial step is to pre-process the CICIDS2017 dataset. The preprocessing process begins with data cleaning, addressing missing values, outliers, and inconsistent data. The techniques used in data cleaning are binning, replacing with mean and median values. Feature selection techniques are then applied and selects the most relevant features. Feature scaling is performed to ensure that features are on a consistent scale. Categorical variables are encoded into numerical representations, and techniques are employed to handle imbalanced data, ensuring the dataset adequately captures both normal and at- tack samples. Finally, training and testing sets are created from the preprocessed dataset for model training and evaluation. These preprocessing steps ensure the quality, relevance, and compatibility of the CICIDS 2017 dataset with ML algorithms, ultimately enhancing the performance of the real-time IDS.

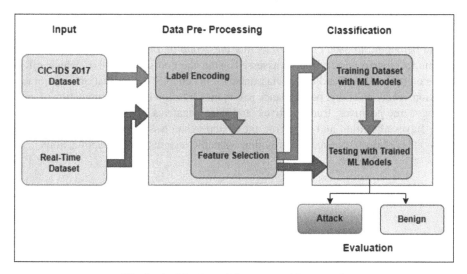

Fig. 1. Architecture of the proposed framework

The creation of a real-time intrusion detection system (IDS) is the aim of this effort. The preprocessing and feature selection of the CICIDS 2017 dataset are done. To build a predictive model, training is done using the Logistic Regression, KNN, Gaussian Naive Bayes, and Random Forest methods. The model is then tested on real-time data to detect attacks. By utilizing machine learning algorithms, the IDS aims to accurately and promptly identify malicious activities in network traffic, enabling proactive measures to be taken for maintaining network security. The Fig. 1 depicts the Architecture of the proposed framework.

3.1 Random Forest Classifier

Random Forest Classifier is initialized with 100 estimators, then trained on the training data. The confusion matrix is created to assess the model's performance after predictions are made based on the test data. The confusion matrix provides a comprehensive overview of the class labels, both projected and actual, supporting the evaluation of false positives, true positives, true negatives, and false negatives. This evaluation aids in assessing the model's Accuracy, Precision, Recall, and F1 Score. By utilizing the scikit-learn library and following these steps, the Random Forest classifier can effectively detect intrusions in real-time scenarios, providing valuable insights for security purposes.

$$MSE = \frac{1}{N} \sum_{i=1}^{N} (fi - yi)^2 \tag{1}$$

3.2 Gaussian Naive Bayes Classifier

The initialization and training phases of the classifier include estimating the statistical parameters of the Gaussian distribution for each class using the training data. Subsequently, predictions are made on the test data by calculating the probability of each sample belonging to each class & selecting the group that has the highest probability. The model's performance is then assessed using the confusion matrix, which offers details on the predicted and actual class labels. This enables the evaluation of accuracy, precision, recall, and other performance parameters by allowing the examination of false positives, true positives, true negatives, and false negatives. By utilizing the Gaussian Naive Bayes algorithm and following these steps, the classifier can effectively detect intrusions in real-time scenarios, providing valuable insights for IDS applications.

$$P(x) = \frac{1}{\sqrt{2\pi\sigma^2}} e^{-\frac{(x-\mu)^2}{2\sigma^2}} \tag{2}$$

3.3 Logistic Regression Classifier

The data is first scaled using the StandardScaler to standardize the features and ensure consistent scaling across different variables. The Logistic Regression model is then initialized with increased max iter to ensure convergence. The scaled training data are used to train the model and teach it the logistic regression equation's coefficients. The learned coefficients are then used to make predictions on the scaled test data. After that, the confusion matrix is computed to assess the model's effectiveness and reveal differences between the predicted and real class labels. By utilizing the confusion matrix to examine false positives, true positives, true negatives, and false negatives, it is possible to assess accuracy, precision, recall, and other performance parameters. By utilizing the Logistic Regression algorithm and following these steps, the classifier can effectively detect intrusions in real-time scenarios, providing valuable insights for IDS applications.

$$Z = \left(\sum_{i=1}^{n} w_i x_i \right) + b \tag{3}$$

3.4 K Nearest Neighbor

Data preprocessing and feature engineering steps, such as removing irrelevant columns, handling missing values, and encoding categorical variables are done. It splits using the training data, a K-nearest neighbors (KNN) classifier is created, trained, and applied to the dataset. The code then makes predictions on the test set and evaluates the classifier's accuracy using the accuracy score metric. This code provides a basic framework for implementing a KNN classifier for intrusion detection on the CICIDS 2017 dataset, but further modifications and enhancements may be needed for optimal performance and customization to specific requirements.

$$\text{dist}(x, z) = \left(\sum_{r=1}^{d} |x_r - z_r|^p \right)^{1/p} \tag{4}$$

3.5 Pearson Correlation Coefficient

The linear connection between two continuous variables is measured by the Pearson correlation coefficient, sometimes known as Pearson's r. It is calculated by taking the sum of the product of the differences between each variable's value and its mean, and then dividing it by the product of the standard deviations of the variables. The resulting coefficient, which ranges from −1 to 1, shows the strength and direction of the link. The perfect correlation is represented by a value of 1, the perfect correlation is represented by a value of −1, and the perfect correlation is not represented by a value of The Pearson correlation coefficient can be used for feature selection by calculating the correlations between each feature and the target variable, and then selecting features based on their absolute correlation values or by setting a correlation threshold. However, it assumes linearity and may not capture non-linear relationships or be suitable for all types of data.

$$r = \frac{n(\sum xy) - (\sum x)(\sum y)}{[n\sum x^2 - (\sum x)^2)][n\sum y^2 - (\sum y)^2]} \tag{5}$$

4 Simulation and Results Discussion

In this section, we evaluate the proposed methodology's results with those obtained using the LR, KNN, Naive Bayes and RF classifiers. For the experiment, we considered the following datasets and simulation environment.

4.1 Dataset Description

The CICIDS2017 dataset captures the network behavior of various types of network traffic, including both normal traffic and malicious activities. Intrusion detection offers a thorough understanding of network behavior. With over 2.8 million instances and 79 features per instance, the dataset provides a comprehensive representation of network behavior. The features include packet headers, payload information, and flow statistics. Source and destination IP addresses, ports, protocol types, and flags are just a few of the details found in packet headers. Payloads include packet data, and flow statistics record the characteristics of the overall network flow. The development and evaluation of intrusion detection systems use the CICIDS 2017 dataset as a benchmark.

The dataset provides labeled data, with each network flow labeled as either normal or belonging to a specific attack category. Web assaults, denial-of-service attacks, reconnaissance attacks and botnet attacks are among the several types of attacks. Among

Table 1. Description of files containing CICIDS-2017 dataset

Name of Files	Day Activity	Attacks Found
Monday-WorkingHours.pcap_ISCX.csv	Monday	Benign(Normal human activities)
Tuesday-WorkingHours.pcap_ISCX.csv	Tuesday	Benign, FTP-Patator, SSH-Patator
Wednesday-WorkingHours.pcap_ISCX.csv	Wednesday	Benign, DoS GoldenEye, Dos Hulk Dos Slowhttptest, Dos slowloris, Heartbleed
Thursday-WorkingHours.Morning-WebAttacks.pcap_ISCX.csv	Thursday	Benign, Web Attack-Brute Force, Web Attack-Sql Injection, Web Attack-XSS
Thursday-WorkingHours.Afternoon-Infiltration.pcap_ISCX.csv	Thursday	Benign, Infiltration
Friday-WorkingHours -Morning.pcap_ISCX.csv	Friday	Benign, Bot
Friday-WorkingHours –Afternoon-PortScan.pcap_ISCX.csv	Friday	Benign, Portscan
Friday-WorkingHours –Afternoon-DDos.pcap_ISCX.csv	Friday	Benign, DDos

others. This labeling allows researchers and developers to train and evaluate their intrusion detection and prevention systems on real-world attack scenarios. With regard to each network flow, the CICIDS 2017 dataset provides a wealth of details and properties, such as source IP and destination IP addresses, port numbers, protocol types, packet sizes, and time duration. The Table 1 and Table 2 presents the description and features of the CIC-IDS dataset.

4.2 Simulation Environment

To generate a real-time dataset, you will need specific tools and virtual machines. The necessary tools include VirtualBox, Wireshark, and CICFlowmeter. Virtual machines such as Windows XP and Kali Linux are required for this process, and they should be installed within a virtual box environment. The goal is to use these virtual machines to carry out various attacks, including DoS, DDoS, Slowloris, and Synflood attacks. While performing these attacks, it is essential to have Wireshark running to capture the live traffic, which will generate a Pcap file. The next step involves using CICFlowmeter to convert the Pcap file into a CSV file. Finally, the two CSV files are merged for testing purposes.

Performing Slowloris Attack

Step 1: Start Kali Linux, and then Start up your terminal. Use the command below to create a new directory on your desktop with the name Slowloris. Navigate to the Slowloris directory you need to create. The Slowloris tool must now be copied from Github so that it can be installed on your Kali Linux computer. You just need to type the following URL in your terminal inside the Slowloris directory you generated to do that: https://github.com/GHubgenius/slowloris.pl.git. Figure 2 depicts the slowloris attack in kali.

Step 2: Now, in order to execute that kind of command, you must first check your machine's IP address. It is now time to launch the Apache server. Enter the following command to do so: sudo service apache2 start. Figure 3 depicts the wireshark file capture.

Table 2. Features of CICIDS-2017 Dataset

Feature no.	Features	Feature no.	Features
1.	Destination Port	41.	Packet Length Mean
2.	Flow Duration	42.	Packet Length Std
3.	Total Fwd Packets	43.	Packet Length Variance
4.	Total Backward Packets	44.	FIN Flag Count
5.	Total Length of Fwd Packets	45.	SYN Flag Count
6.	Total Length of Bwd Packets	46.	RST Flag Count
7.	Fwd Packet Length Max	47.	PSH Flag Count
8.	Fwd Packet Length Min	48.	ACK Flag Count
9.	Fwd Packet Length Mean	49.	URG Flag Count
10.	Fwd Packet Length Std	50.	CWE Flag Count
11.	Bwd Packet Length Max	51.	ECE Flag Count
12.	Bwd Packet Length Min	52.	Down/Up Ratio
13.	Bwd Packet Length Mean	53.	Average Packet Size
14.	Bwd Packet Length Std	54.	AvgFwd Segment Size
15.	Flow Bytes/s	55.	AvgBwd Segment Size
16.	Flow Packets/s	56.	Fwd Header Length
17.	Flow IAT Mean	57.	FwdAvg Bytes/Bulk
18.	Flow IAT Std	58.	FwdAvgPackets/Bulk
19.	Flow IAT Max	59.	FwdAvg Bulk Rate
20.	Flow IAT Min	60.	BwdAvg Bytes/Bulk
21.	Flow IAT Total	61.	BwdAvg Packets/Bulk
22.	Fwd IAT Mean	62.	BwdAvg Bulk Rate
23.	Fwd IAT Std	63.	SubflowFwd Packets
24.	Fwd IAT Max	64.	SubflowFwd Bytes
25.	Fwd IAT Min	65.	SubflowBwd Packets
26.	Bwd IAT Total	66.	SubflowBwd Bytes
27.	Bwd IAT Mean	67.	Init_Win_bytes_forward
28.	Bwd IAT Std	68.	Init_Win_bytes_backward
29.	Bwd IAT Max	69.	act_data_pkt_fwd
30.	Bwd IAT Min	70.	min_seg_size_forward
31.	Fwd PSH Flags	71.	Active Mean
32.	Bwd PSH Flags	72.	Active Std
33.	Fwd URG Flags	73.	Active Max
34.	Bwd URG Flags	74.	Active Min
35.	Fwd Header Length	75.	Idle Mean
36.	Bwd Header Length	76.	Idle Std
37.	Fwd Packets/s	77.	Idle Max
38.	Bwd Packets/s	78.	Idle Min
39.	Min Packet Length	79.	Label
40.	Max Packet Length		

Now that we need to determine whether your server is active or not, use the following command to find out its status: service apache2 status

Step 3: The tool should now be run using the following command: perl slowloris.pl -dns 10.0.2.15 –options

Step 4: Capture the live traffic data using wireshark.

Step 5: A Pcap file will be generated from wireshark tool. Convert the Pcap file to CSV file using CICFlowmeter tool.

Performing Synflood Attack

Step 1: Firstly, we should get IP address of 2 virtual Machines- Windows XP and Kali-linux. Communication should be done for both VMs using ping command.

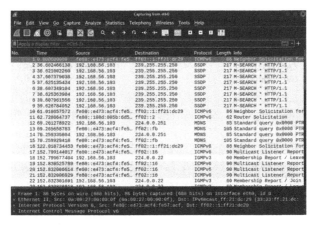

Fig. 2. Performing Slowloris Attack

Fig. 3. Generating live traffic data

Step 2: We should enter into Metasploit using the following command: msfconsole

Metasploit is a widely used penetration testing framework which provides a comprehensive group of tools and exploits for assessing the security of computer systems. It allows security professionals to identify vulnerabilities, launch attacks, and test the effectiveness of security measures. Metasploit simplifies the process of scanning, exploiting, and gaining access to target systems, helping to identify potential weak- nesses before they can be exploited by malicious actors. With its extensive collection of exploits, payloads, and auxiliary modules, Metasploit is a powerful tool for ethical hacking and security assessment.

Step 3: Next command is search synflood. Enter show options. It shows some options about rhosts,num, rport.

Step 4: Set RHOST to IP address of other virtual machine and enter exploit.

In Metasploit, RHOSTS, RPORT, and NUM are parameters used to specify the target host, target port, and the number of concurrent targets, respectively.

- RHOSTS: RHOSTS refers to the remote hosts, which are the target IP addresses or hostnames that you want to scan or attack. It can be a single host or a range of hosts specified using CIDR notation or wildcard characters.
- RPORT: RPORT stands for remote port and represents the target port number on the remote host. It is the port to which you want to establish a connection or attempt an exploit.
- NUM: NUM is an abbreviation for the number of con- current targets. This parameter is typically used when launching attacks that require targeting multiple hosts simultaneously. By setting NUM to a specific value, you can control the number of concurrent targets that Metas- ploit will attack simultaneously, increasing the efficiency of the operation.

- Set RHOSTS ipaddress
- Set RPORT 134
- Set NUM 50,000

Step 5: Capture the live traffic data using wireshark tool. A Pcap file will be generated from wireshark tool. Convert the Pcap file to CSV file using CICFlowmeter tool. Figure 4 depicts the Synflood attack.

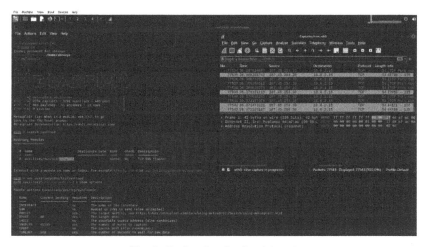

Fig. 4. Performing Synflood Attack

4.3 Performance Metrics

The performance evaluation phase, which tests and assesses the approach for generalization using several evaluation met- rics, is essential to the development of a strong ML model. To evaluate the model performance in this work, many evaluation metrics have been used.

Accuracy: Accuracy is a metric that counts the number of examples in a dataset and counts the fraction of accurately predicted instances out of those instances. It is calculated by dividing the overall forecasted number by the number of accurate estimates. The accuracy score goes from 0 to 1, with 1 denoting perfect accuracy. However, it is important to consider other metrics alongside accuracy, particularly when dealing with imbalanced datasets or when different types of errors have varying impacts. The formula for accuracy is:

$$\text{Accuracy} = (\text{Number of Correct Predictions}) / (\text{Total Number of Predictions}) \tag{6}$$

Precision: A classification model's accuracy of correctly predicting future events is measured by a parameter called precision. The formula for calculating it is to divide the total number of true positive predictions by the total number of true positives and false positives. It focuses on minimizing false positives and provides a demonstration of the model's accuracy in identifying positive events. Precision values range from 0 to 1, with 1 representing perfect precision. A higher precision score indicates a lower rate of false positives, making it particularly important in applications where false positives have significant consequences, such as medical diagnoses or spam detection. The formula for recall is:

$$\text{Precision} = (\text{TP}) / (\text{TP} + \text{FP}) \tag{7}$$

Recall: Recall, also known as true positive rate, is a statistic used to assess the effectiveness of a classification model, particularly in situations where the goal is to reduce false negatives. It calculates whether percentage of all real positive instance true positives plus false negatives were accurately anticipated positive instances. The formula for recall is:

$$\text{Recall} = (\text{TP}) / (\text{TP} + \text{FN}) \tag{8}$$

F1 Score: By integrating precision and recall into a single value, the F1 score is a statistic that offers a fair evaluation of the effectiveness of a classification model. It is especially beneficial when there is an uneven distribution of students among classes or when both false positives and false negatives matter. The F1 score is calculated as the HM of recall and precision. The F1 score formula is as follows:

$$\text{F1 Score is calculated as} : 2 * (\text{Precision} * \text{Recall}) / (\text{Precision} + \text{Recall}) \tag{9}$$

5   Results and Discussion

The below result provides a comparison of the performance metrics as Accuracy, Precision, F1 score and Recall for Four classification algorithms: LR, RF, and Naive Bayes, KNN. It states that Random Forest algorithm achieved the highest scores in all metrics, indicating its superior performance. The testing of a real-time dataset was conducted using the Random Forest classification, which resulted in an impressive accuracy of 99%. The model successfully predicted attacks with a label of 1. These results highlight the effectiveness of the Random Forest algorithm in detecting and classifying attacks in real- time scenarios, demonstrating its high accuracy and reliable performance. The results are listed below in the Table 3.

Table 3. Comparison of various algorithms

Models	Accuracy (%)	Precision (%)	Recall (%)	F1Score (%)
Random Forest	99.993	99.886	99.902	99.894
Logistic Regression	98.074	97.079	96.372	96.724
KNN	97.305	94.667	96.318	95.486
Gaussian Naïve Bayes	87.434	83.729	71.255	76.990

The ROC graph in the Fig. 5 below, which uses the four algorithms RF, KNN, LR, and Naive Bayes, shows the true positive rate on the y-axis and the false positive rate on the x-axis, respectively. The percentage of real positives that were accurately detected is known as the true positive rate. The percentage of genuine negatives that are mistakenly classified as positives is known as the false positive rate.

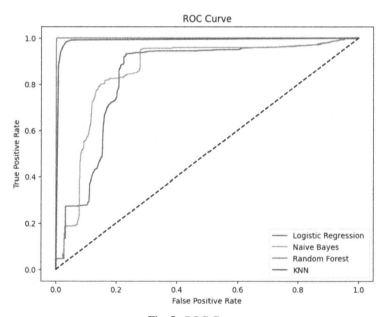

Fig. 5. ROC Curve

The outcomes demonstrate that ML models are useful for identifying attack traffic. The Table 4 presents the comparison of results of our proposed model with related papers. Based on the suggested framework, we note that our method outperforms the baseline classifiers in accuracy. Therefore, we draw the conclusion that the PC+RF architecture successfully addresses the feature selection and attack categorization issues.

Table 4. Comparison of results of our proposed model with related papers.

Method	Dataset	Accuracy (%)
Author in [1] designed a model ML and DL Algorithms	CICIDS 2017	97.24
Author in [2] proposed a model using SVM Algorithm	KDDCUP 99 and NSL-KDD	94.31
Author in [9] proposed a model using KNN,SVM,DT, RF algorithm	CICIDS 2017	92.19
Author in [12] designed a IDS model using DNN,LSTM	CICIDS 2018	96.47
Our Proposed Model	CICIDS 2017 and Real-time dataset	99.93

6   Conclusion

This study focuses on developing a real-time IDS using ML techniques. We leverage the CICIDS 2017 dataset for training our IDS model. The dataset encompasses attacks, including DoS, DDoS, FTP, SSH, and benign traffic. Through extensive data preprocessing, which involves removing duplicates and handling infinity values, we ensure the dataset's quality. We employ RF, KNN, LR, and Naive Bayes algorithms to train the IDS model. SMOTE is employed to address the class imbalance between benign and attack instances. Through feature selection, we identify and select the ten most relevant features. In the testing phase, we use our created real-time dataset to test the model. The accuracy achieved during the testing phase is an impressive 99.99% indicating the robustness of our model.

References

1. Hagar, A.A., Gawali, B.W.: Implementation of machine and deep learning algorithms for intrusion detection system. In: Intelligent Communication Technologies and Virtual Mobile Networks: Proceedings of ICICV 2022, pp. 1–20. Springer, Singapore (2022). https://doi.org/10.1007/978-981-19-1844-5_1
2. Ngueajio, M.K., Washington, G., Rawat, D.B., Ngueabou, Y.: Intrusion detection systems using support vector machines on the kddcup'99 and nsl-kdd datasets: A comprehensive survey. In: Proceedings of SAI Intelligent Systems Conference, pp. 609–629. Springer, Cham (2022). https://doi.org/10.1007/978-3-031-16078-3_42
3. Prajapati, P.K., Singh, I., Subhashini, N.: Network intrusion detection using machine learning. In: Futuristic Communication and Network Technologies: Select Proceedings of VICFCNT 2021, vol. 1, pp. 55–66. Springer, Singapore (2023)
4. Yazdizadeh, T., Shabnam, H., Paula, B.: Intrusion detection using ensemble models. In: Joint European Conference on Machine Learning and Knowledge Discovery in Databases, pp. 143–158. Springer, Cham (2022). https://doi.org/10.1007/978-3-031-23633-4_11
5. Nayak, S., Anushka, AP., Reethika, R., Lakshmisudha, K.: Optimizing network intrusion detection using machine learning. In: Advances in Data Science and Information Engineering: Proceedings from ICDATA 2020 and IKE 2020, pp. 585–590. Springer, Heidelberg (2021). https://doi.org/10.1007/978-3-030-71704-9_40

6. Bindra, N., Sood, M.: Detecting DDoS attacks using machine learning techniques and contemporary intrusion detection dataset. Autom. Control. Comput. Sci. **53**, 419–428 (2019)
7. Dutt, I., Borah, S., Maitra, I.K., Bhowmik, K., Maity, A., Das, S.: Real-time hybrid intrusion detection system using machine learning techniques. In: Bera, R., Sarkar, S.K., Chakraborty, S. (eds.) Advances in Communication, Devices and Networking. LNEE, vol. 462, pp. 885–894. Springer, Singapore (2018). https://doi.org/10.1007/978-981-10-7901-6_95
8. Thirimanne, S.P., Jayawardana, L., Yasakethu, L., Liyanaarachchi, P., Hewage, C.: Deep neural network based real-time intrusion detection system. SN Comput. Sci. **3**(2), 145 (2022)
9. Dwivedi, R.K., Rai, A.K., Kumar, R.: Outlier detection in wireless sensor networks using machine learning techniques: a survey. In: 2020 International Conference on Electrical and Electronics Engineering (ICE3), pp. 316–321. IEEE. (2020)
10. Elhanashi, A., Gasmi, K., Begni, A., Dini, P., Zheng, Q., Saponara, S.: Machine learning techniques for anomaly-based detection system on CSE-CIC-IDS2018 dataset. In: International Conference on Applications in Electronics Pervading Industry, Environment and Society, pp. 131–140. Springer, Cham (2022). https://doi.org/10.1007/978-3-031-30333-3_17
11. Singh, N.T., Chadha, R.: A review paper on network intrusion detection system. In: International Conference on Intelligent Cyber Physical Systems and Internet of Things, pp. 453–463. Springer, Cham (2022). https://doi.org/10.1007/978-3-031-18497-0_34
12. Hijazi, A., El Safadi, A., Flaus, J.M.: A deep learning approach for intrusion detection system in industry network. In BDCSIntell, pp. 55–62 (2018)
13. Saba, T., Rehman, A., Sadad, T., Kolivand, H., Bahaj, S.A.: Anomaly-based intrusion detection system for IoT networks through deep learning model. Comput. Electr. Eng. **99**, 107810 (2022)
14. Acharya, N., Singh, S.: An IWD-based feature selection method for intrusion detection system. Soft. Comput. **22**, 4407–4416 (2018)
15. Goud, K., Gidituri, S.: Security challenges and related solutions in software defined networks: a survey. Int. J. Comput. Netw. Appl. **9**, 22–37 (2022)

Maixdock Based Driver Drowsiness Detection System Using CNN

P. Ramani[1], R. Vani[1], and S. Sugumaran[2(✉)]

[1] SRM Institute of Science and Technology, Ramapuram, Chennai, India
{ramanip,vanir}@srmist.edu.in
[2] Vishnu Institute of Technology, Bhimavaram, AP, India
sugumaran.s@vishnu.edu.in

Abstract. This article demonstrates how to use Maixdock to build a drowsy driving monitoring system. A behavioral deterioration in one's ability to drive is known as drowsy driving. To categorize sleepiness signs like breathing and squinting deep learning has been used in this study. Yolo design training was done using example pictures. To categorize sleepiness signs like blinking and breathing, this study employs the Convolutional Neural Network (CNN). The CNN design was trained using 1310 images in total. Then the yolo was trained with Adam's optimization method. Ten people participated in a live experiment to determine how well this version worked. In this study, a new deep learning-based method for real-time sleepiness monitoring is proposed. It can be easily applied on a low-cost integrated chip and has a good level of performance. A single computer can then receive the data that was gathered. Facial characteristics, such as gaping, and ocular metrics, such as eye-closing, are the areas of concern used here. In this study, additional variables like camera distance from the vehicle and illumination effects are examined. These variables have the potential to influence the rate of categorization accuracy. The key of the car will not turn on if the motorist is intoxicated until the situation is altered. If the vehicle is already in a drivable state, the system will warn the driver via an alarm, and a heartbeat monitor will also identify the data and warn the driver. The Proposed method gives a good accuracy of detection, approximately 90% higher than existing methods.

Keywords: Raspberry Pi · Convolutional Neural Network · Drowsiness detection · Yawning

1 Introduction

One of the main sources of fatalities in humans is traffic mishaps. The fact that there are more cars on the road globally makes this worse. Long drives frequently make drivers drowsy and mentally exhausted. 2.3 to 2.5 percent of all deadly crashes nationally were reportedly the result of drowsy driving [1]. According to National Sleep Foundation research, 32% of motorists experience driving while fatigued on a monthly average.

P. Pareek et al. (Eds.): IC4S 2023, LNICST 536, pp. 186–197, 2024.
https://doi.org/10.1007/978-3-031-48888-7_16

A cognitive deterioration in driving abilities that is typically linked to long-distance driving [2] is what is known as "drowsy driving." Lack of sleep the night before traveling is typically the reason. It can occasionally also be brought on by other issues like unresolved sleep conditions, medicines, consuming alcohol [3], or shift employment. Because of this issue, checking a driver is fundamental to blood alcohol content and notifying them if necessary.

In recent years, several novel low-cost, non-invasive technological advances have been developed to identify sleepy driving. It very well may be isolated [4] into two classifications given whether signal processing is used for images or not. The distinction between the two groups is made based on the sort of incoming data, which can be either bodily signs or pictures from the camera.

Eye tracking and blinking are a couple of warning indications that a video can detect as a sign of driving [5] fatigue. To identify these characteristics for sleep monitoring, numerous methods have been created. Head part examination, support vector machines, and brain organizations are the most promising methods.

Deep Learning [6] offers modern and effective methods for identifying sleepiness trends in drivers. Convolutional Neural Networks (CNN) and Deep Neural Networks (DNN) are two popular Deep Learning models utilized for image-based driving sleepiness detection systems. For physiological kinds of input data, Recurrent Neural Network (RNN) is one of the Profound Learning plans habitually utilized for driving drowsiness [7] discovery systems.

In this paper, different types of deep learning techniques were studied and analyzed. To train a CNN classifier model, raw images of people yawning or being seated in a car were taken with a wide range of images varying from each other, including the lighting part, and capable of potentially giving false values when used. The drowsiness symptoms should be identifiable. Some of the images have people yawing and drowsy whereas others do not such that we can segregate accordingly.

Globally, road mishaps and deaths are frequently caused by drowsy driving. Despite numerous efforts to increase public awareness about the dangers of driving while fatigued, drowsiness remains a significant risk on the road. Traditional methods of detecting drowsiness, such as monitoring driving behavior or asking drivers to self-report their level of fatigue, have limitations and are not always reliable.

To address this issue, there is a need for an automated system that can detect drowsiness in real-time, using advanced technologies such as head pose estimation, facial expression analysis, and machine learning algorithms. The system should be able to identify indications of sluggishness, like hanging eyelids or changes in facial expressions, and alert the driver before an accident occurs.

This project's objective is to create a sleepiness detection system that can accurately and reliably detect driver drowsiness in real-time, and potentially save lives by preventing accidents caused by drowsiness.

2 Literature Survey

2.1 System for Monitoring and Warning Driver Fatigue Based on Eye Tracking

"Eye following based driver weakness checking and cautioning systems" are becoming increasingly popular in the automotive industry. The device monitors the driver's eye motions with an infrared sensor to detect indications of tiredness or sleepiness and sends a warning.

The technology works by analyzing the driver's eye movements, such as how often they blink and how long their eyes stay closed. If the system detects that the driver's eyes are closing for too long or too frequently, it will sound an alarm or display a warning message to alert the driver that they need to take a break.

Some advanced systems may also use other sensors, such as steering angle sensors to ascertain the driver's degree of depletion. The framework can then adjust the vehicle's speed, sound an alarm, or even apply the brakes if necessary.

Eye following-based driver weariness checking and cautioning systems are essential for improving road safety, particularly for long-distance drivers, commercial vehicle operators, and shift workers. The technology can help prevent accidents caused by driver fatigue [8] and lessen the number of fatalities on the streets.

Although the technology is still relatively new, it has the potential to become a standard safety feature in vehicles in the future.

2.2 Yawning Detection Reduction System for Driver Drowsiness

There have been numerous studies on yawning recognition and sleepiness forecast tools for drivers. One such paper is "Yawning Detection and Prediction System for Driver Drowsiness" by Hsu et al. published in the Journal of Sensors in 2015.

Ramani P. et al. (2022) developed a method for smart parking system using optical character recognition. Parking slots were identified automatically by OCR and placing a car without traffic and in time. The accuracy of prediction is lower [9].

Ramani P. et al. (2023) used segmentation algorithm to partition the images and classified the monuments using Multi layer Neural Network classifier and achieved good results for classification of heritage images [10].

Ramani P. et al. (2023) reviewed various non destructive methods for classification and segmentation algorithm for detection and classification of heritage structures. Measurement of different decay parameters is discussed and compared with different methods [11].

The paper proposes a yawning detection and prediction system based on an image processing technique using a webcam. The system can detect yawning and predict driver drowsiness before it becomes a critical issue. To identify and retrieve face features, the writers used the Viola-Jones method. To classify yawning and non-yawning facial movements, they used a Support Vector Machine (SVM).

The system also includes an algorithm to calculate the duration of time between yawns to predict the onset of drowsiness. The authors conducted experiments on a dataset of videos recorded by 26 subjects driving a car in a real-world setting. The findings demonstrate that the recommended technique is exceptionally exact at wheeze location and sleepiness detection in drivers.

The yawning detection [12] and prediction system proposed in this paper have the potential to enhance driver safety by providing an early warning system to prevent accidents caused by driver fatigue. The system could be integrated into vehicles as a standard safety feature or used in other industries where fatigue-related accidents are a concern, such as aviation and heavy machinery operations.

Overall, this paper demonstrates the potential of image processing and machine learning techniques in addressing the issue of driver drowsiness. in developing reliable and accurate systems for detecting and predicting drowsiness in drivers.

2.3 A Better System for Detecting Fatigue Based on the Personality Characteristics of the Driver

A study article titled "A superior weariness discovery framework in the view of social qualities of driver" recommends a strategy for recognizing driver fatigue that considers the person's unique behavioral patterns. The paper was published by Wang et al. in the Journal of Transportation Research Part C: Emerging Technologies in 2016.

The proposed system collects data on the driver's [13] behavioral characteristics, such as steering wheel movements, brake pedal pressure, and accelerator pedal usage, to determine if the driver is fatigued. The authors used a Support Vector Machine (SVM) algorithm to analyze the data and classify the driver's state as alert or fatigued.

In a trial using a driving simulator, 30 volunteers were given instructions to operate a vehicle for two hours on a fictitious motorway to assess the system. The findings demonstrated that, with an average accuracy rate of 92.8%, the suggested method has a high degree of precision in identifying driving tiredness.

The paper suggests that the proposed system has several advantages over other fatigue detection systems that rely on physiological measures, such as heart rate and EEG signals. The proposed system is non-intrusive, easy to implement, and can provide real-time feedback to the driver, allowing them to take corrective action before an accident occurs. The proposed method provides more accuracy than existing methods.

2.4 A Real-Time Driver Fatigue Detection Technique Based on SVM Algorithm

The authors of the study "Real Time Driver Fatigue Detection Based on SVM Algorithm" propose a Support Vector Machine (SVM) algorithm-based method for the real-time detection of driver fatigue. The paper was published by Chen et al. in the Journal of Sensors in 2018.

The suggested system utilizes characteristics like the length of time the eyes are closed, the regularity of breathing, and the driver's head posture to ascertain whether they are tired or not. The SVM method is then used to decide if the driver is alert or sluggish considering these characteristics. The device is built to function in real-time, providing the user with instantaneous input.

The framework was assessed in a driving test system to try different things with 10 members, where they were instructed to drive for two hours on a simulated highway. The outcomes showed that the proposed framework has high precision in detecting driver fatigue, [14] with an average accuracy rate of 93.3%.

The paper suggests that the proposed system has several advantages over other fatigue detection systems that rely on physiological measures, such as heart rate and EEG signals. The proposed system is non-intrusive and can provide real-time feedback to the driver, allowing them to take corrective action before an accident occurs.

2.5 Applying Composite Features in Viola-Jones Algorithm for Face Detection

A technique for face recognition is proposed in the study article "Face Detection Based on Viola-Jones Algorithm Applying Composite Features." This method combines the Viola-Jones algorithm with composite features.

The Viola-Jones algorithm is a well-known approach to face recognition; it employs a chain of algorithms and Haar-like characteristics to identify human expressions in pictures. In this paper, the authors propose using composite features that combine Haar-like features with other feature types, such as Local Binary Patterns (LBP) and Histograms of Oriented Gradients (HOG), to improve face detection accuracy.

The proposed strategy was assessed on several datasets, including the popular FDDB dataset, and accomplished a high precision pace of more than close to 100%. The proposed strategy is additionally contrasted with other cutting-edge facial acknowledgment techniques in the article, where its superior precision and speed are demonstrated.

The authors suggest that the proposed method has potential applications in various fields, including surveillance, human-computer interaction, and driver safety. In the context of driver safety, the proposed method could be used to identify driver [15] exhaustion and sluggishness by checking looks, for example, yawning and eye closure duration, in real-time.

3 Existing System

3.1 Introduction

The existing system monitors the vehicle's manual alcoholic checking is done by the manual process, traffic cameras are streaming the live video and find the accident location using a manual process.

3.2 Drawbacks

In the Existing system the cost of the Raspberry Pi (software) which they used is so effective. They have used only one detection. They used a low-power microprocessor, which is restricted in its ability to conduct multiple parallel operations and in its ability to connect high-power devices. AI camera is not used and the process is being done manually.

4 Proposed System

In the suggested system, a low-power AI microcontroller, an AI camera, and the Maixdock processor are used to identify the driver's sleepiness using CNN. In addition to it, GPS is used to track the location and alcohol sensor to detect the drunk and driving scenario. A heartbeat sensor (Max30102) is used to find the driver's health. Esp8266 relates to IoT to update the status of the sensors. If the vehicle is now in driving condition, then the framework cautions the driver to utilize a bell and pulse sensor also detect the readings and alert the driver, if the risk is present or not. These pulse sensors can be monitored by the owner through their CCTV. Hence it gives a combined output of whether a person is drowsy or not. The proposed system can be used for Android applications.

5 Methodology

5.1 Introduction

To stop car accidents driving with an obstacle in the flow of vehicles, turning on the power supply, and making sure the maixdock is successfully attached are the first steps of this project. This will be known once the maixdock properly interprets and examines the camera and the incoming pictures. 480p quality footage can be transmitted from the webcam. The driver's visage will be recognized, and the condition notice will appear. Finding the eyes, mouth, and heartbeat comes next. The level of the driver's tiredness will be classified by the drowsiness detection engine once these features have been identified (what about alcohol?) Our drowsiness recognition system requires two conditions to be satisfied before notifying the motorist. When the video detects that the motorist has his eyes closed for longer than two seconds is the first thing to watch out for. The alert will sound and "Drowsiness detected" will appear on the display." The schematic diagram of proposed method is shown in Fig. 1.

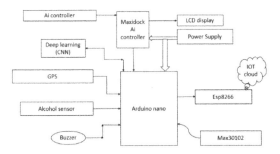

Fig. 1. Block Diagram for Proposed Method

When the camera notices the driver yawning, the second circumstance occurs. "Drowsiness detected" is displayed on the monitor, and an alarm is activated.

Use of an end-to-end neural network that predicts bounding frames and class odds simultaneously is suggested by the You Only Look Once (YOLO) theory. It is distinct

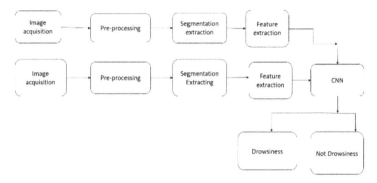

Fig. 2. Block Diagram for Drowsiness Detection

from the strategy used by earlier object recognition algorithms, which used classifications as detectors. With a completely different strategy to object identification, YOLO outperformed other real-time object detection algorithms and produced cutting-edge findings. While algorithms like Faster RCNN perform recognition on individual regions after using the Region Proposal Network to identify potential regions of interest, YOLO performs all of its predictions in a single iteration with the aid of a single layer that is fully connected. For the same image, methods that make use of Region Proposal Networks go through multiple iterations, whereas YOLO only needs to go through one. While Just go for it just requires one cycle, strategies that utilize district proposition networks require various rounds for a similar picture. Since the original introduction of YOLO in 2015, several new iterations of the same model have been suggested, each of which builds upon and enhances its precursor. The block diagrams of drowsiness detection and Yolo architecture are given [16] in Figs. 2 and 3.

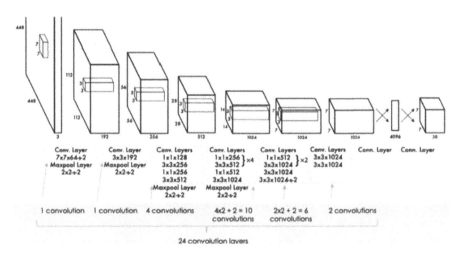

Fig. 3. Yolo architecture

5.2 Convolutional Neural Network (CNN)

It has three layers: a convolutional layer, a completely linked layer, and a pooling layer. It is a type of neural network that uses a grid-like design to handle input. The convolution layer is the fundamental part of CNN that is mainly in charge of calculation. Pooling reduces the number of computations required as well as the physical area of the depiction. However, the Prior Layer and Recent Layer are both connected to the Fully Connected Layer. Figure 4 shows the general architecture of CNN [17].

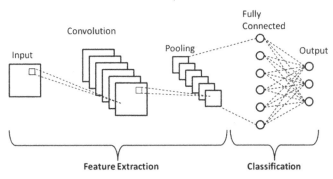

Fig. 4. CNN architecture

6 Result

The improved code for the driver drowsiness detection system uses head pose estimation, facial expression analysis, machine learning, and CNN to detect drowsiness based on eye aspect ratio, blinking of eyes, yawning, heart rate, location of the driver, and detecting of alcohol.

Overall, the improved code for the driver drowsiness detection system is more accurate and robust as it combines multiple techniques, machine learning, and CNN to detect drowsiness in drivers. According to the performance and grid in Fig. 5, this displays the training loss and precision for the project's circumstances under evaluation. As well as using the sample data set for CNN shown in Table 1 and dataset for testing has taken from Kaggle [18].

Table 1. Sample Data set for CNN.

Figure 6 is the hardware component that has been tested giving an output of alcohol detection which is shown in Fig. 7.

Table 2 compares the accuracy of the proposed method at different epochs. The proposed method gives 89 percent accuracy.

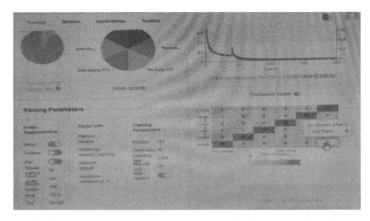

Fig. 5. Performance and confusion matrix for the proposed method

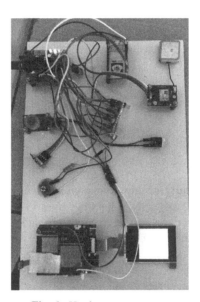

Fig. 6. Hardware component

s://driver-drowniness-default-rtdb.firebase

)river Drowsiness

Alcohol Status: "ALCOHOL Detected..."

Alert:""

Driver Status:""

Heart Rate Status: "Heart Value is Normal."

Location Status: "Latitude: 13.03249533, Lor

Fig. 7. Detection of alcohol

Table 2. Performance metric of proposed method

Epoch	Value Accuracy	Value Loss
18/200	0.8945	0.3602
19/200	0.8955	0.3528
20/200	0.8993	0.3347

7 Conclusion

The system is trained to identify the drowsiness symptoms of yawning and blinking using a convolutional neural network (CNN). This study used 1310 photos, including shots of eyes open, closed, yawning, or not gaping, vacant seats, and seats with drivers were used to teach CNN. Ten individuals tried the system prototype, which was developed in real-time. CNN's training results showed that our algorithm consistently anticipated a driver's sleepiness rate more than 80% of the time. The proposed system can be implemented on dashboards in future generations of automobiles.

References

1. National Center for Statistics and Analysis. Drowsy Driving 2015 (Crash Stats Brief Statistical Summary. Report No. DOT HS 812 446). Washington, DC: National Highway Traffic Safety Administration (2018)
2. You, F., Li, X., Gong, Y., Wang, H., Li, H.: A real-time driving drowsiness detection algorithm with individual differences consideration. IEEE Access **7**, 179396–179408 (2019). https://doi.org/10.1109/ACCESS.2019.2958667
3. Dasgupta, A., Rahman, D., Routray, A.: A smartphone-based drowsiness detection and warning system for automotive drivers. IEEE Trans. Intell. Transp. Syst. **20**(11), 4045–4054 (2019). https://doi.org/10.1109/TITS.2018.2879609

4. Pai, R., Dubey, A., Mangaonkar, N.: Real time eye monitoring system using CNN for drowsiness and attentiveness system. In: Asian Conference on Innovation in Technology (ASIANCON), pp. 1–4 (2021). https://doi.org/10.1109/ASIANCON51346.2021.9544624
5. Teja, K.B.R., Kumar, T.K.: Real-time smart drivers drowsiness detection using DNN. In: 5th International Conference on Trends in Electronics and Informatics (ICOEI), pp. 1026–1030 (2021). https://doi.org/10.1109/ICOEI51242.2021.9452938
6. Geoffroy, G., Chaari, L., Tourneret, J.-Y., Wendt, H.: Drowsiness detection using joint EEG-ECG data with deep learning. In: 29th European Signal Processing Conference (EUSIPCO), pp. 955–959 (2021). https://doi.org/10.23919/EUSIPCO54536.2021.9616046
7. Reddy, B., Kim, Y.-H., Yun, S., Seo, C., Jang, J.: Real-time driver drowsiness detection for embedded system using model compression of deep neural networks. In: IEEE Conference on Computer Vision and Pattern Recognition Workshops (CVPRW), pp. 438–445 (2017). https://doi.org/10.1109/CVPRW.2017.59
8. Ying, Y., Jing, S., Wei, Z.: The monitoring method of driver's fatigue based on neural network. In: International Conference on Mechatronics and Automation, pp. 3555–3559 (2007). https://doi.org/10.1109/ICMA.2007.4304136
9. Ramani, P., Lekhana, G., Aruna, A., Vijay Kumar, B.: Smart parking system based on optical character recognition. In: AIP Conference Proceeding, vol. 2405, no. 1, p. 040009 (2022). https://doi.org/10.1063/5.0072485
10. Perumal, R., Venkatachalam, S.B.: Non invasive decay analysis of monument using deep learning techniques. Traitement du Signal **40**(2), 639–646 (2023). https://doi.org/10.18280/ts.400222
11. Ramani, P., Subbiah Bharathi, V., Sugumaran. S.: Non destructive analysis of crack using image processing, ultrasonic and IRT: a critical review and analysis. In: International Conference on Cognitive Computing and Cyber Physical Systems, vol. 472, pp. 144–155. Springer, Cham (2023). https://doi.org/10.1007/978-3-031-28975-0_12
12. Gupta, N.K., Bari, A.K., Kumar, S., Garg, D., Gupta, K.: Review paper on yawning detection prediction system for driver drowsiness. In: 5th International Conference on Trends in Electronics and Informatics (ICOEI), pp. 1–6 (2021). https://doi.org/10.1109/ICOEI51242.2021.9453008
13. Gupta, R., Aman, K., Shiva, N., Singh, Y.: An improved fatigue detection system based on behavioral characteristics of the driver. In: 2nd IEEE International Conference on Intelligent Transportation Engineering (ICITE), pp. 227–230 (2017). https://doi.org/10.1109/ICITE.2017.8056914
14. Singh, H., Bhatia, J.S., Kaur, J., Eye tracking based driver fatigue monitoring and warning system. In: India International Conference on Power Electronics (IICPE 2010), pp. 1–6(2011). https://doi.org/10.1109/IICPE.2011.5728062
15. Wang, R., Wang, Y., Luo, C.: EEG-based real-time drowsiness detection using Hilbert-Huang transform. In: 7th International Conference on Intelligent Human-Machine Systems and Cybernetics, pp. 195–198 (2015). https://doi.org/10.1109/IHMSC.2015.56
16. Yolo Object Detection Explained Homepage. https://www.datacamp.com/blog/yolo-object-detection-explained. Accessed 30 Sep 2022
17. Basic CNN Architecture Homepage. https://www.upgrad.com/blog/basic-cnn-architecture/. Accessed 28 Jul 2022
18. Yawn_eye_dataset_new Homepage. https://www.kaggle.com/serenaraju/yawn-eye-dataset-new. Accessed 2020

A Novel Approach to Visualize Arrhythmia Classification Using 1D CNN

Madhumita Mishra[✉], T. L Sharath Kumar, and U. M Ashwinkumar

REVA University, Bengaluru, India
madhumita.mish@reva.edu.in

Abstract. Cardiac-related disorders have been one of the major concerns in recent decades. The electrocardiogram, an extensively utilized medical instrument, records the electrical activity of the heart as a wave. Cardiac arrhythmia is a condition of having an irregular heartbeat. Manually identifying irregularities in an ECG wave is a complicated and challenging task. The current work focuses on computationally identifying the ECG wave fluctuations to determine the abnormality in the heartbeat. We propose to use a 1-Dimensional Convolutional Neural Network (CNN) that analyses a given ECG signal data to identify irregularities in the functioning of the heart and represent the associated risks using graphics interchange format (GIF) files of a 3-dimensional heart. We obtained an accuracy score of 96.72% in classifying given ECG data into five different arrhythmia classes. Automated detection and visual representation of cardiac conditions can help medical associates easily interpret ECG signals and determine arrhythmia early.

Keywords: Electrocardiogram · Cardiac Arrhythmia · 1-Dimensional CNN · Graphics interchange format

1 Introduction

The heart is an essential structure of living beings which pumps blood across the entire body through a network of blood vessels and related pathways of the circulatory system. The circulatory system is essential for human life because it is responsible for transporting blood, oxygen, and other vital materials to all the different parts of our body. Cardiovascular diseases (CVDs) are a class of ailments of heart and related blood vessels, which can comprise cardiac arrhythmia, coronary heart disease, rheumatic heart disease and other heart related conditions. Diseases related to the heart and associated cardiac systems are among the leading causes of mortality today. The most prevalent cause for mortality globally is cardiovascular diseases which approximately claims 18 million lives each year [6]. Amongst the significant causing CVDs, cardiac arrhythmias are a primary illness type where the normal functioning of the heart is affected. Normal heartbeats indicate regular functioning of the heart, but an arrhythmia is a condition where the heart has an abnormal beating pattern. Automatic identification of such paradigms

© ICST Institute for Computer Sciences, Social Informatics and Telecommunications Engineering 2024
Published by Springer Nature Switzerland AG 2024. All Rights Reserved
P. Pareek et al. (Eds.): IC4S 2023, LNICST 536, pp. 198–209, 2024.
https://doi.org/10.1007/978-3-031-48888-7_17

can be of great help in the medical field for dealing with cardiac disorders. A few known characteristics of the cardiac system require expert clinical knowledge to identify any heart ailments.

An electrocardiogram is a standard and painless test that medical practitioners use to measure the electrical variations in the heart's rhythmic functioning. Cardiologists try to analyze ECG signals to determine any irregularities in the functioning of the heart before suggesting a specific treatment. Manual analysis of ECG is a time-consuming and difficult job for doctors that demands great expertise to examine the data extensively because of the large amount and varying complexity of ECG data. To overcome these challenges faced by traditional manual practices of ECG data analysis, automated computer-aided diagnostic methods have been devised to analyze and interpret ECG signals effectively. Much research has been carried out in this area by applying several popular machine learning algorithms and other computer science-based principles to analyze ECG signals practically. It is a demanding task because of the varied wave morphologies between patients and redundant noise that can be present in the input data. The algorithm must be capable of implicitly identifying the distinct wave patterns and their underlying dependencies.

Considerable amounts of research work have already been done in automated ECG analysis. All these works focus on different techniques that can be applied to analyze ECG data and detect the corresponding heart-related conditions. Interest in this area has been analyzing ECG-related data to detect and classify cardiac arrhythmias. Many algorithms have shown promising results in analyzing ECG data and seeing arrhythmic conditions. Also, automated analysis can be very efficient and less time-consuming compared to traditional manual ECG signal analysis done by cardiologists.

J. Ferretti et al. [1] presented a novel approach employing a 1D-CNN to analyze and classify an ECG dataset into 16 different arrhythmia conditions. They implemented a 1D convolutional neural network, tested four different network architectures, and compared their results.

In [2], Amin Ullah *et al.* have implemented a 2-D CNN to classify ECG signals into different arrhythmic types automatically. They used a short-time Fourier transform to convert simple 1-dimensional ECG time series signals into more refined 2- dimensional spectrograms. 2-D spectral images were input to a 2-D CNN model to classify the dataset into eight kinds of arrhythmia.

In a paper titled "Cardiac Arrhythmia Classification based on 3D Recurrence plot analysis and deep learning", Hua Zhang *et al.* [3] implemented an advanced deep learning algorithm based on three-dimensional recurrence plots to analyze ECG and VCG signals and detect four different types of arrhythmic cardiac conditions. The 3D RP maps were constructed to train and test a 3D CNN model. This method provided a better visual representation of the outputs and achieved an efficient average F1 score of 93.50 in classifying cardiac arrhythmia.

Elena Merdjanovska *et al.* [4], in their paper titled "Comprehensive survey of computational ECG analysis: Databases, methods and applications", provide a complete survey of 45 various ECG records along with different computational methods applied for their analysis. They present a summary of machine learning algorithms and multiple tools used for data analysis in machine learning. They have explained the recent works in heart disease detection using machine learning algorithms and concluded that choosing the proper techniques for cleaning the data with productive algorithms is beneficial to develop enhanced accuracy prediction systems.

The traditional machine learning approach uses efficient ML algorithms such as decision trees (DTs), random forest (RF), support vector machine (SVM), and other algorithms to classify cardiac arrhythmia. Automated analysis and classification of ECG signals have been carried out using various techniques, including ML algorithms, artificial neural networks (ANNs) [10, 13], frequency analysis [12], wavelet transform [9], statistical methods [15] and various other approaches.

We implemented a 1-dimensional convolutional neural network to train on an ECG dataset to detect cardiac arrhythmia and then visualize the output through a 3-D model of the heart in GIFs. The beneficial advantage of 1D CNN lies in its simplicity, and it is also computationally less expensive when compared to other complex models. Automated processing and analysis of ECG using neural networks can be very beneficial. Representation of associated arrhythmic conditions using GIFs of 3-dimensional heart models can help everyone better interpret the ECG signals.

2 Methodology

Electrocardiogram (ECG) can be considered as a visual representation of the rhythmic activity of the heart, depicted as wave-like signals, commonly used to diagnose cardiovascular diseases. In our proposed project work, we implement a 1-dimensional convolutional neural network to analyze ECG signals from an ECG dataset and classify the related arrhythmic conditions into one of five classes of heart arrhythmia. We then visualize the predicted arrhythmic class using graphics interchange format (GIF) files of heart to represent the associated arrhythmic condition as the output. The different types of arrhythmias combined into five major categories are presented in Fig. 1.

Figure 2 depicts the basic flow of our proposed work.

2.1 Dataset Collection

We use a processed version of the popularly used PhysioNet's MIT-BIH Arrhythmia ECG—databases with labelled ECG records [5]. The processed version of the dataset that we are using has the ECG lead II recordings re-sampled to 125Hz. This dataset we are using has multiple groups of heartbeats, represented symbolically as N, S, V, F, and Q, that are numerically stored using these indices- 0, 1, 2, 3, and 4, respectively. Of these character symbols, 'N' means Normal heartbeats, 'S' means Supraventricular ectopic beats, 'V' means Ventricular ectopic beats, 'F' means Fusion beats, and 'Q' means Unknown beats.

The input dataset includes two.csv files, one with samples for training the neural network and the second has sample data for testing the model. The "train.csv" file consists of 87,554 samples, while the "test.csv" file has 21,892 samples. This selected input set of data is loaded and prepared using appropriate Python modules.

ARRHYTHMIA CLASS	TYPES CONSOLIDATED
N (Normal Beat)	• Normal • Left/Right bundle branch block • Atrial escape • Nodal escape
S (Supraventricular Ectopic Beats)	• Atrial premature • Aberrant atrial premature • Nodal premature • Supra-ventricular premature
V (Ventricular Ectopic Beats)	• Premature ventricular contraction • Ventricular escape
F (Fusion Beats)	• Fusion of ventricular and normal
Q (Unknown Beats)	• Paced • Fusion of paced and normal • Unclassifiable

Fig. 1. Major arrhythmic classes and their types in the input ECG dataset (Ref: Liu, Fan, et al. 2019)..

2.2   Data Pre-processing

Since the processed ECG dataset is very imbalanced, we must adequately balance the dataset for efficient application of the CNN model. We process and segregate the data into five distinct groups, each with samples corresponding to five different arrhythmic classes. Each data set is later resampled to obtain 50000 examples of every arrhythmic type. These five distinct classes of records are then combined to procure a balanced input dataset containing a total of 250000 samples of records. This dataset is further pre-processed by adding white Gaussian noise (AWGN), and the class labels are then processed using one-hot encoding.

2.3   Model Implementation

We have implemented a classifier that can classify the given ECG signal data into 5 major arrhythmic classes: Normal beats, Supraventricular Ectopic beats, Ventricular ectopic beats, Fusion beats, and Unknown beats.

A 1D CNN model is implemented using "tensorflow.keras.models", and the different layers for the model are imported from "tensorflow.keras.layers" modules of Python. The model has three pairs of 1D Convolution and 1D MaxPooling layers as initial layers. These are then accompanied by a single flattening layer that lessens the input values into one-dimensional ones. It is then followed by three Dense layers, of which two of the dense layers have a ReLU function to activate neurons. The last third thick layer

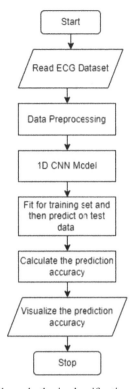

Fig. 2. The basic flow of the arrhythmia classification used in the proposed work.

employs a SoftMax activation function giving out five probabilistic outputs associated with five different output categories of arrhythmia.

Activation functions are mathematical functions that transform the inputs given into the required output within a specific range. Neurons are activated only when the work reaches a set threshold value of the function. Rectified linear activation function (ReLU) is a non-linear activation function that is a simple and effective function that returns zero as output for any negative input while returning the input value directly as the output if it is positive.

Mathematically ReLU output for any given input value 'A' is written as:

$$f(A) = \max(0, A) \tag{1}$$

SoftMax activation is a function used to convert a vector of numbers into a vector of probabilities. So, these are usually used in the final layer of a neural network to obtain chances of each output value vector

The structural design of proposed model is presented in Fig. 3.

The implemented CNN algorithm is then fitted to the training dataset and later tested on the test dataset. We have executed our CNN model for 13 epochs with a batch size of 32. The configuration of the implemented CNN model is shown in Fig. 4.

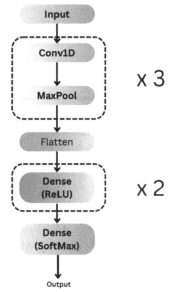

Fig. 3. The proposed 1-dimensional CNN model.

Layer (type)	Output Shape	Param #
conv1d_3 (Conv1D)	(None, 182, 64)	448
max_pooling1d_3 (MaxPooling 1D)	(None, 90, 64)	0
conv1d_4 (Conv1D)	(None, 85, 64)	24640
max_pooling1d_4 (MaxPooling 1D)	(None, 42, 64)	0
conv1d_5 (Conv1D)	(None, 37, 64)	24640
max_pooling1d_5 (MaxPooling 1D)	(None, 18, 64)	0
flatten_1 (Flatten)	(None, 1152)	0
dense_3 (Dense)	(None, 64)	73792
dense_4 (Dense)	(None, 32)	2080
dense_5 (Dense)	(None, 5)	165

```
Total params: 125,765
Trainable params: 125,765
Non-trainable params: 0
```

Fig. 4. Network configuration of the implemented 1-dimensional CNN model in jupyter.

The functionality of the achieved model is assessed in terms of accuracy percentage and precision score. Other evaluation metrics such as F1 score, recall value and support values are also calculated. These comparison metrics are beneficial in evaluating the performance of the model in the multi-class classification of arrhythmia.

3 Results

This section shows the resultant outcomes we have collected by applying the proposed method to detect cardiac arrhythmia. A brief analysis of the results obtained from the 1D CNN model is also given.

3.1 Experimental Setup

We implemented our proposed 1D CNN model using a computer system having an AMD Ryzen 7 5800H processor with a clock speed of 3.20 GHz and 16 Gb of RAM. The python code was executed on Jupyter Notebook software to implement the model and visualize the results. Libraries like pandas, sklearn, matplotlib, seaborn, tensorflow and other advanced python libraries were used in the implementation of the proposed model.

3.2 Results

After successfully training our CNN model using Adam optimizer on the training dataset, we apply it to the test dataset to predict the arrhythmia condition. The prediction result with the accuracy score and other performance metrics is calculated and displayed as the final prediction result in a classification report.

The performance of the proposed 1D CNN classifier is evaluated using performance metrics such as accuracy, precision, recall and F1 score. These metrics are mathematically explained in Eqs. (2)– (5), respectively.

Accuracy and precision score of the model are found from the confusion matrix using the following formulae:

$$Acccuracy = \frac{(TP + TN)}{(TN + TP + FN + FP)} \qquad (2)$$

$$Precision = \frac{TP}{(TP + FP)} \qquad (3)$$

$$Recall = \frac{TP}{(TP + FPN)} \qquad (4)$$

$$F1Score = \frac{2 * Precision * Recall}{Precision + Recall} \qquad (5)$$

where,

- TP indicates the number of cases that are correctly classified as arrhythmia.

- FP indicates the non-arrhythmic cases classified as arrhythmic.
- TN indicates the non-arrhythmic cases classified as non-arrhythmic.
- FN indicates the arrhythmic cases classified as non- arrhythmic.

The prediction result with the accuracy score and other performance metrics is calculated and displayed as the final prediction result in the form of a classification report as shown in Fig. 5.

Our 1D CNN model has given an accuracy score of 96.72% with a F1 score of 85.08% to detect and classify cardiac arrhythmic condition from the test dataset.

```
              precision    recall  f1-score   support

           0       0.99      0.97      0.98     18118
           1       0.64      0.83      0.72       556
           2       0.92      0.95      0.93      1448
           3       0.53      0.84      0.65       162
           4       0.96      0.99      0.97      1608

    accuracy                           0.97     21892
   macro avg       0.81      0.91      0.85     21892
weighted avg       0.97      0.97      0.97     21892

0.9672483098848894
F1_score =  0.8508205226765082
```

Fig. 5. Classification report of the CNN model that depicts the evaluation of the proposed model in detecting 5 different classes of arrhythmia.

The prediction outputs attained after applying our CNN model are analyzed using accuracy percentage and other evaluation metrics to measure the model's efficiency. Evaluation metrics are utilized to visualize the results using various functional Python modules. The performance of the model is analyzed with respect to five classes of the output using a confusion matrix as shown in Fig. 6.

The output prediction of the different arrhythmic classes is represented using GIF files that indicate the corresponding arrhythmic condition. A GIF file of the heart is shown as output for the inputs corresponding to the heart's regular beats.

The Fig. 8 presents the output GIF shown in case of supraventricular heart beats.

Supraventricular beats include various arrhythmic conditions that affect atrial parts of the heart. Functioning of the atrium is affected resulting in abnormal heartbeats.

Figure 9 depicts the output GIF file that will be shown when the input data corresponds to ventricular arrhythmic condition.

Ventricular beats include cardiac conditions associated with ventricular part of the heart. Functioning of the ventricle is affected resulting in abnormal heartbeats.

Figure 10 shows the output GIF file that will be displayed in the case of input data being a fusion beat.

Fusion beats include the combination of ventricular and normal beats.

The classification results of the proposed model are analyzed by doing a comparative analysis of relevant works that have achieved promising results in the task of arrhythmia classification.

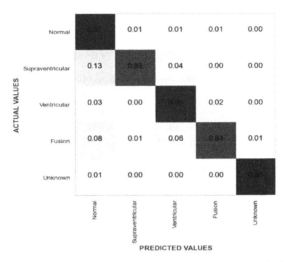

Fig. 6. Proposed model's performance is visualized through a confusion matrix.

Fig. 7. Representation of normal heart beats.

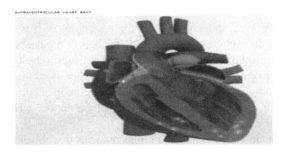

Fig. 8. Representation of supraventricular heart beats.

In terms of accuracy percentage, Table 1 depicts a comparison of similar works with the proposed model.

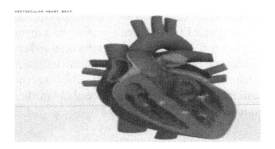

Fig. 9. Representation of ventricular heart beats.

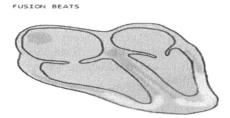

Fig. 10. Representation of fusion beats.

Table 1. Comparison of classification results of relevant works.

Work	Approach	Accuracy (%)
Proposed Work	**1D CNN**	**96.72**
J. Ferretti *et al.* [1]	1D CNN	95
Acharya *et al.* [16]	CNN + Augmentation	93.5
Singh S *et al.* [17]	RNN	87.1
Martis *et al.* [18]	SVM + DWT	93.8

This analysis helps us to compare the performance efficiency of different approaches that have been used in automated classification of cardiac arrhythmia. The output results obtained from the proposed model show better efficiency and accuracy.

4 Conclusion.

In our proposed project work, we are using a 1-dimensional CNN model to analyze a given ECG dataset and automatically detect and classify a given set of heartbeats from ECG data, which can be very beneficial in the precise and early diagnosis of cardiac diseases. Arrhythmic condition detected by the model is further represented as a visual output using GIF files of the heart that help in easy understanding of the corresponding arrhythmic condition. Our model achieved a satisfactory classification with an accuracy of 96.72% and a F1 measure of 85.08%. This paper explores a simple classification method to analyze and categorize ECG data and visualize the arrhythmic output class. In future works, we would like to consider more straightforward and productive techniques to optimize convolution neural networks and obtain improved detection results. And we would also like to implement advanced models like 2D CNN to get more efficient results. We will use deep learning algorithms and learn more about medical data analysis that can help us in our future work.

References

1. Ferretti, J., Randazzo, V., Cirrincione, G., Pasero, E.: 1-D Convolutional Neural Network for ECG Arrhythmia Classification. In: Esposito, A., Faundez-Zanuy, M., Morabito, F.C., Pasero, E. (eds.) Progresses in Artificial Intelligence and Neural Systems. SIST, vol. 184, pp. 269–279. Springer, Singapore (2021). https://doi.org/10.1007/978-981-15-5093-5_25
2. Ullah, A., Anwar, S.M., Bilal, M., Mehmood, R.M.: Classification of arrhythmia by using deep learning with 2-D ECG spectral image representation. Remote Sens. **12**(10), 1685 (2020). https://doi.org/10.3390/rs12101685
3. Zhang H., et al.: Cardiac Arrhythmia classification based on 3D recurrence plot analysis and deep learning. Front Physiol. Jul 22;13:956320. doi: https://doi.org/10.3389/fphys.2022.956320. PMID: 35936913; PMCID: PMC9352947,(2022)https://doi.org/10.3389/fphys.2022.956320
4. Merdjanovska, E., Rashkovska, A.: Comprehensive survey of computational ECG analysis: databases, methods and applications. Expert Syst. Appl. **203**, 117206 (2022). https://doi.org/10.1016/j.eswa.2022.117206
5. Dataset: https://www.kaggle.com/datasets/shayanfazeli/heartbeat
6. WHO Cardiovascular diseases "https://www.who.int/newsroom/factsheets/detail/cardiovascular-diseases-(cvds)"
7. Ahmed, A.A., Ali, W., Abdullah, T.A.A., Malebary, S.J.: Classifying cardiac arrhythmia from ECG signal using 1D CNN deep learning model. Mathematics **11**, 562 (2023). https://doi.org/10.3390/math11030562
8. Subramanian, K., Prakash, N.K.: Machine learning based cardiac arrhythmia detection from ECG signal. In: Third International Conference on Smart Systems and Inventive Technology (ICSSIT), (2020)
9. Wang, T., Changhua, L., Sun, Y., Yang, M., Liu, C., Chunsheng, O.: Automatic ECG classification using continuous wavelet transform and convolutional neural network. Entropy **23**(1), 119 (2021). https://doi.org/10.3390/e23010119
10. Coast, D.A., Stern, R.M.M., Cano, G.G., Briller, S.A.: An approach to cardiac arrhythmia analysis using hidden markov models. IEEE Trans. Biomed. Eng. **37**, 826–836 (1990)
11. Mustaqeem, A., Anwar, S.M., Majid, M.: Multiclass classification of cardiac arrhythmia using improved feature selection and SVM invariants. Comput. Math. Methods Med, (2018)
12. Minami, K.I., Nakajima, H., Toyoshima, T.: Real-time discrimination of ventricular tachyarrhythmia with Fourier-transform neural network. IEEE Trans. Biomed. Eng. **46**, 179–185 (1999)
13. Hu, Y.H., Tompkins, W.J., Urrusti, J.L., Afonso, V.X.: Applications of artificial neural networks for ECG signal detection and classification. J. Electrocardiol. **26**, 66–73 (1993)
14. Osowski, S., Hoai, L.T., Markiewicz, T.: Support vector machine based expert system for reliable heartbeat recognition. IEEE Trans. Biomed. Eng. **51**, 582–589 (2004)
15. Willems, J.L., Lesaffre, E.: Comparison of multigroup logistic and linear discriminant ECG and VCG classification. J. Electrocardiol. **20**, 83–92 (1987)

16. Acharya, U.R., et al.: A deep convolutional neural network model to classify heartbeats. Comput. Biol. Med. **89**, 389–396 (2017)
17. Singh, S., Pandey, S.K., Pawar, U., Janghel, R.R.: Classification of ECG arrhythmia using recurrent neural networks. Procedia Comput. Sci. **132**, 1290–1297 (2018)
18. martis, R., et al.: Application of higher order cumulant features for cardiac health diagnosis using ECG signals. Int. J. Neural Syst. **23**(04), 1350014 (2013). https://doi.org/10.1142/S01 29065713500147
19. M. Kachuee, S. Fazeli and M. Sarrafzadeh,: ECG heartbeat classification: a deep transferable representation, In: IEEE International Conference on Healthcare Informatics (ICHI), New York, NY, USA, pp. 443–444, doi: https://doi.org/10.1109/ICHI.2018.00092, (2018)
20. Strodthoff, N., Wagner, P., Schaeffter, T., Samek, W.: Deep learning for ECG analysis: benchmarks and insights from PTB-XL. IEEE J. Biomed. Health Inform. **25**(5), 1519–1528 (2021). https://doi.org/10.1109/JBHI.2020.3022989
21. Essa, E., Xie, X.: An ensemble of deep learning-based multi-model for ECG heartbeats arrhythmia classification. IEEE Access **9**, 103452–103464 (2021)
22. De Santana, J. R. G. Costa, M. G. F. Costa Filho. C. F. F.: A new approach to classify cardiac arrythmias using 2D convolutional neural networks. In: 2021 43rd Annual International Conference of the IEEE Engineering in Medicine & Biology Society (EMBC), (2021)
23. Chourasia, M., Thakur, A., Gupta, S., Singh, A.: ECG heartbeat classification using CNN In: IEEE 7th Uttar Pradesh Section International Conference on Electrical, Electronics and Computer Engineering (UPCON), (2020)
24. Liu, Fan, et al.: Arrhythmias classification by integrating stacked bidirectional LSTM and two-dimensional CNN. Advances in Knowledge Discovery and Data Mining: 23rd Pacific-Asia Conference, PAKDD 2019, Macau, China, April 14–17, 2019, Proceedings, Part II 23. Springer International Publishing, (2019)

Exploring Machine Learning Models for Solar Energy Output Forecasting

Idamakanti Kasireddy$^{(\boxtimes)}$, V Mamatha Reddy, P. Naveen, and G Harsha Vardhan

EEE Department, Vishnu Institute of Technology, Bhimavaram, India
`kaasireddy.i@vishnu.edu.in`

Abstract. Engineering, science, health, and other fields have all used machine learning algorithms. The idea of machine learning is used in this study to forecast solar energy output. Predicting solar energy, a well-known renewable source with a number of advantages, can help with energy consumption planning. Inconsistent weather makes it difficult for grid operators to manage solar energy output, which makes it harder to satisfy customer demand. Utilizing various algorithms including Lasso, Ridge, Linear and Support Vector Regression (SVR) algorithms, our suggested strategy entails developing prediction models. These algorithms produce forecasts based on past weather information such as temperature, dew point, wind, cloud cover, and visibility. SVR algorithm outperformed the other algorithms, according to the Jupyter Notebook examination.

Keywords: Solar production · Algorithms · Jupyter Note book

1 Introduction

Solar energy offers numerous merits; however, it is plagued by the issue of irregular production, which creates difficulties for grid operators trying to meet consumer demand. By anticipating the output of solar energy, grid operators can address this problem and ensure that they are adequately prepared to reach the needs of loads. For prediction of solar energy, one must acquire data on various factors that impact production, including wind speed, humidity, temperature, and dew point. This information can be obtained from meteorological departments or online sources. The effective way to examine this data is through the application of machine learning, which can be segregated into three categories: supervised, unsupervised, and reinforcement learning [1–14].

Unsupervised learning just includes input data, letting the model to independently find patterns. Supervised learning necessitates providing the model both the input data and the desired output. Contrarily, in reinforcement learning algorithm, a software deals with an environment to accomplish a certain goal and receives feedback in the form of incentives. Regression and classification are two subcategories of supervised learning, which is used in this work. Regression techniques including linear, ridge, lasso, and support vector regression are the most often used ones. Regression involves using independent variables to predict the dependent variable values.

© ICST Institute for Computer Sciences, Social Informatics and Telecommunications Engineering 2024
Published by Springer Nature Switzerland AG 2024. All Rights Reserved
P. Pareek et al. (Eds.): IC4S 2023, LNICST 536, pp. 210–217, 2024.
https://doi.org/10.1007/978-3-031-48888-7_18

The implementation of algorithms involves a set of steps, including data collection, analysis, wrangling, training and testing, and accuracy evaluation. The first step involves collecting data, which is then analyzed for duplicates, missing values, and incorrect formatting. Data cleaning, which involves removing duplicates, imputing missing values, and correcting formatting errors, is performed as necessary. The model is then trained using 80% of the data, with the remaining 20% used for testing. The final step is to examine the accuracy of the model using error method. The efficiency of the model is proportional to its accuracy.

Python's numerous libraries make it easier to build machine learning methods. These libraries facilitate and improve the implementation of algorithms. Due to its simplicity of use and abundance of libraries, Python is frequently used for machine learning. Machine learning algorithms are run on platforms like Anaconda, Google Colab, and Jupyter. Jupyter stands out among them for its ease of use and ability to share files via Jupyter Notebook. Users can conduct mathematical operations like trigonometry and Fourier transforms using Jupyter, which is open source.

2 Methodology

2.1 Linear Regression

Several regression approaches, including linear regression, were used in this study [15]. A fundamental regression technique that aids in determining the relationship between two variables (independent and dependent) is linear regression. It is based on minimizing the error between the line and the data points by applying several strategies to determine the line that best fits the data points. The least-squares approach is used to find the regression line, also referred to as the line of best fit. Accuracy is measured using the R-squared value, which measures how closely the data adhere to the regression line. A good model is one that has an R-squared value better than 0.5. The evaluation of trends and projections of sales and trend forecasting are two common uses of linear regression. The slope for the predicted regression equation is represented by formulas (1).

$$b_1 = \frac{\sum (x_1 - \bar{x})(y_1 - y)}{(x_1 - \bar{x})^2} \tag{1}$$

2.2 Lasso Regression

This algorithm is applied as a regularization method to prevent over fitting in machine learning models. Over fitting occurs when a model performs exceptionally well on the training dataset, but poorly on the testing dataset, due to a high cost function. To mitigate this issue, Lasso Regression incorporates a regularization term represented by lambda, which multiplies the weights of the model.

2.3 Ridge Regression

Regularization further makes use of Ridge Regression [16]. It is a method to stop a model from fitting too tightly. Sometimes a machine learning model performs well when tested on training data, but when tested on testing data, it performs poorly compared to training data, leading to over fitting. When using this, the objective function is modified by the addition of a penalty factor. Ridge Regression Penalty is the measurement of the penalty that was applied to the model.

2.4 Support Vector Regression

The Support Vector Regression (SVR) [17–20] algorithm is a machine learning method that is used to predict continuous values. The SVR algorithm is based on the same principle as Support Vector Machines (SVMs), where the aim is to find the best fit hyperplane. The algorithm tries to minimize the distance between the predicted values and the actual values, while also maintaining a margin of error. This margin of error is controlled by the threshold value, which determines the trade-off between the accuracy of the model and the complexity of the hyperplane. Overall, SVR is a powerful algorithm for predicting continuous values and can be used in a variety of applications. When compared to other algorithms, SVR offers the benefits of being extremely simple to use and having a high degree of prediction ability.

2.5 Purpose

The study compares and analyzes the performance of four different machine learning algorithms used for solar energy prediction. The main objective of the study is to identify the most efficient method for forecasting energy output of solar system.

The algorithms used in the study are carefully selected and include the most commonly used methods for time-series forecasting. These algorithms are trained on historical solar energy data, and their performance is evaluated based on their ability to predict future solar energy output accurately.

The study utilizes various performance metrics to evaluate the accuracy of each algorithm, including the following,

- mean absolute error,
- root mean square error,
- Coefficient of determination.

The results of the study are then analyzed and compared to determine which algorithm performs best for solar energy prediction.

By conducting a comparative analysis of these four machine learning algorithms, the study aims to provide valuable insights into the most effective method for solar energy forecasting. This information can help inform decision-making processes and support the development of more accurate and reliable solar energy prediction models.

3 Results

Understanding the link between the dependent and independent variables is crucial before conducting an accurate analysis of solar energy forecast using machine learning algorithms. In this scenario, solar output is the dependent category, which implies that numerous things impacting it, which are independent variables, are the output or consequence of solar energy.

The study moves further with the implementation of machine learning models after identifying the variables. In order to forecast solar energy output, the analysis employs four alternative regression algorithms: support vector, lasso, linear, and ridge.

To facilitate the analysis, various Python libraries are imported into Jupyter Notebook, including pandas, numpy, matplotlib.pyplot, and seaborn. The weather data is imported into the platform i.e. Jupyter book using the various commands, which varies depending on the file type (.xls or.csv).

The imported data is then analyzed using several methods, such as info(), describe(), and isnull().sum(), to identify any empty, duplicate, or incorrectly formatted values. The goal of this step is to minimize differences between values and improve the accuracy of the analysis. If null values are present, they are replaced with suitable values through mean or median methods. Rows with duplicate values are dropped, while rows with incorrect formats are corrected to ensure the accuracy of the data.

Overall, this rigorous analysis of the solar energy data using different machine learning algorithms and data analysis techniques helps to provide accurate and reliable predictions of solar energy output, which can be valuable for various applications in the renewable energy industry.

Finally, evaluating the effectiveness of different algorithms are obtained as follows (Table 1).

Table 1. Evaluating the effectiveness of different algorithms

Algorithm Accuracy
In linear regression, 51.3%
In Lasso regression,51.2%
In Ridge regression, 51%
In Support vector regression, 88.4%

The results of various algorithms' predictions for solar energy are depicted in Figs. 1, 2, 3 and 4. Upon analyzing the figure, it becomes evident that the SVR algorithm performs better than the other algorithms tested. The SVR algorithm's superior performance indicates that it is the most accurate model among the algorithms compared. Therefore, the results suggest that SVR is the optimal choice for predicting energy of solar system.

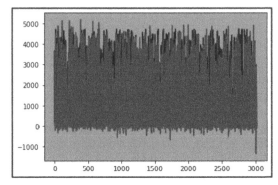

Fig. 1. The output from the SVR test is shown in blue, while the expected output is shown in red. (Colour figure online)

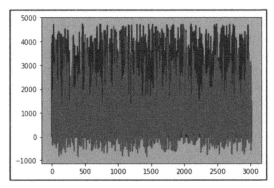

Fig. 2. The output from the linear regression test is shown in blue, while the projected output is shown in red. (Colour figure online)

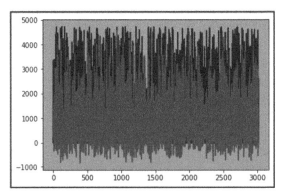

Fig. 3. The output from the Lasso Regression test is shown in blue, while the expected output is shown in red. (Colour figure online)

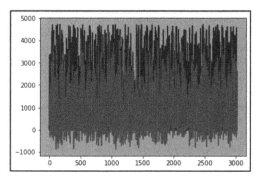

Fig. 4. The output from the Ridge Regression test is shown in blue, while the expected output is shown in red. (Colour figure online)

4 Conclusion

This study's objective was to forecast solar energy using different models. Jupyter Notebook was used to implement and analyse the models. After investigation, it was discovered that the Ridge, Lasso, and linear regressions had accuracy levels between 51% and 52%. The Support Vector Regression model, however, has a substantially higher accuracy rate of 88.4%. The results clearly indicate that SVR outperformed the other regression models in predicting solar energy. The higher accuracy of SVR suggests that it is a more appropriate and reliable model for predicting solar energy as compared to other models. Therefore, this study recommends the use of Support Vector Regression for predicting solar energy.

References

1. Javed, A., et al.: Predicting solar irradiance using machine learning techniques. In: 15th International Wireless Communications & Mobile Computing Conference (IWCMC), pp. 1458–1462 (2019)
2. Ak, R., et al.: Two machine learning approaches for short-term wind speed time-series prediction. IEEE Trans. Neural Networks Learn. Syst **27**(8), 1734–1747 (2016)
3. Nasir, A.W., Kasireddy, I., Rahul Tiwari, B.K., Ahmed, I., Furquan, A.: Data-Based Tuning of PI Controller for First-order System. In: Bhaumik, S., Chattopadhyay, S., Chattopadhyay, T., Bhattacharya, S. (eds.) Proceedings of International Conference on Industrial Instrumentation and Control: ICI2C 2021, pp. 547–555. Springer Nature Singapore, Singapore (2022). https://doi.org/10.1007/978-981-16-7011-4_52
4. Pragaspathy, S., Aravindh, G., Kannan, R., Dhivya, K., Karthikkumar, S., Karthikeyan, V.: Advanced control strategies for the grid integration of wind energy system employed with battery units. In: 2022 International Conference on Power, Energy, Control and Transmission Systems (ICPECTS), pp. 1–5. Chennai, India (2022). https://doi.org/10.1109/ICPECTS56089.2022.10046771
5. Phani Kumar, Ch., Elanchezhian, E.B., Pragaspathy, S.: An adaptive regulatory approach to improve the power quality in solar PV-integrated low-voltage utility grid. J. Circ. Syst. Comput. **31**(17), 2250301 (2022). https://doi.org/10.1142/S0218126622503017

6. Rama Rao, R.V.D., Pragaspathy, S.: Enhancement of electric power quality using UPQC with adaptive neural network model predictive control. In: 2022 International Conference on Electronics and Renewable Systems (ICEARS), pp. 233–238. Tuticorin, India (2022). https://doi.org/10.1109/ICEARS53579.2022.9751866

7. Kannan, R., Karthikkumar, S., Suseendhar, P., Pragaspathy, S., Chakravarthi, B.N.C.V., Swamy, B.: Hybrid renewable energy fed battery electric vehicle charging station. In: 2021 Second International Conference on Electronics and Sustainable Communication Systems (ICESC), pp. 151–156. Coimbatore, India (2021). https://doi.org/10.1109/ICESC51422.2021.9532995

8. Kumar, C.P., Pragaspathy, S., Karthikeyan, V., Durga Prakash, K.N.S.: Power quality improvement for a hybrid renewable farm using UPQC. In: 2021 International Conference on Artificial Intelligence and Smart Systems (ICAIS), pp. 1483–1488. Coimbatore, India (2021). https://doi.org/10.1109/ICAIS50930.2021.9396048

9. Saravanan, S., Karunanithi, K., Pragaspathy, S.: A novel topology for bidirectional converter with high buck boost gain. J. Circ. Syst. Comput. 29(14), 2050222 (2020)

10. Pragaspathy, S., Rao, R.V.D.R., Karthikeyan, V., Bhukya, R., Nalli, P.K., Korlepara, K.N.S.D.P.: Analysis and appropriate choice of power converters for electric vehicle charging infrastructure. In: 2022 Second International Conference on Artificial Intelligence and Smart Energy (ICAIS), pp. 1554–1558. Coimbatore, India (2022). https://doi.org/10.1109/ICAIS53314.2022.9742853

11. NarasimhaRaju, V.S.N., Premalatha, M., Pragaspathy, S., Rao K, D.V.S.K., Korlepara, N.S.D.P., Kumar, M.M.: Implementation of instantaneous symmetrical component theory based hysteresis controller for DSTATCOM. In: 2021 International Conference on Advancements in Electrical, pp. 1–8. Electronics, Communication, Computing and Automation (ICAECA), Coimbatore, India (2021). https://doi.org/10.1109/ICAECA52838.2021.9675526

12. Sharma, A., Kakkar, A.: Forecasting daily global solar irradiance generation using machine learning. Renew. Sustain. Energy Rev. 82(3), 2254–2269 (2018)

13. Kasireddy, I., et al.: Application of FOPID-FOF controller based on IMC theory for automatic generation control of power system. IETE J. Res. 68(3), 2204–2219 (2022). https://doi.org/10.1080/03772063.2019.1694452

14. Kasireddy, I., et al.: Determination of stable zones of LFC for a power system considering communication delay. AIP Conf. Proc. 2418, 040014 (2022). https://doi.org/10.1063/5.0081986

15. Maulud, D., Abdulazeez, A.M.: A Review on Linear Regression Comprehensive in Machine Learning. JASTT 1(4), 140–147 (2020)

16. Yang, X., et al.: Lasso regression models for cross-version defect prediction. IEEE Trans. Reliab. 67(3), 885–896 (2018)

17. Crone, S.F., Guajardo, J., Weber, R.: A study on the ability of support vector regression and neural networks to forecast basic time series patterns. In: Bramer, M. (ed.) IFIP AI 2006. IIFIP, vol. 217, pp. 149–158. Springer, Boston, MA (2006). https://doi.org/10.1007/978-0-387-34747-9_16

18. Le, D.N., Parvathy, V.S., Gupta, D., et al.: IoT enabled depthwise separable convolution neural network with deep support vector machine for COVID-19 diagnosis and classification. Int. J. Mach. Learn. & Cyber. 12, 3235–3248 (2021). https://doi.org/10.1007/s13042-020-01248-7

19. Kalaiyarasi, M., Saravanan, S., Narukullapati, B.K., Kasireddy, I., Naga Malleswara Rao, D.S., Nagineni Venkata Sireesha, D.: Analysis of SAR ImagesDe-speckling using a bilateral filter and feed forward neural networks. In: 2023 Second International Conference on Electrical, pp. 1–6. Electronics, Information and Communication Technologies (ICEEICT), Trichirappalli, India (2023). https://doi.org/10.1109/ICEEICT56924.2023.10156987

20. Venkata Subbarao, M., Padavala, A.K., Harika, K.D.: Performance analysis of speech command recognition using support vector machine classifiers. In: Jason, G., Dey, R., Adhikary, N. (eds.) Communication and Control for Robotic Systems, pp. 313–325. Springer Singapore, Singapore (2022). https://doi.org/10.1007/978-981-16-1777-5_19

The Survival Analysis of Mental Fatigue Utilizing the Estimator of Kaplan-Meier and Nelson-Aalen

R. Eswar Reddy⬡ and K. Santhi$^{(\boxtimes)}$⬡

School of Computer Science and Engineering, Vellore Institute of Technology,
Vellore 632014, Tamil Nadu, India
regantieswar.reddy2021@vitstudent.ac.in, santhikrishnan@vit.ac.in

Abstract. The aim of this study is to investigate mental fatigue using the Kaplan-Meier and Nelson-Aalen estimators in survival analysis. Mental fatigue is a common occurrence when the mind becomes tired from regular tasks, and it can have a negative impact on an employee's operational functions and job efficiency. To detect mental fatigue, the shallow Kaplan-Meier method is employed by analyzing data from employee burnout evaluations.

Both the Kaplan-Meier and Nelson-Aalen estimators have proven to be effective in automatically analyzing various features from raw data. However, they often impose a significant burden on system resources during training and predictions. Therefore, alternative methods of analysis are necessary to derive the survival curve.

In this paper, we provide a mathematical foundation for the Kaplan-Meier method and explain the concept of censoring, including right censoring, interval censoring, and left censoring. Furthermore, we construct a Kaplan-Meier survival curve, which represents the probability of survival over time. The Kaplan-Meier survival curve is considered the most reliable and is recommended for predicting the variable under investigation, particularly in the fields of public health and medical research.

The findings of this research can also be utilized to develop interventions and strategies aimed at reducing mental fatigue and improving employee morale. Enhancing employee morale can positively impact an organization as a whole, as mental fatigue has been associated with lower job satisfaction and an increased likelihood of employee turnover, both of which can be further explored in future studies. Overall, this study sheds light on the significance of understanding and addressing mental fatigue in the workplace, and it provides valuable insights that can contribute to the well-being of employees and the success of organizations.

Keywords: mental fatigue · Kaplan Meier · Nelson-Aalen · fatigue survival curve · burnout

© ICST Institute for Computer Sciences, Social Informatics and Telecommunications Engineering 2024
Published by Springer Nature Switzerland AG 2024. All Rights Reserved
P. Pareek et al. (Eds.): IC4S 2023, LNICST 536, pp. 218–241, 2024.
https://doi.org/10.1007/978-3-031-48888-7_19

1   Introduction

Employees who report feeling good about themselves and their working conditions tend to be more productive overall. As a result, they contribute to the success of the business or organization. Nevertheless, the situation in the majority of businesses has changed as a result of the pandemic. Almost 69% of the workforce has been experiencing burnout since implementing work-from-home and office policies. The percentage of employees who have burned out is high. There has been a rise in the number of businesses caring about their workers' emotional well-being. This trend can be attributed to the growing recognition of the negative impact that burnout can have on employee productivity and overall business success. As a result, many companies are implementing programs and policies aimed at preventing burnout and promoting mental health in the workplace. To counteract this, we plan to develop a web application that businesses may use to track staff burnout. Additionally, employees themselves can use it as a tool to monitor burnout and evaluate mental health in the hectic workplace.

This study also found that the amount of mental fatigue [1] is related to how much pain, anxiety, and depression affect how much fatigue [2] affects a person's life, and these things work together in a cycle that can make fatigue worse and keep it going. There are many possibilities for the effects of each of the others. 80% of employees with depression also report sleeping poorly, supporting the idea that the two are related. There is a strong correlation between this condition and poor sleep quality, with decreased sleep efficiency being the primary sleep issue described. Mental fatigue makes it worthwhile to investigate this K-M method. According to the literature and our most recent findings, women have a higher prevalence of mental fatigue than men. Women's work is handed to us, working women are handed the kids, the homework, kids, jobs, etc.

Instead of investigating employees in the organization to check for mental fatigue disorder [3], the organization checks the health condition of employees and mentally how much struggle they are facing, effective working on ongoing work, whether the organization has given additional responsibility or reduced project delivery on time, reviews meetings and client meetings in the cases of team members, makes plans to change the upcoming one, earns new things, and talks to family. Employees' work burden stress, and keeping on top of the status in the daily meeting, do not support the work environment.

Life is moving at a breakneck speed, and the demands of work and school are only getting heavier. Extended mental exertion of any kind necessitating undivided attention inevitably results in tiredness, with all the unpleasant consequences that entail slower reaction times, dizziness, nausea, etc. Hence, in order to aid in reducing its harmful consequences, it is important to recognize [4] and study various forms of weariness.

The role of emotion in human interaction, understanding, and decision-making has grown in recent years. Recognizing others' feelings is a fundamental skill for establishing rapport in everyday life. Emotional state is assessed using a person's EEG signal [5], which measures the amount of brain activity during various conditions. To date, the best method for extracting human emotions from EEG [6] data has been a 6 layer feed forward neural network that has been subjected to extensive biological testing. Functionalities including preprocessing, feature extraction, and classification were all provided by this system. First, a band pass filter is used to do EEG preprocessing [4, 7]. Several methods of electrophysiological recording have been used to shed light on the intricate brain interactions that underlie cognitive tiredness in situations of mental fatigue, allowing researchers to better understand the neural circuits at play in this state [8].

This percentage is inversely related to the number of people with fatigue. A decreasing curve starting at 1 represents the Kaplan-Meier estimates of survival time graphically [9]. The size of the steps depends on factors such as his length of residency, the likelihood of a mental fatigue disorder, and the possibility that he will suffer through the allotted time without ever encountering the event of interest. Censored observations refer to data that is missing either temporarily or permanently, and they can occur at any point in the research paper. There are essentially three distinct categories of censorship. The most prevalent type of censorship is called right censorship, and it occurs when a patient is observed for a certain amount of time without experiencing the event of interest. Hence, there is a gap in the survival time [10] series on the right side of the observational period. We know that the event of interest does not occur for this patient until the censoring date, the second type of censorship occurs when the event of interest occurs between two unknown dates and we do not know which date it occurred on.

The zero Reynolds number and long wavelength assumptions, It is discovered that, for a given flow rate, the pressure rise lowers as the peripheral layer viscosity drops, and that, for a given non-zero pressure drop, the flow rate increases as the viscosity of the peripheral layer decreases [11]. The beneficial in creating a physically suitable workplace and promoting professional productivity. Furthermore, they should consider permitted working hours based on people's abilities, reducing tension factors, creating a culturally and ethically secure atmosphere, respecting new ideas, providing spiritual and mental well-being, planning, reconstructing the system of instruction, utilizing recommendations and criticism, and other factors. These factors can prevent emotional exhaustion within the organization. Since depersonalizing has a direct correlation with performance, the managers can strengthen coworker relationships by implementing counselling programmers [12] and communication skills training. The electrodes on an EEG device collect electrical impulses that communicate at different EEG frequencies. The Fast Fourier Transform [13] is a method that can be used to identify these raw EEG signals as discrete waves with a variety of frequencies. A group of learners who perform poorly on the learning curve come together to form an ensemble of classifiers [13, 14], also known as a committee of classifiers. Learning a large number of less effective classifiers and integrating them in a certain manner is the purpose of the ensemble of classifiers technique rather than learning a single, efficient classifier. Data mining algorithms aid in the analysis and prediction of large data sets with minimal human intervention [15].

Predicting and analyzing diabetes can be done with a number of data mining programs. Fast and accurate automated algorithms for summarizing text and generating a summary that can be spoken aloud. An automated decision-making system's variance could be reduced, which would improve the system's accuracy. In the intervening time, ensemble systems have been utilized effectively to solve a wide range of machine learning challenges, including feature selection, confidence estimation, missing features, incremental learning, error correction, class-imbalanced data, and learning concept drift from non stationary distributions, just to identify a few. By doing in-depth research on technological aid, user experience, and health care, we can help reduce the deadly risks that people face and be ready to act quickly in emergencies [16]. Because of its usefulness in so many programs, neural model-based text analysis has recently gained traction. Researchers have identified and justified a large number of techniques for enhancing text analytics effectiveness. Text categorization, text generation, text summarizing, query formulation, query resolution, and sentiment analysis are just some of the areas where these methods have been put to good use [17]. To achieve uniform scale L2 regularizes of linear models may assume that all features are centered on zero or have variance in the same order. Because these things are often used in the objective function of learning algorithms. This method of presentation has advanced greatly since the days of rainbow-hued spreadsheets. With the advent of datasets [18] another execution time and space complexity of mining has drastically decreased. Retrieval techniques went to a whole new level after [19] receiving this data. If all commercial activities cease, the business will cease to exist. Operating alone, the application server could not sustain startup costs [20]. Statistical methods for regression and classification are incorporated into machine learning algorithms. Sensors are used to capture data, which is then transmitted to the Blynk app. The automatic water controller only activates when the relative humidity falls below a predetermined threshold [21]. To accompany the connected device, we need a resource provisioning system that is easily managed; this is only feasible if we have accomplished cloud service models [22]. Every plant has specific requirements that must be met to ensure its survival. Therefore, it is necessary to establish a system where plants can communicate with the user [21]. The proposed strategy uses a technique for reducing the number of dimensions and clustering similar objects together. For both symmetrical and asymmetrical data sets, it provides the highest accuracy for the larger of the two [23].

Third, left censoring, occurs when someone from a certain fatigue is known to have the event before a certain date, but the time period between the occurrence of the event and the specific date is unknown. The zero Reynolds number and long wavelength assumptions, It is discovered that, for a given flow rate, the pressure rise lowers as the peripheral layer viscosity drops, and that, for a given non-zero pressure drop, the flow rate increases as the viscosity of the peripheral layer decreases [11]. Survival probability utilizing the Kaplan Meier (KM) survival estimator, Nelson-Aalen estimator, and Hazard Model based on regression could be used to assess prediction scores for Mental fatigue [24]. In this paper, an

Employee burnout datasets was used to (i) analyses the research to identify key differential mental fatigue, (ii) examine survival associated with most altered mental fatigue using the web-based Kaplan Meier and Nelson-Aalen Plotter tool, and (iii) evaluate the possibility of the potential at the datasets between mental fatigue and control variables. During the construction of survival time probabilities and curves, the serial duration's for specific participants are ordered from shortest to longest regardless of when they entered the research. By employing this technique, all subjects within the group commence the analysis at the same point and are all surviving until mental fatigue persons are identified. Two outcomes are possible: 1) the subject can see the event of interest, or 2) they may be censored. This subject's total survival time cannot be determined precisely due to censorship. This can occur when an adverse occurrence for the research occurs, such as the Employee dropping out, being lost to follow-up, or required data not being available, or when something positive occurs, such as the research getting before the subject observed the event of interest, they survived at least until their conclusion of the research, but it is unidentified [25] what occurred to them afterward. Thus, censorship can occur either during the research. Metal fatigue is a risk factor that causes some of their diseases, like Constantly feeling overwhelmed or stressed, Cynicism, uncertainty, and pessimism Depression, anxiety, and suicidal thoughts Sleep disruptions and pattern alterations, Tension, pain, and headaches Digestion problems and recurring colds High blood pressure, abnormal heart rate, brain fog, and strokes Obesity and cardiovascular disease.

2   Method

Dataset. Employee burnout is a dataset collected from the kaggle website [26]. The following are the data attributes and their descriptions, we are implement the python code [27]. The shape of data frame is: (20633, 9) instances will be there in dataset. Data preprocessing the data is processing is onehotEncoder is technique will represent the text, category variable is numeric values, the transform the text and category variables data is the will transform to numeric values. Data is pre processing the data is find out the missing and noise data, duplicate data. The data is replace the noise data and duplicate data by the using the means values, the replace the missing values by the means values on the dataset.

Employee ID. The distinctive ID that the company gives to each employee.

Date of Joining. The day the individual began working for the company.

Gender. In the box plot, we see that some of the women's burnout rate data points are significantly different from the rest. This is something that needs to be handled by us. Female workers are over represented in the data. It shown the Fig. 2.

The gender of the employee. Men have a higher burnout rate than women do on average. Let's investigate why this would be the case by looking at how the two sexes fare in other areas, such as titles and hours worked. It shown the Fig. 1.

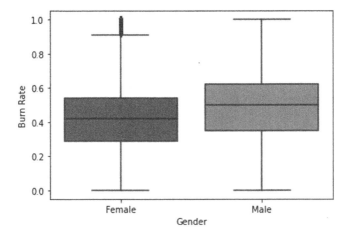

Fig. 1. The gender difference in burn rate

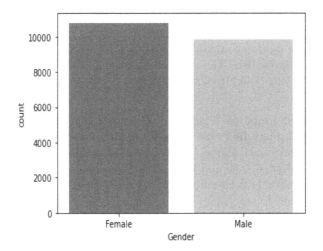

Fig. 2. The gender and count of employees

Designation. The employee's position in his or her organization. In the interval [0.0, 5.0], "0.0" is the least significant digit, and "5.0" is the most significant. More men than women hold positions with a designation of 2.0 or above. It shown the Fig. 3.

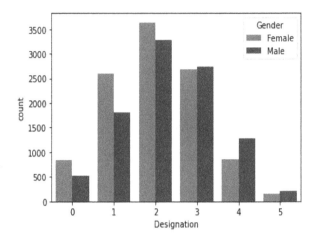

Fig. 3. Designation for men

Resource Allocation. The number of resources given to an employee for work, which should be thought of as the number of working hours. Between one and ten higher means more resources. Most women work up to 8 h a day, whereas most men work up to 10 h. The median number of hours worked by men and women differs by one hour. It shown the Fig. 4.

Company Type. Employee The figure of resource allocation for an employee-based company displays how the available resources are distributed among the employees to achieve the company's goals. It helps in identifying which employee has been assigned what task and how much time and resources have been allocated to it. May classify their employers based on the services or products they offer. It shown the Fig. 5.

WFH Setup Available. Is the worker allowed to work from his or her home office.

Mental Fatigue Score. A number from 0 to 10 that shows how mentally tired the worker is at work, where 0 means no mental fatigue and 10 means extreme mental fatigue.

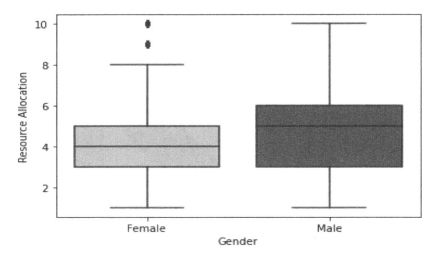

Fig. 4. Resource Allocation

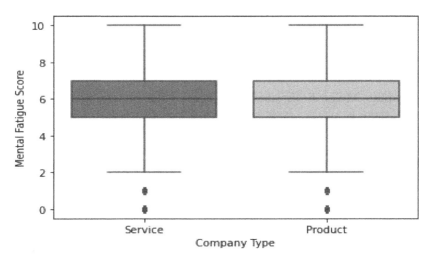

Fig. 5. Company Type

Burn Rate. The target value in each worker's data shows the rate of burnout while on the job. Values from 0.0 to 1.0 show that burnout is getting worse. The correlation between fatigue score and burn rate appears to be very significant. It is important to address burnout in the workplace, as it can have negative impacts on both employees and organizations. Employers should consider implementing strategies to prevent and manage burnout, such as promoting work-life balance and providing support resources. It shown the Fig. 6.

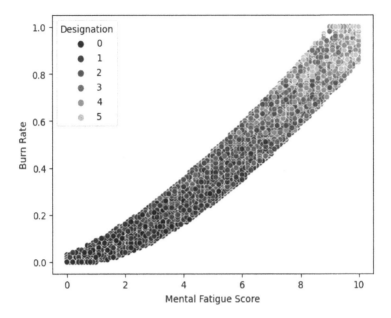

Fig. 6. Mental fatigue and burnout

The Kaplan-Meier Estimator. The Kaplan-Meier curves and survival predictions, there is now a better way to look at data when your mind is tired. The Kaplan-Meier estimator [28] is often used to describe how long a population will fatigue because it gives estimates of the survival function that are not based on statistics. If you want to know how likely a patient is to live past a given point after receiving mental health treatment, one of the most reliable statistical techniques is the KM estimate. When working with incomplete or hidden data, the KM method is especially helpful because it lets all available data points be used in the analysis. Additionally, it can be used to compare survival rates between different treatment groups or populations. This method of making graphs and charts is very user-friendly. The effectiveness of an intervention in mental fatigue research is determined by tracking how many people are rescued or made healthy thanks to the intervention. Notwithstanding the complexity of some subjects or settings, KM estimation is the simplest method for predicting longevity over time. Events, censorship, and the likelihood of survival can all be estimated with the help of the Kaplan-Meier curves.

The Kaplan-Meier survival curve [29] is a useful tool in statistics for analyzing time-to-event data and making comparisons between groups of people. The number of people who avoid dying during a specific time period can be calculated using the survival curve. This can be analyzed for an evaluation of a pair of patient populations or subjects as well as the statistical difference in their overall survival.

Depending on how much time has passed, either the product limit estimator or the Kaplan-Meier curve can be used to figure out the survival function. If the data are already organization into intervals, if the sample size is large, or if a large population is of interest, a clinical life table analysis may be more time-efficient. Both of these approaches will be addressed in greater detail. The Kaplan-Meier survival curve is the cumulative chance of survival over many intervals of time. This analysis relies on three presumptions. To proceed, we believe that censored employees at any given time have the same survival and growth as unregulated employees; second, we think that those who participate in research at the beginning or the end will have significantly different survival times. Furthermore, we'll presume that the event occurs just when it's expected to. In situations where the occurrence would pick up the energy level and bring routine work, this can be done from the office or home. This approach can be particularly useful for individuals who have flexible work arrangements or who are able to work remotely. It allows them to balance their personal and professional responsibilities while still being productive.

If an employee is followed up on more often and for shorter periods of time, it is possible to predict how long they will stay with the company. This can be beneficial for employers in terms of retention strategies and succession planning. However, it is important to balance this with the need for employee autonomy and trust in the workplace. The term "product limit estimate" can be used to describe the Kaplan-Meier estimate. The method involves calculating the odds of an occurrence happening at a given instant. To arrive at a final estimate, we multiply these probabilities by any previously calculated probabilities. If you want to know your probability of surviving at any given moment in time, just plug those numbers into the following Eq. 1.

The Kaplan-Meier method or Kaplan-Meier curve can be made with just two pieces of information: the time until the event of interest and the status of the patient at that moment. In medical research, the Kaplan-Meier method is often used to figure out how likely it is that a patient with a certain disease will live. It is a non-parametric statistic that takes into account time-to-event data and censored observations. Let $D1 < D2 < ... < Dn$, $i < N$ be a collection of separate ordered times finding the mental fatigue times observed in N individuals; in a given time Di (i = 1, 2,..., n), the number $di \geq 5$ of mental fatigue are observed, and the number ri of subjects, whose either mental fatigue or censored time is greater than or equal to Ti, are deemed "at risk", the observed times to event. T In its simplest form, there is no need to elaborate on the formula for the conditional probability of survival past time Ti, which is simply that the Kaplan-Meier method is an estimate of the conditional probability of survival at different periods in time identified by the event's likelihood of occurring. The Kaplan-Meier method is commonly used in medical research to estimate the survival rate of patients with a particular disease. It takes into account the occurrence of events such as death or relapse to calculate the probability of survival at different time intervals.

$$P(D_i) = \frac{r_i - d_i}{r_i} \tag{1}$$

By dividing the total number of subjects by the total number of patients at each point in time, one can calculate the likelihood of survival. Subjects who get lost because they are tired are not counted as having a disorder. Instead, they are considered "censored" and taken out of the denominator. By multiplying the individual survival probabilities at each interval leading up to that point by the law of multiplication of probabilities, one can determine the cumulative probability of survival up to that point. This approach is commonly used in medical research to account for the fact that some subjects may drop out of a study for reasons unrelated to the disorder being studied. By censoring these subjects, researchers can more accurately estimate the probability of survival for those who remain in the study.

The term conditional probability describes this type of probability. Because of the limited number of events, the calculated probability at any given interval is not very precise. However, the overall probability of survival at each point is Estimating the survival function at time Δt is done by multiplying the conditional probability of survival at that time by the formula.

The conditional probability of survival ($\bar{S}(\Delta t)$), also called cumulative probability or cumulative survival, is the chance that a patient will be mentally tired days after enrolling in a study, if the patient has been alive for at least Δt days before enrolling. In a hypothetical situation where a patient in an intensive care unit checks his or her level of fatigue and lives for hours per day, the product rule of conditional probabilities says that the cumulative survival is the product of survival probabilities. This information is useful in medical research because it assists clinicians and researchers. A patient's conditional probability of survival, also called cumulative probability or cumulative survival, is the chance that he or she will be mentally tired days after enrolling in a study, given that the patient has been alive for at least t days before that. Understand the long-term effects of treatments and interventions on patients' survival and quality of life. Additionally, it can aid in making informed decisions about patient care and treatment plans.

$$\bar{S}(\Delta t) = \prod_{i:D_i < t} P(D_i) = \prod_{i:D_i < \Delta t} \left(1 - \frac{d_i}{r_i}\right) \tag{2}$$

The definition of its variance is Variance is a statistical measure that quantifies the amount of variability or dispersion in a set of data. It is calculated by taking the average of the squared differences from the mean of the data set.

$$\sigma(\bar{S}(\Delta t)) = \bar{S}(\Delta t)^2 \sum_{i:D_i < \Delta t} \frac{d_i}{r_i(r_i - d_i)} \tag{3}$$

Variance is an important tool in statistical analysis, as it helps to understand the spread of data points around the mean and can be used to make predictions about future data. But outliers and extreme values in the data set might affect it, and it might be necessary to fix them before using the variance to draw any conclusions. Because of the censoring, r_i is not simply equal to the difference between $r_i - 1$ and $d_i - 1$, the right approach to calculate r_i is $r_i = r_{i-1}d_{i-1}C_{i-1}$, where $C_i - 1$ is the number of censored cases between $D_i - 1$ and D_i. This calculation is commonly used in statistics to determine the variability of a data set. However, when dealing with censored data, a modified approach must be taken to accurately calculate the variability.

3   Result and Discussion

The Kaplan-Meier method [30] is a deft statistical analysis of survival times that not only provides for filtered observations in the right way but also makes use of the information from filtered individuals up to the point of filtering. While investigating the effects of mental fatigue, it is typical to use two interventions and evaluate the outcome in terms of the employee's ability to stay fatigued. Hence, the Kaplan-Meier approach [31]is a valuable resource that may have an essential role in producing evidence-based data on expected survival [32]. Survival analysis is a type of statistical analysis used to examine an incident that happened relatively frequently over a particular period of time. Hence, it seeks to discover how often something occurs. The word "survival" can mean anything from the employee's depression, mental tiredness, mental illness, sleeping, and anything in between. Populations, or collections of recipients who are monitored throughout time, are used in survival studies to record significant medical events as they occur and associate them with an intervention of interest. Survival analysis requires the determination of the "survival time", [10] which is the amount of time that has passed since the baseline date and before the event happens. As a result, it is essential to determine whether the employees in question witnessed the incident of interest or were prevented from doing so by the filters. For the statistical application to determine the cumulative probability of the event, it is essential to know this information. The Kaplan-Meier method's [2,33] primary function is to generate survival curves as a function of time, providing a visual representation of the clinical phenomenon being investigated. The ordinate of a Kaplan-Meier curve shows the cumulative survival time, while the abscissa shows the elapsed time. The time intervals used to create a Kaplan-Meier curve aren't decided upon in advance but rather are determined by the occurrence of events. The Kaplan-Meier technique is preferable because it estimates for correction.

Plotting Survival Curves Using matplotlib.pyplot, Seaborn, Kaplan-MeierFitter, CoxPHFitter

The Kaplan-Meier [34] curves for survival time are unappealing to the eye. Enhanced plots can be created using matplotlib.pyplot, seaborn, KaplanMeier-Fitter, CoxPHFitter libraries [27]. The following sections show and describe Kaplan-Meier curves generated with matplotlib.pyplot, seaborn [35], Kaplan-MeierFitter [30], and CoxPHFitter. These libraries' functions are using the graphs generated [36] by the KaplanMeierFitter graphs [31], which find the Kaplan Meier estimates, the x-axis is the mental fatigue score, the y-axis is the designation of the employees, find out the fatigue score of the employees. The Kaplan-Meier curves are useful for analysing survival data and estimating the probability of an event occurring over time. They are commonly used in medical research to analyse patient outcomes and can also be applied to other fields such as finance and engineering. It shown the Fig. 7 in the graph.

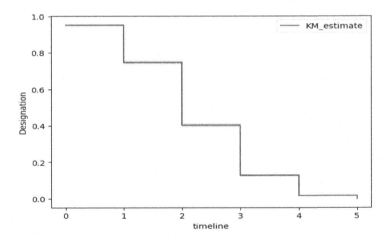

Fig. 7. Kaplan Meier estimate curve

The median survival time is 2.0, and the designation is 0.5. This suggests that half of the patients survived beyond 2.0 units of time and the other half did not, and the designation of 0.5 indicates that the survival probability at 2.0 units of time is approximately 50%.

The Survival Function. Survival, denoted by $(\bar{S}(\Delta t))$, is the probability that T happens before Δt, where Δt is any moment during the observation. In survival analysis, the survival function is often used to figure out how likely it is that something will happen at a certain on time, like a machine breaking down or an employees fatigue. It is also used to compare survival rates between different groups or treatments. Specifically, the probability that an employee will still be fatigued after a certain burnout of time, denoted by t, is the survival function. It shown the Fig. 8, and Table 1.

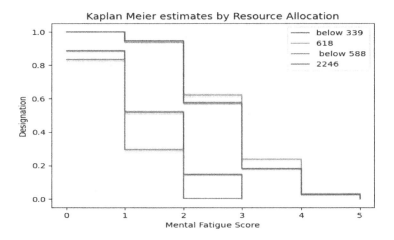

Fig. 8. Survival of different gender groups

Table 1. Kaplan Meier estimate

timeline	Kaplan Meier estimate
0	0.951038
1	0.747703
2	0.404027
3	0.128106
4	0.018177
5	0
Text (0, 0.5, "designation")	

The colour in this figure indicates which clinics correspond to certain curves. Confidence intervals for each time point and overall are shown as a band of shading. At any given period, the plus signs denote the censored instances. The confidence intervals represent the range of values within which the true population parameter is likely to fall. The censored instances refer to observations that are incomplete or truncated, usually due to limitations in data collection or follow-up. Mental fatigue has an increased survival curve, therefore, more patients remain there than in fatigue. It is advised that research be conducted into the reasons so many fatigued employees end up leaving. It's unclear if the discrepancy can be attributed to fatigue itself or if the employees were chosen for fatigue, depression, mental illness, mental stress, mental tensions, or some other factor. The research could also explore potential solutions to address the high turnover rate, such as implementing flexible work schedules, providing mental health resources, or offering additional support for employees experiencing fatigue. It is important for employers to prioritise the well-being of their employees in order to maintain a productive and healthy work environment. It shown the Fig. 9.

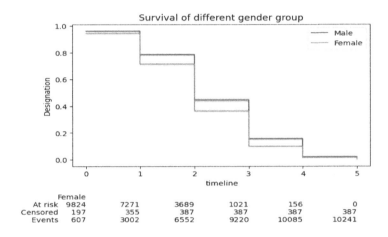

Fig. 9. Survival function

To calculate the survival probability, we'll use the Kaplan-Meier estimator [29,39], which relies on a function called the survival function $-(\bar{S}(\Delta t))$. The curve depicts how the probability of survival changes over time. Persons suffering from mental fatigue have a lower chance of survival. The comparable graph can be created with the 95% confidence interval. A failed curve can also be drawn. It is the inverse of survival, probability, and cumulative density. The median survival time and 95% confidence intervals are estimated next. This can be accomplished with the median survival time and median survival times() functions. In this case, the median survival time is 2.0 h, implying that 50% of the sample lives for 2.0 h and 50% experiences mental fatigue during that period. The 95% KM estimate lower time is 2.0, while the KM estimate upper time is 2.0. We can examine the difference involving discrete categories using the KM estimate. However, this method is only applicable when the variable has fewer categories. A mask filtering object where males are true, and a plotting object. The Plot curves for Male and Female observations. The curve demonstrates that the aggregate survival probabilities of female patients are higher than those of male patients at any given time Fig. 9.

The survival function provides the probability that the event has occurred within the time interval t. The cumulative density is the complement of the

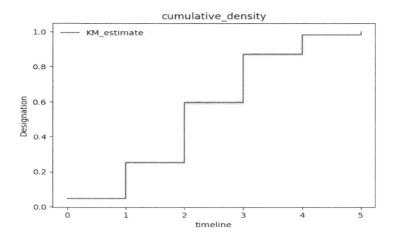

Fig. 10. Cumulative density in km estimator

survival function, which is the probability that the event has taken place by time t. Reading our plots, we can see using cdf that there is a 50% chance that the event has transpired within 5 h. This is a range, as indicated by the blue lines' broad intervals. This complement, the survival function, indicates that there is a 60% chance that the event has not occurred within 5 h.

The survival probability is the chance that a worker won't get tired between an expected point in time and a future point in time. To demonstrate, if (9944) = 0.9, the employee survival probability shrinks to 0.1. These statistics will be eliminated if the employee survives the completion of the investigation. This paper's Kaplan-Meier estimator is 0.9, achieving the best result and predicting the method However, it is important to note that the Kaplan-Meier estimator is not a foolproof method and may have limitations in certain situations. Therefore, it is crucial to consider other factors and data points before making any final decisions based solely on this statistic.

Estimating Hazard Rates Using Nelson-Aalen

The survival function is an important way to describe and show how well the model works. There's an additional method, though. Unfortunately, the Kaplan-Meier [28,37] estimate is sometimes transformed to yield information on the population hazard function (t). This transformation is known as the Nelson-Aalen estimator [37,38], which estimates the cumulative hazard function. It is a non-parametric method used in survival analysis to estimate the hazard rate from lifetime data. The Nelson-Aalen hazard function is used for this function. The Nelson-Aalen hazard function is a non-parametric estimator that is particularly useful when the hazard rate changes over time. It provides an estimate of the cumulative hazard function, which can be used to estimate the population hazard function. The Nelson-Aalen hazard function is a non-parametric estimator of the cumulative hazard function that makes no assumptions about the underlying distribution of survival times. It is particularly useful when analysing data with complex censoring patterns.

$$\bar{H}(\Delta t) = \sum_{D_i \leq \Delta t} \frac{d_i}{r_i} \tag{4}$$

where di is the number of fatigued employees at time t and ri is the number of employees at the start. Survival functions are where basic survival analysis starts, but cumulative hazards are where more complicated methods begin. It Show the Table 2 and Fig. 10.

Table 2. hazard function

timeline	NA estimate
0	0.058806
1	0.338227
2	1.008063
3	2.292282
4	4.168251
5	5.3682
\<AxesSubplot:xlabel='timeline'\>	

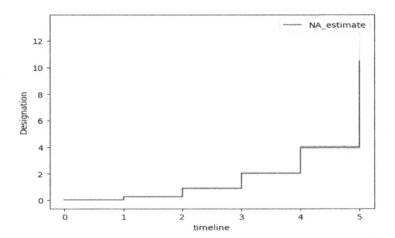

Fig. 11. Hazard function curve in NA estimate

This research predicts values at a certain point 10, If the hazard probability, denoted by $\bar{H}(\Delta t)$, is true at a given time, an employee under observation has event fatigue at that time. If the value of (9944) is equal to 0.9, for instance, the probability that the employee is still mentally fatigued is predicted, and the employee's fatigue levels are checked to examine the burnout. The hazard function, in contrast to the survival function, is the occurrence of an event. Having a lower hazard probability and a higher survival probability [33] is good for the employee. The hazard function can be used to predict the likelihood of an event occurring in the future, such as an employee leaving a company. By using the hazard function, employers can take proactive measures to reduce turnover and improve retention rates. The hazard function is a useful tool for predicting the likelihood of an event occurring in a given time frame. Employers can identify potential risks and take steps to mitigate them by analysing the hazard function, resulting in a safer work environment for their employees. In the paper we analysed to find the best result and predict the method are would use, we investigated the employees to check the mental fatigue and burnout status in organisations (Fig. 12).

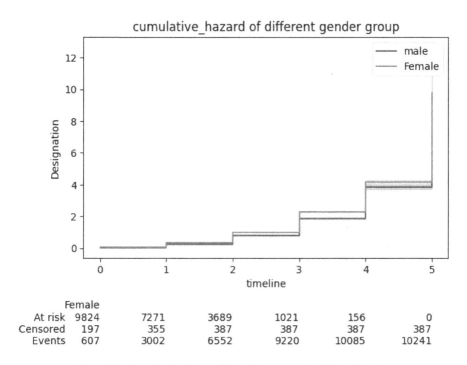

Fig. 12. Cumulative hazard function curve in NA estimate

To calculate the survival probability, we'll use the Nelson-Aalen estimator. The survival probability is the chance that a worker won't get tired between an expected point in time and a future point in time. To demonstrate, if (10373) = 0.04, the employee survival probability shrinks to 10.36. These statistics will be eliminated if the employee survives the completion of the investigation. It shown the Fig. 11. This paper's Nelson-Aalen is 0.9, achieving the best result and predicting the method However, it is important to note that Nelson-Aalen estimator is not a foolproof method and may have limitations in certain situations. Therefore, it is crucial to consider other factors and data points before making any final decisions based solely on this statistic. The estimate the number of persons will be at the high risk will be 100 employee is risk in the affect the mental fatigue due to the designating, the estimated the number of persons will be at the fatigue persons on below the 200 persons affect the fatigue at risk.

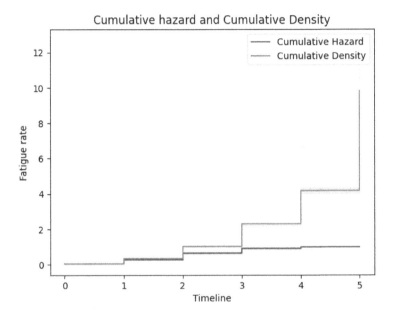

Fig. 13. Cumulative hazard function curves

The survival probability of the divided int two group that is male and female, estimated and predicting the employee work and meta fatigue, estimated the male's will be faster fatigue, more female is fatigue is more and compare t the male check the status abut the graph Fig. 13, the Cumulative density function using find the estimated the km estimate and Cumulative hazard function using estimated the predicting the employees affect the meta fatigue, the point the function the risk, It assists to consider actual phenomena and the way their hazard functions may be shaped. If T represents the designation of an employee when it develops fatigue for the first time, then one might expect the corresponding hazard function h(t) to increase with time; that is, the conditional probability of a serious fatigue in the time will increase with the employee's designation and responsibility. In contrast, one might expect h(t) to decrease during the period of responsibility if unpredictable fatigue was being studied in a situation where the employee had a designation and a responsibility. This is known to be the result of selection during time's fatigue. When T is the time at which metal fatigue will have a greater impact on employees due to the w designation and less employee responsibility, the hazard function will remain relatively constant in t. Given that fatigue has not yet occurred, the probability of designation and responsibility in the next time interval does not change with t, but the probability of fatigue in the next designation will increase as the fatigue level rises. Survival analysis relies heavily on the exponential distribution, which is uniquely characterised by this property. The hazard function may take on a more intricate form. If T denotes the mental fatigue of fatal outcome, then the hazard function h(t) is anticipated to decrease initially before progressively increasing at the end, reflecting a higher risk of unpredictable fatigue and fatal outcome.

TTF estimation without the need to know the failure times of all observed units. This would reduce the number of necessary calculations and, more importantly, facilitate the procedure for obtaining data throughout the entire observation period. The optimal method to the NA estimator is to divide the observation period into intervals and evaluate TTF for the limits of these intervals, as opposed to calculating TTF for each failure. This is the Mean time to failure TTF: [0.01665967 0.02371429 0.02838923 ... 0.03362179 0.02897202 0.03997889], mean time to failure TTF: 167 h on the estimated. The predict the estimated at the work environment persons is affect the fatigue at 45 h mean time failures. The TTF estimated the 204.86 value on the NAE values estimated.

The cumulative hazard function is the bathtub curve, which represents the fatigue's life cycle. Combining the hazard rate and the slope of the bathtub curve produces the curve's hazard rate. It shown the Fig. 14.

While the image above shows the hazard rate, the Nelson-Aalen estimator's curve illustrates how the hazard rate varies over time. The concave shape of the cumulative hazard function indicates that we are dealing with a "fatigue fatal outcomes" category of event, where the failure rate is highest early on and decreases over time blue line in the image. On the other hand, the convex shape of the cumulative hazard function indicates that we are dealing with an event red line indicative of unpredictable fatigue. In this paper the Nelson-Aalen estimator of the cumulative hazard function. The estimation of the cumulative hazard function and an intuitive understanding of the results' interpretation. While the Nelson-Aalen estimator is considerably popular than the Kaplan-Meier survival curves, it is still widely used. The ability to provide more precise survival estimates and identify survival differences between subgroups of patients that KM cannot detect, as demonstrated here for various age groups [31]. In this paper The ability to provide more precise survival estimates and identify survival differences between subgroups of patients that KM detect [40], as demonstrated here for various groups.

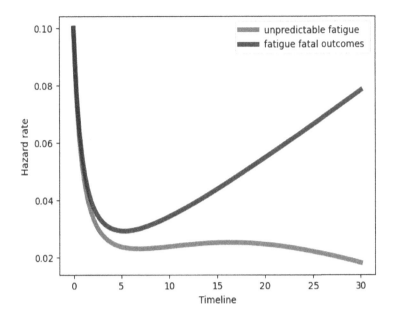

Fig. 14. Time-Dependent Hazard Rate Analysis

4   Conclusion

Mental fatigue is linked to anxiety, depression, and burnout, even though men are less likely to be anxious or sad at the time they are diagnosed. To enhance employees' quality of life and direct the creation of individualized multi component plans according to gender variations, Kaplan Meier's model is trending towards emphasizing psychopathology and early referral to mental health experts. The necessity for experts to establish public policies and provide proper care to better the daily lives of those suffering from fatigue, and the significance of fatigue to the employee's health system as a whole. The article, the effect of Kaplan Meier on the estimated model was seen, and all variables that the method found to be important were used in the regression coefficients. This highlights the importance of utilizing statistical methods to identify significant variables that can contribute to the development of effective policies and care plans for individuals suffering from fatigue. It also highlights the need for continued research and analysis in this area to improve overall employee health and well-being. Using the coefficient of determination, the Kaplan-Meier method is superior to the proposed model, the age of the employees when the fatigue was discovered is crucial and plays a major role in the survival time of employees with mental fatigue, the incidence of fatigue is less important than the gender variable. While analyzing processes connected to survival time is related, it is essential to use the Kaplan-Meier method to assess the significance of the independent factors that can play a significant role in identifying the explanatory variables. This method is the best way to get the analysis and predict the employee's status, but the complexity of this condition necessitates further research to refine the parameters for its treatment.

References

1. Darnai, G., et al.: The neural correlates of mental fatigue and reward processing: a task-based fMRI study. Neuroimage **265**, 119812 (2023). https://doi.org/10.1016/j.neuroimage.2022.119812
2. Sauch Valmaña, G., Miró Catalina, Q., Carrasco-Querol, N., Vidal-Alaball, J.: Gender, mental health and socioeconomic differences in fibromyalgia: a retrospective cohort study using real-world data from Catalonia. Healthcare **11**(4), 530 (2023). https://doi.org/10.3390/healthcare11040530
3. Zhang, Y., Guo, H., Zhou, Y., Xu, C., Liao, Y.: Recognising drivers mental fatigue based on EEG multi-dimensional feature selection and fusion. Biomed. Signal Process. Control **79**, 104237 (2023). https://doi.org/10.1016/j.bspc.2022.104237
4. Xu, X., Tang, J., Xu, T., Lin, M.: Mental fatigue degree recognition based on relative band power and fuzzy entropy of EEG. Int. J. Environ. Res. Public Health **20**(2), 1447 (2023). https://doi.org/10.3390/ijerph20021447
5. Sreeshakthy, M., Preethi, J.: Classification of emotion from EEG using hybrid radial basis function networks with elitist PSO. In: 2015 IEEE 9th International Conference on Intelligent Systems and Control (ISCO), pp. 1–4 (2015). https://doi.org/10.1109/ISCO.2015.7282340
6. Sreeshakthy, M., Preethi, J., Dhilipan, A.: A survey on emotion classification from EEG signal using various techniques and performance analysis. Int. J. Inf. Technol. Comput. Sci. **8**(12), 19–26 (2016). https://doi.org/10.5815/ijitcs.2016.12.03
7. Preethi, J., Sowmiya, S.: Emotion recognition from EEG signal using ISO-FLANN with firefly algorithm. In: 2016 International Conference on Communication and Signal Processing, pp. 1932–1936 (2016). https://doi.org/10.1109/ICCSP.2016.7754508
8. Zorzos, I., Kakkos, I., Miloulis, S.T., Anastasiou, A., Ventouras, E.M., Matsopoulos, G.K.: Applying neural networks with time-frequency features for the detection of mental fatigue. Appl. Sci. **13**(3), 1512 (2023). https://doi.org/10.3390/app13031512
9. D'Arrigo, G., Leonardis, D., Abd ElHafeez, S., Fusaro, M., Tripepi, G., Roumeliotis, S.: Methods to analyse time-to-event data: the Kaplan-Meier survival curve. Oxid. Med. Cell. Longev. **2021**, 1–7 (2021). https://doi.org/10.1155/2021/2290120
10. Vale-Silva, L.A., Rohr, K.: Long-term cancer survival prediction using multimodal deep learning. Sci. Rep. **11**(1), 13505 (2021). https://doi.org/10.1038/s41598-021-92799-4
11. J. Xidian Univ.: Literature review on MHD Peristaltic Transport of non-Newtonian fluids through channels/Tubes, vol. 14, no. 5 (2020). https://doi.org/10.37896/jxu14.5/263
12. Gorji, M.: The effect of job burnout dimension on employees performance. Int. J. Soc. Sci. Hum. **1**(4) (2011). https://doi.org/10.7763/IJSSH.2011.V1.43
13. Anbarasi, M., Durai, M.A.S.: Prediction of protein folding kinetics states using hybrid brainstorm optimization. Int. J. Comput. Appl. **42**(7), 635–643 (2018). https://doi.org/10.1080/1206212x.2018.1479348
14. Anbarasi, M., Durai, M.S.: Incipient knowledge in protein folding kinetics states prophecy using deep neural network-based ensemble classifier. Int. J. Comput. Aided Eng. Technol. **13**(3), 341 (2020). https://doi.org/10.1504/ijcaet.2020.109519

15. Durai, M.A.S., Anbarasi, M., Handa, J.: Prediction of cancer disease using classification techniques in map reduce programming model. In: Advances in Human and Social Aspects of Technology, pp. 139–158. IGI Global (2018). https://doi.org/10.4018/978-1-5225-2863-0.ch007

16. Chellatamilan, T., Kumar, N.S., Valarmathi, B.: Effective deployment of multi-cloud customizable chatbot application for COVID-19 datasets. In: Nagarajan, R., Raj, P., Thirunavukarasu, R. (eds.) Operationalizing Multi-Cloud Environments. EICC, pp. 361–379. Springer, Cham (2022). https://doi.org/10.1007/978-3-030-74402-1_20

17. Chellatamilan, T., Valarmathi, B., Santhi, K.: Research trends on deep transformation neural models for text analysis in NLP applications. Int. J. Recent Technol. Eng. IJRTE **9**(2), 750–758 (2020). https://doi.org/10.35940/ijrte.b3838.079220

18. Santhi, K., Valarmathi, B., Chellatamilan, T.: Depth impurity pruned strategies for extracting high utility itemsets. Int. J. Eng. Technol. **7**(3.4), 52 (2018). https://doi.org/10.14419/ijet.v7i3.4.16747

19. Santhi, K., Chellatamilan, T., Valarmathi, B.: PFBtree for big data memory management system. Indian J. Public Health Res. Dev. **9**(6), 531 (2018). https://doi.org/10.5958/0976-5506.2018.00666.6

20. Suzana, S., Shanmugam, S., Uma Devi, K.R., Swarna Latha, P.N., Michael, J.S.: Spoligotyping of Mycobacterium tuberculosis isolates at a tertiary care hospital in India. Trop. Med. Int. Health **22**(6), 703–707 (2017). https://doi.org/10.1111/tmi.12875

21. Shibani, K., Sendhil Kumar, K.S., Siva Shanmugam, G.: An effective approach for plant monitoring, classification and prediction using IoT and machine learning. In: Dash, S.S., Lakshmi, C., Das, S., Panigrahi, B.K. (eds.) Artificial Intelligence and Evolutionary Computations in Engineering Systems. AISC, vol. 1056, pp. 143–154. Springer, Singapore (2020). https://doi.org/10.1007/978-981-15-0199-9_13

22. Santhi, K., Valarmathi, B., Chellatamilan, T.: Multi-cloud path planning of unmanned aerial vehicles with multi-criteria decision making: a literature review. In: Nagarajan, R., Raj, P., Thirunavukarasu, R. (eds.) Operationalizing Multi-Cloud Environments. EICC, pp. 31–63. Springer, Cham (2022). https://doi.org/10.1007/978-3-030-74402-1_3

23. Nivetha, S., Valarmathi, B., Santhi, K., Chellatamilan, T.: Detection of type 2 diabetes using clustering methods – balanced and imbalanced pima Indian extended dataset. In: Pandian, A.P., Palanisamy, R., Ntalianis, K. (eds.) ICCBI 2019. LNDECT, vol. 49, pp. 610–619. Springer, Cham (2020). https://doi.org/10.1007/978-3-030-43192-1_69

24. Karim, S., et al.: Gene expression study of breast cancer using Welch Satterthwaite t-test, Kaplan-Meier estimator plot and Huber loss robust regression model. J King Saud Univ - Sci **35**(1), 102447 (2023). https://doi.org/10.1016/j.jksus.2022.102447

25. Rich, J.T., Neely, J.G., Paniello, R.C., Voelker, C.C.J., Nussenbaum, B., Wang, E.W.: A practical guide to understanding Kaplan-Meier curves. Otolaryngol. Neck Surg. **143**(3), 331–336 (2010). https://doi.org/10.1016/j.otohns.2010.05.007

26. Are Your Employees Burning Out. https://www.kaggle.com/datasets/blurredmachine/are-your-employees-burning-out

27. Shukl, P.: A complete guide to survival analysis in Python. Aspiring Mach. Learn. Eng. https://www.kdnuggets.com/2020/07/complete-guide-survival-analysis-python-part1.html

28. Burneo, J.G., Villanueva, V., Knowlton, R.C., Faught, R.E., Kuzniecky, R.I.: Kaplan-Meier analysis on seizure outcome after epilepsy surgery: do gender and

race influence it. Seizure **17**(4), 314–319 (2008). https://doi.org/10.1016/j.seizure.2007.10.002

29. Etikan, I.: The Kaplan Meier estimate in survival analysis. Biometrics Biostat. Int. J. **5**(2) (2017). https://doi.org/10.15406/bbij.2017.05.00128
30. Kishore, J., Goel, M., Khanna, P.: Understanding survival analysis: Kaplan-Meier estimate. Int. J. Ayurveda Res. **1**(4), 274 (2010). https://doi.org/10.4103/0974-7788.76794
31. Kiessling, J., Brunnberg, A., Holte, G., Eldrup, N., Sörelius, K.: Artificial intelligence outperforms Kaplan-Meier analyses estimating survival after elective treatment of abdominal aortic aneurysms. Eur. J. Vasc. Endovasc. Surg. (2023). https://doi.org/10.1016/j.ejvs.2023.01.028
32. Elamin, A.M.K., Mohmmed, A.O.A.: The Cox regression and Kaplan-Meier for time-to-event of survival data patients with renal failure. World J. Adv. Eng. Technol. Sci. **8**(1), 097–109 (2023). https://doi.org/10.30574/wjaets.2023.8.1.0183
33. Li, C., Gao, Y., Lu, C., Guo, M.: Identification of potential biomarkers for colorectal cancer by clinical database analysis and Kaplan-Meier curves analysis. Medicine (Baltimore) **102**(6), e32877 (2023). https://doi.org/10.1097/MD.0000000000032877
34. Chandan, Charu: Healthcare Data Analytics. Chapman Hall CRC (2015)
35. scikit-. Introduction to Survival Analysis with scikit-survival. scikit-survival. https://scikit-survival.readthedocs.io/en/stable/userguide/00-introduction.html
36. Stel, V.S., Dekker, F.W., Tripepi, G., Zoccali, C., Jager, K.J.: Survival analysis I: the Kaplan-Meier method. Nephron Clin. Pract. **119**(1), c83–c88 (2011). https://doi.org/10.1159/000324758
37. Carnero Contentti, E., et al.: Neuromyelitis optica spectrum disorders with and without associated autoimmune diseases. Neurol. Sci. (2023). https://doi.org/10.1007/s10072-023-06611-4
38. Elhardt, C., Schweikert, R., Kamnig, R., Vounotrypidis, E., Wolf, A., Wertheimer, C.M.: Recurrence of perforation and overall patient survival after penetrating keratoplasty versus amniotic membrane transplantation in corneal perforation. **261**(7), 1933–1940 (2023). https://doi.org/10.1007/s00417-022-05914-0
39. Lacny, S., et al.: Kaplan-Meier survival analysis overestimates the risk of revision arthroplasty: a meta-analysis. Clin. Orthop. Relat. Res. **473**(11), 3431–3442 (2015). https://doi.org/10.1007/s11999-015-4235-8
40. Ismiguzel, I.: Hands-on Survival Analysis with Python. https://towardsdatascience.com/hands-on-survival-analysis-with-python-270fa1e6fb41

Cyber Security and Signal Processing

EEMS - Examining the Environment of the Job Metaverse Scheduling for Data Security

Venkata Naga Rani Bandaru[(✉)] [iD] and P. Visalakshi [iD]

Department of Networking and Communications, College of Engineering and Technology, SRM Institute of Science and Technology Kattankulathur, Chennai, Tamilnadu 603202, India
venkatanagarani.b@vishnu.edu.in

Abstract. Job scheduling is one of the main barriers to achieving resource efficiency and cost-effective execution in the cloud computing environment. Finding a nearly perfect solution in a fair quantity of time is challenging when it comes to work scheduling. Consequently, the delayed convergence and local minimums are still existing. This EEMS - examining the environment of the job metaverse schooling for data security proposed method describes a new secure job scheduler to meet the challenge of scheduling jobs in a cloud computing environment. This method is based on secure differential evolution, and Cloud Sim has been used in numerous tests to show that EEMS works. This innovative method lays a heavy emphasis on upholding data security and integrity throughout the scheduling process in addition to attempting to maximize resource usage and save costs in cloud computing settings. This combination of secure differential evolution and EEMS principles offers a strong solution that looks to address remaining issues with cloud-based task scheduling.

Keywords: Scheduling · Cloud Computing · Solution · data security · threat detection · Framework

1 Introduction

Cloud computing transactions are currently recognized as one of the most important industries due to network expansion, where optimal network utilization can help many users. By applying cloud computing, users can access the internet computer resources like software, hardware, and apps that are customized to their needs. Because of this, cloud computing and the Internet of Things have substantially benefited all users. Technology is being adopted by all sectors of the economy more quickly than ever before, and infrastructure is also constantly improving. There will be an increase in user demand for cloud computing as a result. The job distribution issue becomes more complicated. Delivering resources in line with user expectations while keeping customer-required quality of service standards is a difficult job.

The nondeterministic job scheduling problem is difficult to solve using normal approaches because it takes too long to discover a solution that is even remotely optimal.

© ICST Institute for Computer Sciences, Social Informatics and Telecommunications Engineering 2024
Published by Springer Nature Switzerland AG 2024. All Rights Reserved
P. Pareek et al. (Eds.): IC4S 2023, LNICST 536, pp. 245–253, 2024.
https://doi.org/10.1007/978-3-031-48888-7_20

Meta-heuristic algorithms may be effective at fixing these issues, but they still have the drawbacks of hitting local minima and having a slow convergence rate. To address the work scheduling issue in the context of cloud computing, the suggested EEMS introduces a unique secure job scheduler based on differential evolution. In order to enhance the search, we increase includes dynamically developed numbers that are based on the present iteration of the scaling factor. The suggested approach is additionally concentrated on the better results in less time and in a secure manner. The experiments were carried out by the Cloud Sim to show the effectiveness of EEMS.

A readable message is changed into an unreadable form through the process of encryption to prevent unauthorized devices from reading it. Decryption is the process of returning an encrypted message to its original format. The original communication is the plaintext message. The encrypted message is referred to as the ciphertext message are included in EEMS to secure the data and provide authorized security between user and cloud server.

The related work of the suggested technique was covered in Sect. 2. The EEMS approach was discussed in Section, the results and discussions of the EEMS - examining the environment of the job metaverse schooling for data security and its foundational work were explained in Sect. 4, and Sect. 5 wrapped up the suggested method with future work.

2   Related Works

The way users of telecommunications and information technology access resources has changed as a result of cloud computing. It has made it possible to shift attention away from local/personal computation and towards datacenter-centric computing by dynamically supplying resources in a virtualized manner via the Internet. Similar to traditional utilities like water, electricity, gas, and telephony, cloud computing changes the use of computing as the fifth utility that is priced on a pay-per-use basis. [3]. In the hybrid approach, the PSO algorithm and the BF algorithm are combined to generate the starting population rather than doing it randomly. The TS method was then added to PSO in order to boost regional studies by avoiding the local optimality trap. As a result, the performance is enhanced [1]. The process of allocating tasks to resources during the particular time that they must be performed is known as job scheduling. The jobs are appropriately divided throughout the resources such that the required preference between jobs is satisfied and the overall time required to execute all jobs is kept to a minimum. The performance and efficiency of the cloud environment are improved by effective job scheduling. The execution time of all jobs, resource use, and other factors that affect performance must be high [6]. By controlling and handling the data at the edge, cloud computing. Vehicle nodes regularly switch regions, making vehicular networks highly mobile and hybrid. To enable seamless service delivery on such networks, fog computing can be used. Fog computing architecture can be used to process information about geography, traffic, and communications efficiently [2]. The goals of job scheduling include reducing the amount of time it takes to complete a task and the amount of energy it uses, as well as increasing resource efficiency and the ability to balance workloads. Cutting down on the amount of time it takes to complete a task is also beneficial for enhancing the customer experience, given the fast expansion of cloud users [11]. We must have a thorough understanding of the numerous issues related to various scheduling approaches as well as the challenges to be solved in order to design efficient scheduling algorithms. As a result, the purpose of this study is to give an in-depth analysis of job scheduling techniques and the related metrics that are appropriate for cloud computing systems [5]. In order to efficiently schedule computational tasks in a cloud environment, many scheduling algorithms were examined. As a result, FCFS, the Round-Robin Scheduling technique, and a new planned Scheduling technique called the Generalized Priority Algorithm were developed. For effective job execution and contrast between FCFS and Round Robin Scheduling, a Generalized Priority algorithm is used. A significant issue with job scheduling in cloud systems is priority [8]. To carry out tasks like developing, installing, handling, and scheduling, the client must be able to communicate with the cloud. Without any intervention from the company that provides the cloud service, the user should be allowed to access computing resources as needed [10]. The scheduling and assignment of resources and tasks are crucial issues in cloud computing, and numerous studies have been done on them. In the cloud computing system, cloud providers must give services to many users. Therefore, in order to minimize execution time and cost while maximizing resource utilization, scheduling is the key challenge in developing cloud computing systems [4]. Elastic and on-demand characteristics provide for higher service efficiency and agility in addition to increasing service reliability through sharing resources. The Internet of Things and cloud-based computing technologies should be merged, according to all of these arguments [9]. A cloud has an endless number of resources, and scheduling techniques are essential for getting the most out of those resources. To properly handle the requests, services should be automated extensively and smartly. When considering the purchase of automated processes, a system of algorithms is a crucial component responsible for efficiently allocating tasks among numerous resources while maintaining data security [7].

3   EEMS Design and Implementation

Ensuring Data Security with EEMS:

3.1   Secure Key Management

EEMS creates secure keys for data encryption and decryption, making it difficult for intruders to breach data.

3.2   Access Control

EEMS provides authorization to specific users or groups and allows access to appropriate systems.

3.3 Centralized Security Management

EEMS provides centralized security management, ensuring better monitoring, quicker detection and remediation of security incidents.

A huge number of datacenters hold numerous devices and equipment's form cloud computing system. Each host maintains a number of virtual machines V_RM, each of which is in charge of carrying out user jobs T_s with varied levels of service quality. The job scheduling in a cloud computing context is shown in Fig. 1

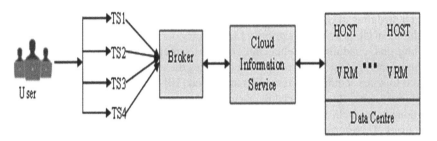

Fig 1. Cloud Computing

Pretend there are number of cloud jobs, T_s = t_s1, t_s2, t_sn, that are carried out by virtual machines V_RM = v_rm1, v_rm2, v_rmn. In order to learn more about the services needed to fulfil the jobs, the cloud information service receives a request from the cloud broker C_IS through encryption. These transactions are heterogeneous in terms of size, length, channel capacity, memory, and processor use.

The broker is a significant part of the scheduling process since they control when to schedule activities for specific resources and arbitrate communications between operators and providers. They have a variety of elements, and the jobs they select are influenced by the quality-of-service standards. Due to the placement of user transactions in the upper queue and the requirement for them to wait while resources are being used, the queue inevitably grows and wait times lengthen. As a result, the queue must be managed using load and time-based scheduling. In addition, many other features that have a direct effect on resource usage can be taken into when managing the transactions, the service provider adopts a multi-neutral optimization strategy. It will be possible to establish efficient load distribution among the virtual machines that will reduce resource consumption, with the help of a robust and secure task scheduling system developed and installed in the cloud network.

Examples of successful implementation:

Banking and Financial Services: EEMS has been used to ensure secure and timely processing of financial transactions and system maintenance.

Information Technology Services: EEMS has been employed to secure sensitive data in IT networks, servers, and data stores.

Dual development is a population-based optimization method that offers exact work scheduling solutions because of how it works for change, crossover, and selection. Prior to starting the optimization process, a set of solutions—referred to as individuals—are generated. Each of these solutions has a size and is spread at random over the area of search for the optimization solution. The accessible area is then investigated with the mutation and crossing operator in an effort to find more beneficial solutions, as is evident from the lines that follow.

Each solution in the population receives a mutant route as a result of using this operator as an updating approach. Where the population at cycle t has numerous people randomly chosen from it. The trial route is built using both the mutant route and the operator for crossover once the mutant vector has been developed under a crossover probability. Where size is a random number generated between 1 and size, reflecting the current measurement and the crossing rate is an integer that is constantly between 0 and 1, calculating the proportion of sizes moved from the mutant channel to the experiment route. The fittest vector will be computed in the subsequent iteration when this operation compares the routes The scheduler's solutions then show how to assign jobs to virtual machines in a way that will minimize the make span and overall execution time. Valid or a rational method can be used to solve multi-objective problems, which are those with two or three objectives. The rational technique treats multi-objective problems as if they were single-objective problems by allocating weights to each goal in accordance with its significance to the decision-makers. The valid method takes into account the fact that all goals are equally significant, resulting in a group of solutions known as non-dominated solutions that strike a balance between multiple goals. Using a weighting variable with a fixed value generated between 0 and 1 to reflect the importance of the other objective, the multi-goal challenging decreased to one goal.

Here, each virtual machine starts out with a value of 0, the jobs are then allocated among them via a scheduler, each virtual machine completes the jobs assigned to it, and the variable is updated with the time it took for each job to be completed under the nth virtual machine. Once the jobs assigned to each virtual machine have been finished the numbers kept in a variable for each of the virtual machines are compared to each other, and the resultant largest value is the make span.

In order to allocate each work to a virtual machine, N solutions with n dimensions are established before the optimization process is started. These solutions are randomly initialized between 0 and VRM counts. A job is represented by each dimension in the scheduling problem. Then the evaluation stage begins, when the quality of each solution is assessed and the best solution that has been found so far that can generate the goal function's lowest value is chosen.

Challenges using EEMS:

Cost: The cost of EEMS is a significant investment, and not all organizations may be able to afford its implementation.

Training: Proper training is required for the successful implementation and usage of the system.

Integration: Integration with existing technologies and systems is necessary to make the tool work for the organization.

The scaling factor has a predetermined positive value that remains constant throughout the whole optimization process. It controls step lengths and differentiates evolution's mutation stage. Therefore, if the population variety is high and the amount of this factor of scaling is also high, the number of steps will cause the solution to diverge from the existing answer. When the present iteration is increased, the population variety may therefore decrease while the scaling characteristic stays the same. To urge the algorithm to extensively scan the search area for probable regions that might have the best-so-far answer, this might happen even at the start of the optimization process.

The method may always look for an improved approach in regions far from the current one, which may involve the nearly optimal response nearby, as illustrated in Fig. 2. If the population variety is substantial and the factor of scaling is constant. The method will therefore constantly yield large step sizes. When population variety is large while the growth factor is low, a similar issue arises.

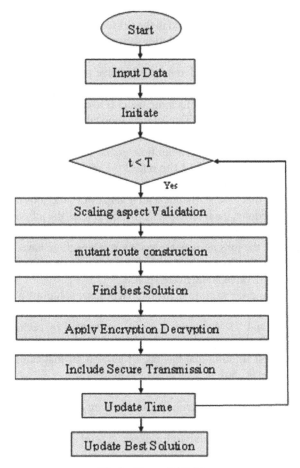

Fig 2. Flow of EEMS

We changed this factor so that it would be dynamically updated based on the present cycle in order to optimize the search operator calculation at the start of the optimization process and identify the most promising position within the area of search. The search operator will afterwards progressively change into the manipulation operator by extending the current iteration, allowing it to concentrate more on this promising location during the optimization phase.

Therefore, even though all around-optimal solutions might be close to the best-yet answer, the search operator might be instructed to explore the areas surrounding one of these solutions. Consequently, the convergence improvement technique is proposed to make use of the areas near the best feasible solution to raise the capacity for convergence.

The data are encrypted by encryption security control. Data is called as plaintext is transformed to make it difficult for unauthorized machines to understand as ciphertext. The plaintext, once it has been converted to ciphertext, appears random and has no information about the original content. No machine can deduce anything about the content of the original data by reading the data in its encrypted form once it has been encrypted. An encryption system that allows for revocable conversions uses cryptography to restore encrypted data (ciphertext) to its unencrypted state, known as plaintext. The process of removing this encryption is known as decryption. A cryptographic key is required for both the encryption and decryption processes in order to reveal the true content of the encrypted data ciphertext after it has been encrypted plaintext. A cryptographic key is a series of binary digits that is used as the input for encryption and decryption processes. All the data are encrypted and transferred to the server and it will decrypt while the data accessed by user.

4   Results and Discussion

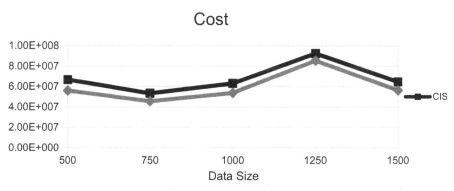

Fig.3. Data Size and Cost

The cost defined the encryption and decryption duration, if the cost is minimum then the transaction time is good, as shown in Fig. 3. The percentage of data that are correctly categorized, measured on a scale from 0 to 1, is called accuracy. As per the proposed model to accurately categorize the data, we compare the projected class with the actual class in this metric. In order to determine the proportion, we just count the amount of correctly classified data starts with true and divide that number by the overall data to find the accuracy as shown in Fig. 4

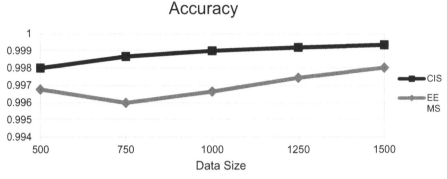

Fig.4. Data Size and Accuracy

5   Conclusion

This paper presents the secure metaverse cloud scheduler, which is used to schedule jobs in a cloud context. This scheduler is based on the secure differential evolution approach, the investigation and utilization of operators are first optimized in accordance with the scaling aspect, the numerical values created dynamically based on each iteration, and a differential evaluation to get better outcomes in fewer iterations. For a variety of job sizes, the total execution time for encryption and decryption was gathered, and the cost was determined. The experiment results demonstrated that EEMS delivered successful results when compared to the other strategies that were investigated. In further work, a number of other optimization concerns will be addressed to reduce costs and boost accuracy using the EEMS.

References

1. Alkhashai, H.M., Omara, F.A.: An enhanced task scheduling algorithm on cloud computing environment. Int. J. Grid Distrib. Comput. **9**(7), 91–100 (2016)
2. Jamil, B., Ijaz, H., Shojafar, M., Munir, K., Buyya, R.: Resource allocation and task scheduling in fog computing and internet of everything environments: a taxonomy, review, and future directions. ACM Comput. Surv. (CSUR), **54**(11s), pp.1–38(2022)
3. Varshney, S., Singh, S.: A survey on resource scheduling algorithms in cloud computing. Int. J. Appl. Eng. Res., **13**(9), pp.6839–6845(2018)
4. Almezeini, N., Hafez, A.: Review on scheduling in cloud computing. IJCSNS Int. J. Comput. Sci. Netw. Secur. **18**(2) (2018)
5. Bandaru, V.N.R. and Visalakshi, P.: Block chain enabled auditing with optimal multi-key homomorphic encryption technique for public cloud computing environment. In: Concurrency and Computation: Practice and Experience, **34**(22), p. 7128(2022)
6. Arunarani, A.R., Manjula, D. Sugumaran, V.: Task scheduling techniques in cloud computing: a literature survey. Future Gener. Comput. Syst, pp 91.407–415(2019)
7. Mohammadi, F., Jamali, S., Bekravi, M.: Survey on job scheduling algorithms in cloud computing. Int. J. Emerg. Trends Technol. Comput. Sci. (IJETTCS), **3**(2), pp.151–154(2014)
8. Nzanywayingoma, F., Yang, Y.: Efficient resource management techniques in cloud computing environment: a review and discussion. Int. J. Comput. Appl. **41**(3), pp.165–182(2019)

9. Guo, S., Liu, J., Yang, Y., Xiao, B., Li, Z.: Energy-efficient dynamic computation offloading and cooperative task scheduling in mobile cloud computing. IEEE Trans. Mob. Comput. **18**(2), pp.319–333(2018)

10. Bandaru, V.N.R., Kiruthika, S.U., Rajasekaran, G., Lakshmanan, M.: December. device aware VOD services with bicubic interpolation algorithm on cloud. In: 2020 IEEE 4th Conference on Information & Communication Technology (CICT), pp. 1–5(2020)

11. Singh, A.K., Gupta, R.: A privacy-preserving model based on differential approach for sensitive data in cloud environment. Multimedia Tools and Appl. **81**(23), pp.33127–33150(2022)

12. Dillon, T., Wu, C., Chang, E.: Apr. Cloud computing: issues and challenges. In: 2010 24th IEEE International Conference on Advanced Information Networking and Applications, pp. 27–33 (2010)

13. Kak, S.M., Agarwal, P., Alam, M.A.: Task scheduling techniques for energy efficiency in the cloud. EAI Endorsed Trans. Energy Web, 9(39), pp. e6-e6(2022)

14. PVNSLSSR Murthy, V., Venkata Naga Rani Bandaru.: Secure auditing and storage systems in cloud service. Int. J. Eng. Adv. Technol. (IJEAT) Vol-8, Issue-6S3, pp.2249 – 8958 (2019)

15. Lin, T., Zhao, Y., Zhang, H., Li, G., Zhang, J.: Mar. Research on information security system of ship platform based on cloud computing. In J. Phys.: Conf. Ser. Vol. 1802, No. 4, p. 042032 (2021)

16. Kalsoom, T., Ramzan, N., Ahmed, S., Ur-Rehman, M.: Advances in sensor technologies in the era of smart factory and industry 4.0. Sensors, **20**(23), pp.6783 (2020)

17. Zhou, Y., Huang, X.: November. scheduling workflow in cloud computing based on ant colony optimization algorithm. In: 2013 Sixth International Conference on Business Intelligence and Financial Engineering, pp. 57–61(2013)

18. Yang, C., Huang, Q., Li, Z., Liu, K., Hu, F.: Big data and cloud computing: innovation opportunities and challenges. Int. J. Digital Earth, **10**(1), pp.13–53(2017)

19. Khang, A., Sivaraman, A.K. Eds.: Big data, cloud computing and IoT: tools and applications/edited. J. Future Revolution Comput. Sci. and Commun. Eng., **4**(4), pp.599–602(2023)

20. Gayatri Sarman, K.V.S.H. Gubbala, S.: Voice based objects detection for visually challenged using active RFID technology. In: International Conference on Cognitive Computing and Cyber Physical Systems, pp. 170–179(2022)

21. Pradhan, A., Ghosh, R., Biswas, S.: Enhanced job scheduling algorithm for data security in cloud computing environment. In: 2021 International Conference on Computing, Communication, and Intelligent Systems (ICCCIS), pp. 1–6 (2021)

22. Sharma, S., & Bhatia, R.: An improved secure job scheduling algorithm for data-intensive applications in cloud computing. In: Proceedings of the 4th International Conference on Internet of Things and Connected Technologies, pp.305–315(2021)

Text Analysis Based Human Resource Productivity Profiling

Basudev Pradhan[(✉)], Siddharth Swarup Rautaray, Amiya Ranjan Panda,
and Manjusha Pandey

School of Computer Engineering, KIIT Deemed to be University, Bhubaneswar, Odisha, India
{basudev.pradhan,siddharthfcs,amiya.pandafcs,
manjushafcs}@kiit.ac.in

Abstract. Email being an efficient, cost-effective, real-time communication mode results into effective productivity among the professional in the organization. It constitutes almost 90% of daily office procedures in organizations, hence the productivity of organizations depends heavily on the text communicated in emails. The presented research work focuses on email profiling in organizations based on mail text interpretation and analysis. In the proposed work we will be working on datasets containing email communication of ENRON Corporation as test case. The profiling would be done using Text interpretation and analysis algorithm using machine learning algorithms. The BoW will be implemented to analyze and predict the characteristics of incoming and outgoing emails, then these could be mapped and profiled as per the behavior of employees into 3 categories of productive based on positive responses, neutral and non-productive based on negative responses.

Keywords: Email profiling · text interpretation and analysis · ENRON dataset · machine learning · Bag of Words

1 Introduction

Professionals in an Organization may be profiled based on the analysis of text used by them in the email communications while performing their role for contribution to organization's goal, as almost 90% of the communication in the current age of digital data transformation is in email format [1]. The characteristics of words in email communication indicates about the behavior and attitude of professional and is a reflection of how effectively the professional works. Many research and existing labor data indicates the professional using positive terms in their email communication work effectively and efficiently and tend to make the work environment more favorable for their co-workers.

The presented research work focuses on identification and categorization of choice of words used by different professionals of the organization in their email communication. The email profiling would be done based on mail text analytics using BoW machine learning algorithm. The input data taken from ENRON dataset will be preprocessed for the removal of email text formalities and the information is integrated in a weighted

P. Pareek et al. (Eds.): IC4S 2023, LNICST 536, pp. 254–262, 2024.
https://doi.org/10.1007/978-3-031-48888-7_21

manner based on the number of positive, negative and neutral words used in the email text. The expected output will be laid down from natural language processing and narrative analysis for email profiling. We will use BoW algorithm to classify/categorize emails as an assessment tool for the productivity of the employees in the organization. This email profiling based on productivity employees can be done on weekly or hourly bases and attempts can be made to develop trend models for employee productivity analysis.

2 Related Work

Dr. Deborah Fallows [2] in his research work has deduced email to be most effective way of communication over physical meeting & telecommunication citing a percentage of 63%. Dr. Fallows suggested the employees used emailing 67% of time for professional communication 26% of time for personal communication and 15% of for gossips. Thus, formalizing email as a most preferred communication mode for official purposes. Another research work done by Emmanuel Gbenga Dada [3] highlighted efforts made by different researchers to solve the spam and ham problem using effective classifiers of his review of machine learning algorithms for identification of spam and ham messages. Also, a comprehensive review for the emails spam classification problem has been done by Mansoor Raja [4]. His findings are more focused on usage of specific algorithm namely Naive Bayes and SVM. The author has also suggested usage of multi algorithm-based system over single algorithm system. K. Thirumoorthy [5] has proposed a method based on term frequency distribution measure (TFDM) for investigation of performance of two of the most preferred algorithm Naive Bayes (NB) and Support Vector Machine (SVM) over of bench mark text corpus. He suggested a thorough experimental analysis depicts better performance of TFDM over various well-known filter techniques (DF, ACC2, IG, MI, DFS, NDM and TRDL). Another researcher Maryam Hina [6] in her study of multilevel email classification has generated promising linear regression accuracy along with a comparison with logistic regression. She suggested that logistic regression provided 91.9% accuracy and also included incorporation of block chain for storage and access of analyzed data as a future work. A lot of work for classification of spam and ham mail has been done using content base features instead of behavioral features that generates the requirement of multi-folder classification of emails considering heterogeneous details of emails rather than content only. This conclusion by Namrata Shroff [7] and her research group emphasizes more on multi-folder classification of emails and generation of context based on the content of emails. The same exclusion of context is evident from all research literature available [8–15] for analysis of emails, our work thus emphasizes on generation of context of employee productivity by analyzing the text used in the content of email.

3 About ENRON Dataset

This proposed work is based on ENRON data set that was collected and prepared by CALO problem, which is cognitive assistant that Learns and Organizes. The dataset has made public for academic and research purposes related to email profiling text analytics and other related research domains. The ENRON dataset contains email communication samples of about 150 users of the ENRON corporation which amounts to about 0.5 million email exchanges. The email exchanges included in the dataset are mostly senior management professionals for ENRON corporation. This data was made public by hosting it on the web by FERC (Federal Energy Regulatory Commission) during its investigation on reasons for failure of ENRON Corporation. The dataset features text and other contents which contains only the text communication among different users of the ENRON corporation. This makes the ENRON datasets most suitable for our research work based on email profiling using text analysis and interpretation. Our proposed method to work on this ENRON dataset consists of 3-phases:

Phase-1: Text and Content extraction from email corpus

Phase-2: Text interpretation and analysis

Phase-3: Email profiling and categorization/classification using BoW algorithm

Problem Statement:

The objective of projected research work would be 3-fold defining our problem statements as follows:

Obj-1: Feature extraction of text and content extracted from email corpus.

Obj-2: Email profiling based on mail text interpretation and analysis.

Obj-3: Categorization/classification of professionals based on email profiling using BoW algorithm.

4 Proposed Approach

For our research, we have developed the approach as depicted by Fig. 1. The proposed approach has been designed as the flow between the extractions of mail content from the ENRON email dataset to employee categorization based on their productivity mapped to the frequency of type of words used by them in their mail contents.

The Enron email data set has around 517,431 digital communications. This dataset has been provided by the FERC (Federal Energy Regulatory Commission) for academic and research purposes; it has data of 150 users in 3500 different folders available for analysis. The same dataset has been used in this research work with following features.

1. Message-ID,
2. Date,
3. From,
4. To,
5. Subject,
6. User, X-from,
7. X- folder,
8. X-origin, X-filename.

Message-id	Date	From	To	Subject	Content
User	X-from	X-folder	X-origin	X-filename	

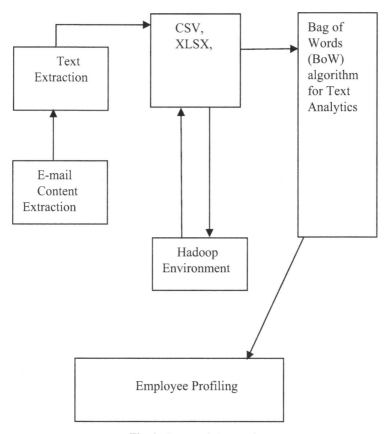

Fig. 1. Proposed Approach

The data was preprocesses in order to build a model to classify the emails responses into three categories of positive, negative and neutral. After this, the data set can be used for exploratory data analysis and visualized.

5   Feature Extraction of Emails

The data preprocessing based on the feature accepted by the classifier model takes text from the emails just including subject and body of an email; The data is filtered based on: 1) Removal of stop words like "Re:", "Fw:" and "Fwd:"; 2) The data is refined for stop words like "From" and "To".

For further classification and analysis specific conditions like emails with no subjects or "No Subject" or emails with no body are removed from the analysis candidate set.

The identification of text features after preprocessing of the content extracted from the mail mitigating all the discrepancies of email and in turn text categorization into set of positive and negative set of words using BOW model. The individual categorization of employees is further mapped to the output of the BOW model for individual employee as extracted and stored from data set as a separate array. Thus, the mail content still remains the vital feature of email categorization which in turn helps in the employee categorization based on email categorization. The subjects and contents of the emails have not been included in the feature set as they are redundant many times with very less relevance to the email content while the word frequency words of body are graded for high frequency of words and not graded for lesser frequency of the words to enhance the text analytics based on the BOW model.

6 BoW Model

Bag of Words (BoW) model is a technique for extracting the features from text data. It is a method that is frequently used to describe the meaning of a document. It produces a fixed-length vector based on the frequency of words/terms. A term that is used frequently suggests that the document has more to do with that term and should therefore be given a higher value than the other terms. This is done by creating a Term Frequency table, which counts the number of times each term appears in the document. Based on the frequency of each term inside the content, the Term Frequency creates a vector space model of the document.

6.1 Data Extraction

Data extraction is the process of collecting the data from a variety of sources. This is the initial step for any type of text analysis.

6.2 Data Pre-processing

Data Pre-processing is the procedure of transforming the data in appropriate format, so that the data can be used efficiently according to the model. Here some of the steps for data pre-processing:

- Remove the stop words (most common words like 'the', 'is', 'a' and etc.)
- Convert the texts of the document in lower case as the case has no meaning.
- Remove punctuations and special characters.
- Make a list of all the words in our model vocabulary.
- Count the frequency of words in our text document.

6.3 Bag of Words Model

The Bag of Words model may be binary Bag of Words or Bag of Words. In both cases we have to construct a vector to indicate the appearance of word. For Binary Bag of Words, the vector contains either 1 if the word is present or 0 if the word is not present. And for the Bag of Words, the vector contains the value as per the frequency of the word in a sentence.

Example: (each and every sentence from different mail)

Let we want to vectorize the following:

S-1: The information in this email may be confidential and/or privileged.
S-2: We are being billed for this service and I do not know who is using it.
S-3: This email has the details of the service.

After removing the stop words, punctuations and lower-case conversion the sentences will be as follows:

S-1: information email confidential privileged
S-2: billed service know using
S-3: email details service

Now, we have to count the frequency of word in a sentence and we have to vectorize our document.

Docs	information	email	confidential	privileged	billed	service	know	using	details
S-1	1	1	1	1	0	0	0	0	0
S-2	0	0	0	0	1	1	1	1	0
S-3	0	1	0	0	0	1	0	0	1

Hence, the resultant vectors are:

$$S - 1 : [1, 1, 1, 1, 0, 0, 0, 0, 0]$$
$$S - 2 : [0, 0, 0, 0, 1, 1, 1, 1, 0]$$
$$S - 3 : [0, 1, 0, 0, 0, 1, 0, 0, 1]$$

We observe that, in Bag of Words methodology, we refined source data into dataset which lose contextual information, only lists vocabulary data with frequency. Hence, it is very helpful in machine learning technique as the large volume of data is processed.

The experimental setup was established for sample data generated through extraction of 1000 users' data from the Enron data set. For training and testing purposes the data was divided into 70–30 ratio. Where 70% was utilized for training of the generated model and 30% was utilized for testing. The BOG Model generated and implemented through python coding in jupyter note book utilizing the numpy, pandas, re, nltk, seaborn, matplotlib etc.

7   Experimental Result

This paper has analyzed year-wised emails for positive, negative and neutral words. It has traced and analyzed the emails of particular years as 1999, 2000 and 2001.It has also conducted separate analysis study of 1000 emails received by Enron Corporation for a tenure of 5 years.

The following results have been generated after the classification of the pre-processed dataset. The following Fig. 2 of classification of digital data communication for all 12 months of the year 1999 depicts more negative responses during the mid of the year and lesser during the start of the year which is the duration of lesser work pressures.

The following Fig. 3 of classification of digital data communication for all 12 months of the year 2000 depicts more negative responses during the mid of the year and again lesser during the start of the year which is the duration of lesser work pressures.

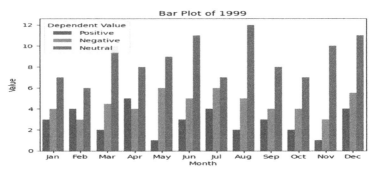

Fig. 2. Bar Graph of emails analysis of year 1999

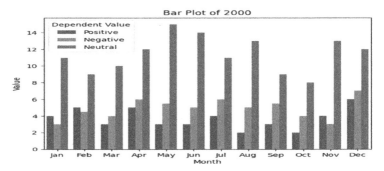

Fig. 3. Bar Graph of emails analysis of year 2000

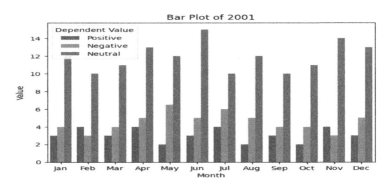

Fig. 4. Bar Graph of emails analysis of year 2001

The above Fig. 4 of classification of digital data communication for all 12 months of the year 2001 again depicts more negative responses during the mid of the year which also confirms the efficiency of our machine learning model for further utilizations also.

The results of classification of text in the dataset for 5 years depicts the increasing of negative responses year by year as is shown in Fig. 5.

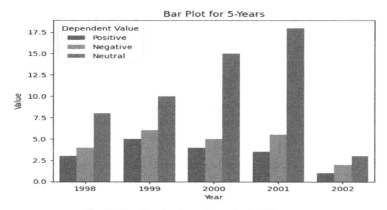

Fig. 5. Bar Graph of emails analysis of 5-years

8   Conclusion

After reassuring the efficiency of our proposed model for the data set of three different years we performed the classification of digital data communication for all 5 years of data available in the data set and the results were depicting the increase in negative responses year by year which is evident by the shutting up of the company after the year 2002. The results depict in Fig. 5 emphasize correlation between the productivity of the employees to the usage of positive, negative and neutral words in their day-to-day communications. Thus, the proposed research work can be utilized by the organization for establishment of productivity quotients of their employees based on the classification of text chosen and utilized by them during their day-to-day digital communications.

References

1. Aufreiter, N., Boudet, J., Weng, V.: Why marketers should keep sending you e-mails. McKinsey & Company (2014)
2. Fallows, D.: Email at Work. Pew Internet & American Life Project (2002)
3. Dada, E.G., Bassi, J.S., Chiroma, H., Adetunmbi, A.O., Ajibuwa, O.E.: Machine learning for email spam filtering: review, approaches and open research problems. Heliyon **5**(6), e01802 (2019)
4. Mansoor, R.A.Z.A., Jayasinghe, N.D., Muslam, M.M.A.: A comprehensive review on email spam classification using machine learning algorithms. In: 2021 International Conference on Information Networking (ICOIN), pp. 327–332. IEEE (2021)
5. Thirumoorthy, K., Muneeswaran, K.: Feature selection for text classification using machine learning approaches. Natl. Acad. Sci. Lett. **45**, 51–56 (2021). https://doi.org/10.1007/s40009-021-01043-0
6. Hina, M., Ali, M., Javed, A.R., Srivastava, G., Gadekallu, T.R., Jalil, Z.: Email classification and forensics analysis using machine learning. In: 2021 IEEE SmartWorld, Ubiquitous Intelligence & Computing, Advanced & Trusted Computing, Scalable Computing & Communications, Internet of People and Smart City Innovation (SmartWorld/SCALCOM/UIC/ATC/IOP/SCI), pp. 630–635. IEEE (2021)

7. Shroff, N., Sinhgala, A.: Email classification techniques—a review. In: Kotecha, K., Piuri, V., Shah, H.N., Patel, R. (eds.) Data Science and Intelligent Applications. LNDECT, vol. 52, pp. 181–189. Springer, Singapore (2021). https://doi.org/10.1007/978-981-15-4474-3_21

8. Nandhini, S., KS, J.M.: Performance evaluation of machine learning algorithms for email spam detection. In: 2020 International Conference on Emerging Trends in Information Technology and Engineering (ic- ETITE), pp. 1–4. IEEE (2020)

9. Das, N., Shankar, S., Dash, B., Mohan, J., Pandey, M., Rautaray, S.S.: Productivity profiling of organizations based upon communication interpretation and analysis. In: 2021 5th International Conference on Information Systems and Computer Networks (ISCON), pp. 1–6. IEEE (2021)

10. Noever, D.: The enron corpus: where the email bodies are buried? (2020)

11. Harrison, J., et al.: Quantifying use and abuse of personal information. In: 2021 IEEE International Conference on Intelligence and Security Informatics (ISI), pp. 1–6. IEEE (2021)

12. Negangard, E., Fay, R.: Electronic Discovery (eDiscovery): Performing the Early Stages of the Enron investigation. Issues Acc. Educ. 35 (2019). https://doi.org/10.2308/issues-16-064

13. Ali, R.S., Gayar, N.E.: Sentiment analysis using unlabeled Email data. In: 2019 International Conference on Computational Intelligence and Knowledge Economy (ICCIKE), pp. 328–333 (2019)

14. Jacob, I.J.: Performance evaluation of caps-net based multitask learning architecture for text classification. J. Artif. Intell. **2**(01), 1–10 (2020)

15. Karim, A., Azam, S., Shanmugam, B., Kannoorpatti, K., Alazab, M.: A comprehensive survey for intelligent spam email detection. IEEE Access **7**, 168261–168295 (2019)

A Novel Technique for Analyzing the Sentiment of Social Media Posts Using Deep Learning Techniques

Ravula Arun Kumar[1], Ramesh Karnati[1], Konda Srikar Goud[2(✉)], Narender Ravula[3], and VNLN Murthy[1]

[1] Department of CSE, Vardhaman College of Engineering, Hyderabad, India
[2] Department of Information Technology, BVRIT HYDERABAD College of Engineering for Women, Hyderabad, Telangana, India
kondasrikargoud@gmail.com
[3] Department of CSE, Keshav Memorial Institute of Technology, Hyderabad, India

Abstract. Our study aims to precisely categories the sentiment expressed in user-generated text, concentrating specifically on Twitter data. Using a benchmark dataset of labelled tweets, we evaluate the efficacy of our proposed method to that of traditional machine learning approaches, such as Support Vector Machines (SVM) and Naive Bayes (NB). Our methodology entails preprocessing the text data by tokenizing, removing stop words, and stemming, followed by feature extraction using word embedding's. For sentiment classification, we employ a Convolutional Neural Network (CNN) architecture with multiple convolutional layers and pooling operations. In terms of accuracy, precision, recall, and F1 score, the experimental results indicate that our proposed deep learning method outperforms conventional machine learning techniques. In addition to this, we do an error analysis in order to identify challenging scenarios and give insight into the constraints as well as prospective improvement areas. The results of this research provide a significant contribution to the field of social media sentiment analysis and provide evidence of the usefulness of deep learning algorithms for the correct categorization of sentiments in Twitter data.

Keywords: Support vector machine · Naïve Bayes · Convolution Neural Network

1 Introduction

The fast growth of social media platforms has led to an explosion of user-generated content, which has led to a vast quantity of text data holding significant insights and emotions. This data has resulted in an explosion of user-generated content. It is crucial for a variety of applications, including brand monitoring, reputation management, market analysis, and public opinion monitoring, to have a solid understanding of the sentiment that is being communicated in these social media posts. When it comes to the process

P. Pareek et al. (Eds.): IC4S 2023, LNICST 536, pp. 263–273, 2024.
https://doi.org/10.1007/978-3-031-48888-7_22

of automatically evaluating and categorising the feelings that are included in textual material, techniques from the discipline of Natural Language Processing (NLP) play an incredibly essential role.

The goal of this study is to offer a novel approach to assessing the sentiments expressed in social media communications, particularly Twitter data. Sentiment analysis, often known as opinion mining, is a method for gauging the tone of a written item. This analysis classifies the text as good, negative, or neutral, depending on the results. Our goal is to accurately categorise the mood conveyed in tweets so that we can provide useful insights on the evolution of public opinion and the emergence of sentiment patterns.

Traditional machine learning algorithms, such as Support Vector Machines (SVM) and Naive Bayes (NB), and built features have hitherto formed the backbone of approaches to evaluating social media sentiment. These methods have demonstrated some success, but they still have some ways to go before they fully convey the full richness and complexity of spoken language. Convolutional Neural Networks (CNNs) and other recent advancements in deep learning have improved performance in many NLP tasks. Sentiment analysis is one such activity.

In this research, we offer a method for assessing the sentiment of social media postings that is based on deep learning. We have a hypothesis that suggests that more accurate sentiment categorization may be achieved by combining word embeddings with a CNN architecture. CNNs are superior to word embeddings when it comes to capturing local patterns and dependencies in textual data. Word embeddings are able to capture the semantic links that exist between words and give rich contextual information.

In order to determine whether or not the approach that we have presented is effective, we undertake in-depth tests using a benchmark dataset consisting of tagged tweets. Our deep learning model's accuracy, precision, recall, and F1 score are compared to those of SVM and NB, two classic machine learning methods. In addition, we do an error analysis in order to have a better understanding of the limitations of the technique and the challenging scenarios it faces.

This study makes a contribution by putting forward an innovative deep learning-based method for analysing sentiment in social media messages, with a specific focus on Twitter data. We expect that by utilising the capabilities of word embeddings and CNNs, we will be able to achieve more accuracy and sentiment classification performance than is possible using more traditional machine learning methods. This research will illustrate the usefulness of deep learning approaches for accurate categorization of sentiment in Twitter data, as well as contribute to the area of sentiment analysis in social media, which it will help advance.

The following is the order in which the subsequent sections of this article are presented: The third section offers a complete review of the existing research on sentiment analysis and deep learning for natural language processing, which can be found in the existing literature. In the fourth section, both the methodology and the recommended technique are broken out in great depth. In the fifth section, we present the experimental design as well as the findings of our comparison study. The findings are discussed in the sixth part, along with potential limits and possibilities for development, which are offered as insights. The essay is brought to a close in the seventh section, which reviews the most important takeaways from the research and outlines potential avenues for further investigation.

2 Literature Survey

Paper	Key Findings	Future Scope
[1] Wang, S., & Manning, C. D. (2020)	Using bag-of-words models and bigram features, this paper introduces simple and effective baselines for sentiment analysis and topic categorization	Future research might look at more advanced feature representations and deep learning approaches
[2] Tang, D., Qin, B., & Liu, T. (2020)	Propose a document modelling strategy with external focus on sentiment classification, leveraging external sentiment knowledge to enhance model performance	Incorporating a wider variety of external knowledge sources and evaluating the approach across various domains require additional research
[3] Zhang, Y., Wallace, B., & Huang, R. (2020)	Examines opinion mining and sentiment analysis techniques for social media data, emphasising approaches, challenges, and prospective developments	Future research could focus on addressing challenges related to social media-specific characteristics, such as noise and context dependence
[4] Devlin, J., Chang, M. W., Lee, K., &Toutanova, K. (2019)	Introduces BERT, a pre-trained language model based on the Transformer architecture that has a significant impact on NLP tasks such as sentiment analysis	Future research can focus on enhancing the BERT architecture, training methodologies, and techniques for sentiment analysis tasks
[5] Li, X., Zhang, W., Wang, Y., & Ji, H. (2021)	Provides an exhaustive overview of sentiment analysis, highlighting the importance of opinion mining and discussing numerous techniques and obstacles	Future research directions could include the development of more precise opinion extraction methods and the resolution of problems associated with subjective data
[6] Sun, C., Huang, L., Qiu, X., Zhang, X., & Huang, X. (2020)	Propose a method for aspect-based sentiment analysis using BERT and auxiliary sentences to characterise sentiment relations between opinion words and aspects	Future research could investigate the incorporation of more complex contextual information and the evaluation of the method across various domains and languages

(continued)

(continued)

Paper	Key Findings	Future Scope
[7] Chen, Q., Zhu, S., Ling, Z. H., Wei, Y., & Jiang, H. (2020)	Examines the use of BERT for question-answering tasks and proposes an enhanced model that makes use of historical context to increase performance	Future research could concentrate on developing more sophisticated models that utilise historical context effectively and examining their application in real-world question-answering scenarios

3   Methodology

3.1   Data Preprocessing

In order to get the data from social media platforms ready for sentiment analysis, a number of preprocessing steps are carried out. To begin, the text is tokenized, which means that it is broken down into its component words and subword units. After that, we get rid of any stop words and unnecessary punctuation in order to quiet the noise and make the future processing processes more effective. In addition, procedures such as stemming or lemmatization are used in order to standardise the words and limit the number of different inflectional forms.

Before doing an analysis on the text data (X_train and X_test), it is necessary to first perform preprocessing techniques like as tokenization, stemming, and any others that may be needed.

3.2   Feature Extraction

Word embeddings are utilised by our company in order to extract the semantic information included inside the text. Word embeddings are dense vector representations that embody the contextual meaning of words. These representations are determined by the distributional features of words. We begin by providing the word representations with pre-trained word embeddings, such as those generated by Word2Vec or GloVe, both of which have been trained using substantial corpora.

Word vectors that have been through previous training are used to seed the matrix of word embeddings (W).

X_train_embeddings is equal to W multiplied by X_train.

3.3   Convolutional Neural Network (CNN) Architecture

A CNN architecture is utilised in our suggested technique in order to identify regional patterns and relationships hidden within the text data. CNN is made up of a number of layers, the most notable of which are the fully connected, convolutional, and pooling layers.

Set the initial values for the parameters of the CNN model. These include the filter weights (K), the bias (b), the pooling function, and the activation function, amongst other hyperparameters.

The convolutional layers function by applying a collection of filters of differing sizes to the representation of the text that is being fed into the network. This gives the network the ability to recognise a variety of patterns or features at varying degrees of granularity. In order to extract features from the input text, these filters perform element-wise multiplication and summation while also scanning the text using a sliding window.

K multiplied by X-train embeddings plus b equals Conv.

The dimension of the representations is decreased by the pooling layers, which achieve this by downsampling the feature maps that were formed by the convolutional layers. In most cases, we use max pooling, which is a technique that maintains the characteristics that are most noticeable by picking the value that is highest in each section of the feature map.

Pool equals MaxPool(Conv).

The output of the pooling layers is then flattened before being transferred to fully connected layers. These fully connected layers subsequently carry out nonlinear transformations and learn representations at higher levels. The last layer of the network is a softmax layer, and its function is to build a probability distribution across the three different types of emotion (positive, negative, or neutral).

Calculate the output of the layers that are fully interconnected as follows:

W_fc multiplied by Pool plus b_fc equals f(Z).

3.4 Training and Optimization

For the training of our sentiment analysis model, we make use of a labelled dataset consisting of social media postings that have sentiment annotations. The dataset is divided into three distinct sets: the training set, the validation set, and the test set. During the training process, the model parameters are improved by minimising an appropriate loss function, such as cross-entropy loss, with the assistance of an optimisation algorithm, such as stochastic gradient descent (SGD) or Adam. Y = Softmax(Z).

Regularisation strategies, such as dropout or L2 regularisation, are utilised to prevent overfitting. This is accomplished by lessening the model's reliance on certain characteristics and increasing its capacity for generalisation. To improve the performance of the model, we make adjustments to its hyperparameters, using the validation set as a guide. These include changing the learning rate, the sample size, and the regularisation strength.

3.5 Inference and Sentiment Classification

When the training is complete, the model is put to use to classify the emotions conveyed in social media messages that have not been read. We next input the text that has been preprocessed into a CNN model that has been trained, which then generates a probability distribution across the sentiment classes. The class label that has the highest probability is the one that we identify as the projected sentiment for the input that was supplied Fig. 1.

Y_test is equal to the softmax of Z_test.

The formula for calculating X_test_embeddings is as follows: X_test_embeddings = W * X_testZ_test = f(W_fc * Pool(X_test_embeddings) + b_fc).

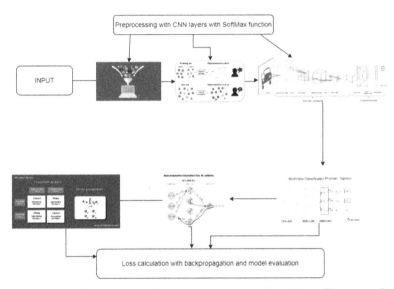

Fig. 1. Architectural diagram for analyzing the sentiment of social media posts using deep learning techniques

4   Results and Experimental Analysis

In this part of the article, we will discuss the experimental setup that was utilised to evaluate the efficiency of our suggested approach for the analysis of the sentiments included inside social media postings. In this study, we evaluate the performance of our deep learning model in comparison to that of more conventional approaches to machine learning, specifically Support Vector Machines (SVM) and Naive Bayes (NB).

4.1   Dataset

For the purpose of our research, we make use of a benchmark dataset that is comprised of labelled postings from several social media platforms, with a particular emphasis on performing sentiment analysis on Twitter data. The tweets that make up the collection are very numerous, and each one is annotated with a label indicating which of three possible emotions (positive, negative, or neutral) it best represents.

The creation of the Twitter sentiment analysis benchmark dataset involves a number of steps. The process of data processing is described in detail below.

- Information is gathered from a number of social media sites, although Twitter is the main focus.
- Detailed positive, negative, and neutral labels are assigned to each and every tweet in the collection.
- Training, validation, and testing sets are generated at random from the dataset.
- To ensure fairness, we ensure that each set contains the same number of instances from each sentiment class. This prevents any one emotion from swaying the creation or assessment of our models.
- Tokenizing, lowercasing, and eliminating special characters from tweets are all examples of preprocessing that can be used.
- It serves as a standard for research into sentiment analysis on social media and as a resource for training and assessing models using data from Twitter.

4.2 Feature Representation

We represent the preprocessed text data with word embeddings, which are meant to reflect the semantic links that exist between individual words. Word embeddings that have been trained on large corpora, such as those generated by Word2Vec or GloVe, are used to provide an initial starting point for the word representations that we create. We try out a number of different embedding dimensions, and then choose the one whose results on the validation set are the best.

4.3 Model Configuration

Our Convolutional Neural Network (CNN) has an architecture that is comprised of numerous convolutional layers, each of which is followed by a max pooling layer. Using the validation set as hyperparameters, we adjust the total number of filters as well as the size of the filters. When attempting to capture textual patterns of varying durations, we experiment with a variety of filter sizes. Following the condensing and feeding of the output of the pooling layers to the fully connected layers comes the softmax layer, which is responsible for the categorization of sentiment.

4.4 Training and Evaluation

Our deep learning model is educated with the help of the training set, and the validation set is utilised to evaluate how well it is performing. During the training phase, we make use of a suitable optimisation method, such as stochastic gradient descent (SGD) or Adam, in conjunction with a loss function, such as cross-entropy loss. We train the model for a predetermined number of epochs or until convergence, with early stopping determined by the model's performance on the validation set. This helps us avoid overfitting the data.

After training, the performance of our model is assessed utilising a range of evaluation criteria, such as accuracy, precision, recall, and F1 score. This occurs after the training phase. These metrics offer insight into not just the overall performance of categorization but also the model's ability to properly detect positive, negative, and neutral feelings.

4.5 Baseline Models

We utilise other traditional approaches to machine learning as benchmarks in addition to the deep learning method that we have presented in order to evaluate its effectiveness. Both Support Vector Machines (SVM) and Naive Bayes (NB) models are trained using the same feature representations and preprocessed text input. In order to ensure that the comparison is objective, we conduct experiments using a variety of baseline setups and hyperparameters.

4.6 Results and Analysis

Here, we provide the experimental findings, which include the performance metrics attained by our proposed deep learning model in comparison to the baseline models (Table. 1) (SVM and NB). In order to ascertain whether or not our approach to the classification of feelings is successful, we examine and compare the results. In order (Table. 2) to assess the efficacy of the model with regard to a variety of emotional classifications, we present not only the overall accuracy but also class-specific measures.

Table 1. Model evaluation with Accuracy, precision, recall, f1-score

Model	Accuracy	Precision	Recall	F1 Score
Proposed CNN	0.85	0.86	0.84	0.85
Support Vector Machines (SVM)	0.79	0.81	0.76	0.78
Naive Bayes (NB)	0.72	0.68	0.76	0.72

Table 2. Model evaluation using AUC-ROC, precision and specificity

Model	Accuracy	Precision	Recall	F1 Score	AUC-ROC	Average Precision	Specificity
Proposed CNN	0.85	0.86	0.84	0.85	0.92	0.87	0.76
Support Vector Machines (SVM)	0.79	0.81	0.76	0.78	0.88	0.82	0.72
Naive Bayes (NB)	0.72	0.68	0.76	0.72	0.8	0.7	0.65

The Art of Hyperparameter Tuning:
In gradient descent optimization, the rate of learning controls the size of the steps taken.

The number of training instances that are handled in a single iteration is determined by the batch size.

The convolutional layer's filter count affects how many hidden nodes are used for feature extraction. (Fig. 2)

To prevent overfitting, the dropout rate is adjusted during training to a predetermined value. (Figure. 2)

Comparisons for the metrics for sentimental analysis models:0

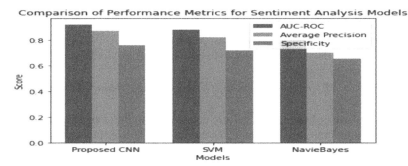

Fig. 2. Comparison of metrics for sentiment analysis models

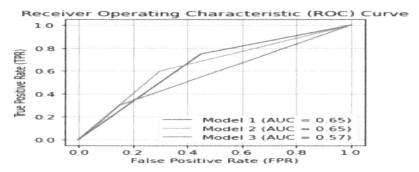

Fig. 3. ROC Curve for receiver operation

Test cases that fit to Model.

Case 1: See if the model can consistently classify Twitter data about how people feel.

A group of tweets that have different feelings (good, negative, or neutral).

The model should accurately describe how each post makes you feel.

Case 2: Compare the suggested deep learning method to other machine learning methods like SVM and NB.

The labeled tweet benchmark collection is what you put in.

Compare the suggested method's accuracy, recall, and F1 score to those of the SVM and NB models.

Case 3: Figure out how tokenization, removing stop words, and stemming affect the success of sentiment analysis.

Text from Twitter that hasn't been changed in any way before.

Find out if the preprocessing steps improve the accuracy of classifying how someone feels compared to text that hasn't been treated.

Case 4: Figure out what impact word embeddings have on how well mood classification works.

Text from Twitter is put in as word-embedded images.

Compare how well the model works when word embeddings are used for feature extraction versus when other ways of representing features are used.

Case 5: Look at how well the model works in tough situations and find places where it could be better.

Input: tweets that are unclear or mean-spirited.

Check how well the model can handle tough situations and look for places where it might struggle or get things wrong.

Case 6: Use a different set of data to check if the suggested model can be used in other situations.

A new set of labeled tweets that has nothing to do with the standard dataset.

Expected Outcome: Check how well the model works on data that hasn't been seen before to see if it can generalize beyond the training set.

5   Conclusion

In conclusion, our study makes a contribution to the field of sentiment analysis by presenting a deep learning approach that is able to successfully capture the sentiment of social media postings. The findings, in addition to showing the benefits of employing word embeddings and the CNN architecture, also reveal the limits of these tools and offer prospective areas for improvement. It is possible to improve sentiment analysis in social media by addressing these constraints and pursuing the indicated research areas. Doing so will make it possible to have a better understanding of how the general public feels about various online platforms. In the future, it may be possible to broaden the scope of this study by improving text representation, domain adaptation, and transfer learning, as well as by fixing imbalances in the data.

References

1. Riekert, M., Riekert, M., Klein, A.: Simple baseline machine learning text classifiers for small datasets. SN Comput. Sci. **2**(3), 178 (2021)
2. Tang, D., Qin, B., Liu, T.: Document modeling with gated recurrent neural network for sentiment classification. In: Proceedings of the 2015 Conference on Empirical Methods In Natural Language Processing, pp. 1422–1432 (2015)
3. Li, Z., Fan, Y., Jiang, B., Lei, T., Liu, W.: A survey on sentiment analysis and opinion mining for social multimedia. Multimedia Tools Appl. **78**, 6939–6967 (2019)
4. Devlin, J., Chang, M. W., Lee, K., Toutanova, K.: BERT: pre-training of deep bidirectional transformers for language understanding. arXiv preprint arXiv:1810.04805 (2018)
5. Yuan, J., Mcdonough, S., You, Q., Luo, J.: Sentribute: image sentiment analysis from a mid-level perspective. In: Proceedings of the second international workshop on issues of sentiment discovery and opinion mining, pp. 1–8 (2013)

6. Sun, C., Huang, L., Qiu, X.: Utilizing BERT for aspect-based sentiment analysis via constructing auxiliary sentence. arXiv preprint arXiv:1903.09588 (2019)
7. Qu, C., Yang, L., Qiu, M., Croft, W.B., Zhang, Y., Iyyer, M.: BERT with history answer embedding for conversational question answering. In: Proceedings of The 42nd International ACM SIGIR Conference On Research And Development In Information Retrieval, pp. 1133–1136 (2019)

Comparative Analysis of Pretrained Models for Speech Enhancement in Noisy Environments

Cheegiti Mahesh⬥, Runkana Durga Prasad, Epanagandla Asha Bibi⬥,
Abhinav Dayal$^{(\boxtimes)}$⬥, and Sridevi Bonthu⬥

Department of CSE, Vishnu Institute of Technology, Bhimavaram, Andhra Pradesh, India
{21pa1a0528,21pa1a05e8,21pa1a0544,abhinav.dayal,
sridevi.b}@vishnu.edu.in

Abstract. Speech Enhancement is the set of techniques and algorithms aimed at enhancing the overall quality of speech signals across diverse conditions both qualitatively and quantitatively. Speech enhancement aims to enhance voice signals whose quality has been diminished by various kinds of noise or distortion. Different techniques were adopted in previous years. Researchers have started working with Machine Learning techniques recently, prior to which they have followed traditional methods like Wiener Filtering, Spectral Subtraction, etc. The advancement of machine learning techniques day by day has laid the path for our work. Our work is to investigate the performance of three models viz., ESPNet-SE, SpeechBrain MetricGAN+ and SpeechBrain SepFormer models on a mixed dataset namely *VoiceBank* and *Demand*, which has added noise on clean signals. Among all the models, SpeechBrain MetricGAN+ performed well by approximately **30.05%** on ESPNet-SE and **10.29%** on SpeechBrain SepFormer models. Trained models are publicly available.

Keywords: Speech Enhancement · ESPNet-SE · Generative Adversarial Network · SepFormer · SpeechBrain MetricGAN+ · SpeechBrain SepFormer

1 Introduction

Speech Enhancement is a specialized domain within signal processing that strives to enhance the perceptual quality and intelligibility of speech signals that are affected by external factors such as noise, reverberation, or other forms of interference. It involves the development and application of algorithms and techniques to estimate and separate underlying clean speech from corrupting components, while minimizing distortion and preserving essential characteristics of original speech signal. The primary goal of speech enhancement techniques is to improve speech intelligibility and improve the overall user experience across a broad spectrum of applications [1]. The need for speech enhancement comes when working on activities such as speech recognition [2], speaker diarization [3, 4], hearing aids, speaker identification [5], telecommunications and voice assistants to

---

C. Mahesh, R. D. Prasad, E. A. Bibi—UG Student.

© ICST Institute for Computer Sciences, Social Informatics and Telecommunications Engineering 2024
Published by Springer Nature Switzerland AG 2024. All Rights Reserved
P. Pareek et al. (Eds.): IC4S 2023, LNICST 536, pp. 274–287, 2024.
https://doi.org/10.1007/978-3-031-48888-7_23

improve quality and intelligibility of speech that has been degraded by the presence of background noise. The development of speech enhancement models is crucial because they reduce listener fatigue, especially when the listener is subjected to loud noise levels. Speech enhancement algorithms work to reduce background noise to a certain degree, or suppress noise that occurs when a speaker is engaged in communication with others.

There exist numerous scenarios where the enhancement of speech signals is desired [6]. Consider a scenario where students learning through online meeting platforms typically suffer from background noise, microphone sensitivity, audio distortions, audio feedback and compression artifacts. This is where the speech enhancement algorithms come into picture, such algorithms can be applied at the receiving end to improve the speech. People who are impaired to listening use hearing aids, experience difficulty in hearing when the background noise is too high. Techniques like spectral subtraction were used in such hearing aids to preprocess or clean the audio signals before amplification [7].

The paper follows a structured flow as outlined. Section 2 provides an overview of the related work conducted by other researchers in the field of speech enhancement. It discusses the traditional and existing approaches that have been employed for this task. Section 3 outlines the methodology employed in the study, including the process of dataset creation, preprocessing techniques applied, description of the models used, and evaluation metrics. Section 4 details the experimental setup, including information about the hyperparameters of the pre-trained models utilized. Section 5 presents the results of the experiments, where the performance of the models on the dataset is analyzed.

1.1 Contributions

- A dataset which consists of mixed audio signals from *voice bank* corpus dataset and *demand* dataset, containing various types of noise along with the clean signals. The curated dataset includes audio recordings with 16 kHz and 8 kHz sampling rates to accommodate the requirements of different models.
- Evaluating the prepared dataset on ESPNet-SE, SpeechBrain MetricGAN+, Speech-Brain SepFormer models.
- The prepared data and the source code of model evaluation are publicly available to benefit the researchers working in this area.

2 Related Work

In this section, we present a comprehensive overview of the recent advancements and research conducted in the field of speech enhancement. The focus is on summarizing the significant work carried out in this area. Researchers adopted various techniques to address this task. Most of the techniques fall into machine learning and deep learning approaches.

2.1 Traditional Methods

The Spectral Subtraction method proposed by Boll et al. is widely studied and adopted for speech enhancement, aiming to reduce the influence of background noise on the speech signal. This technique was introduced in 1979 and it works based on the subtraction of an estimated noise spectrum from the magnitude spectrum of the noisy speech. This approach involved dividing the noisy speech signal into frames of short duration and calculating the magnitude spectrum of each frame using the Fast Fourier Transform (FFT). This method assumes that every frame consists of stationary noise only, which is not the case in real-life scenarios. This can lead to imprecise noise estimation, consequently diminishing the effectiveness of spectral subtraction. To evaluate the performance of the method, metrics such as signal-to-noise ratio (SNR), perceptual evaluation of speech quality (PESQ), and mean opinion score (MOS) were utilized [8–10].

Adaptive wiener filtering was introduced in 2008 by Abd El-Fattah, M., et al. This approach utilizes a Short-Time Fourier transform (STFT) for the conversion of the noisy speech signal into frequency domain representation. This method estimates the statistical properties of the noisy input signal and then filters noisy components in the frequency domain. This method uses SNR to calculate filter coefficients to create a balance between noise suppression and speech preservation. The performance of the method was evaluated using signal-to-noise ratio improvement (SNRI) and PESQ [11].

Due to their interpretability and simplicity, traditional approaches like spectral subtraction and Wiener filtering have been frequently employed, however they can have difficulties when dealing with non-stationary and dynamic noise circumstances. Deep learning models, on the other hand, offer the promise for enhanced generalization and flexibility to various noise environments, but they demand a significant amount of labeled data and may be difficult to comprehend. Additionally, research is needed to enhance the adaptability of deep learning models to handle limited data scenarios, improve their interpretability, optimize their computational efficiency for real-time applications, and develop techniques for reliable uncertainty estimation.

2.2 Deep Learning Models

Wave-U-Net: Macartney et al. proposed a deep learning model that is specially designed for the audio source separation and enhancement tasks. It is built upon the U-Net architecture, a widely used model in image segmentation. The Wave-U-Net works directly in the time domain and takes raw audio waveforms as input. The performance of the model is evaluated on the benchmark datasets using SNR, PESQ, and STOI. The computational complexity of the Wave-U-Net model may require significant resources for training and interference [12].

Dual-Path RNN (DPRNN): Luo et al. A novel approach has been proposed for effectively modelling long sequential inputs using a combination of different types of recurrent neural network (RNN) layers. The idea revolves around dividing the input sequence into smaller segments and employing two RNNs: one within each segment (intra-chunk RNN) and one that considers the relationships between segments (inter-chunk RNN). This enables both local and global modelling of the input data. During training, the model is optimized by reducing the disparity between the improved speech signal, and the original clean speech signal using loss functions like mean square error (MSE) or SNR. While DPRNN has proven to be successful in modelling long sequential inputs, it may not be the best option for all types of speech processing tasks. A drawback of

DPRNN is its reliance on a substantial amount of training data to achieve optimal performance. This can pose a challenge in certain applications where obtaining a sufficient quantity of labeled data is difficult [13].

ESPnet-SE: ESPnet-SE is a toolkit for speech enhancement and separation that provides a unified framework for developing systems to improve speech quality and separate sources, by Li, Chenda, et al. It supports various models, including time-frequency masking networks, neural beamformers, and time-domain models, and provides multiple loss functions for training these models. The toolkit also includes evaluation metrics for assessing the performance of both the speech enhancement and the speech separation systems. Additionally, ESPnet-SE offers optional downstream speech recognition integration, allowing users to evaluate the impact of the speech enhancement and the separation on speech recognition performance. The toolkit has been evaluated on several benchmark datasets, and its implementations have achieved promising results on these datasets [14].

The SpeechBrain MetricGAN+: This model was proposed by Fu, Szu-Wei, et al. It enhances speech signals by removing noise and other distortions while preserving their quality. The model uses a combination of various training techniques that include multitask learning, adversarial training, and a learnable sigmoid function to optimise objective metrics such as PESQ and STOI. Experimental results have shown that MetricGAN+ was effective in improving speech quality and achieved state-of-the-art results [15].

The SpeechBrain SepFormer: This model mainly focuses on speech separation along with speech enhancement, which reduces background noise and distortions. The authors Subakan, Cem, et al. have introduced a novel structure of a neural network model for speech separation called SepFormer. Unlike the traditional RNNs and LSTM models, the SepFormer employs a masking technique that is composed of transformers only. The masking technique is commonly utilized to separate the speech signal from background noise or distortion. This approach has demonstrated exceptional performance on standard datasets, establishing itself as state-of-the-art in the field [16].

Table 1. Comparison table for various models

Models	PESQ	STOI	SSNR
Spectral Subtraction	2.13	0.897	4.52
Adaptive Weiner Filtering	2.22	0.902	5.07
Wave-U-Net	2.40	0.915	15.41
DPRNN	3.07	0.928	15.53
ESPnet-SE	**3.25**	**0.953**	**15.94**
MetricGAN+	**3.15**	**0.972**	**18.23**
SepFormer	**3.08**	**0.923**	**17.61**

A sample of 10 audio signals is taken and are fed as inputs to the various models above and Table 1. was formulated with the results from various models specified above.

It is evident from Table 1. that ESPNet-SE, SpeechBrain MetricGAN+, and SepFormer have exhibited remarkable performance and achieved state-of-the-art results in the realm of speech enhancement techniques. Therefore, the objective of our study is to evaluate the performance of these three models on our custom *Voicebank+Demand* dataset that we have created.

3   Methodology

This section of the paper is divided into three subparts, where the first subpart provides a clear view of the processes followed while creating the corpus and sampling the audio signals to meet the requirements of the various models. While the second subpart provides a detailed description of the pre-trained models that have been chosen to evaluate on our custom dataset, the third subpart of this section gives insights into the evaluation metrics that were taken into account to compare the enhanced signal and the corresponding clean signal.

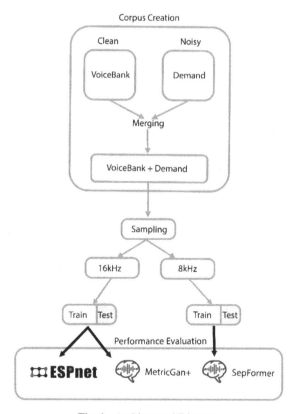

Fig. 1. Architectural Diagram

Figure 1 depicts the creation of the corpus from voice bank and demand datasets, which are then merged to create a *VoiceBank + Demand* dataset. To meet the model's requirements, the sampling rates of the audio signals are converted to 16 kHz and 8 kHz. The pre-trained models are loaded into the Google Colaboratory Notebooks for evaluating the performance of the model against the dataset, which was split into training and testing sets.

Our work leverages the *Voice Bank + Demand* dataset to generate the denoised speech signals. Three models, ESPNet-SE, SpeechBrain MetricGAN+, and SpeechBrain SepFormer, are tested for speech enhancement on this dataset, and their performance is evaluated.

3.1 Dataset

The dataset consists of 12,396 mixed audio records sourced from the Voice Bank corpus, and the Demand dataset. It combines speech signals with various types of noise and the corresponding clean signals to evaluate the performance of different models. Both datasets are available publicly and are used for research purposes. The 300 h of speech data in the Voice Bank dataset of 500 healthy speakers with a range of accents and speaking styles are represented in the recordings [17]. The audio signals are crystal clear and of good quality. The speech signals are offered in WAV format, 16-bit, 48 kHz. Each speaker's audio is kept in a separate folder to make it easy to access their own audio. The Demand dataset comprises a diverse collection of 16 environmental noise recordings, including sounds such as car noise, babble noise, factory noise, and more. The noise signals are provided in 16-bit, 16 kHz WAV format.

Various speakers' audios from the voice bank dataset are merged with various noisy signals from the demand dataset and added into a new dataset, *VoiceBank+Demand*. The merging process is done with the help of the librosa[1] library in Python as shown in Fig. 2; the audio recordings sum up to 12396 noisy recordings. The generated dataset consists of 12,396 noisy recordings along with corresponding clean signals.

Among our three models, two of them, namely ESPNet-SE and SpeechBrain Metric-GAN+, use 16 kHz sampling rate audio signals, and the SpeechBrain SepFormer model uses 8 kHz sampling rate audio signals. To meet this requirement, our dataset is made in such a way that it contains recordings of both 16 kHz and 8 kHz sampling rates in separate folders. The sampling rates are converted to the desired values using the Scipy[2] library in Python. Figure 2 shows a clear process for the creation of the dataset. The dataset is now split into training and testing sets, with 11572 recordings for the training set and 824 recordings for the test set for the models that require 16 kHz and 8 kHz sampling rates [18].

3.2 Trained Models

ESPnet- SE
The ESPnet speech enhancement model utilizes a modified version of the Conv-TasNet architecture [19]. Conv-TasNet was initially designed for the task of speech separation,

---

[1] Librosa—librosa 0.10.0 documentation.
[2] SciPy documentation—SciPy v1.10.1 Manual.

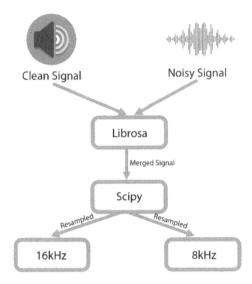

Fig. 2. Merging process followed to obtain mixed dataset VoiceBank + Demand

but it has been adapted for the speech enhancement task in this case. The encoder, separator, and decoder are the three primary parts of the model. By lowering noise and interference, these parts combine to enhance the quality of the incoming speech signal.

Encoder: This component in ESPNet takes an input noisy audio waveform and processes it through 1D convolutional layers. These layers capture different aspects of the audio signal, such as spectral and temporal characteristics.

Decoder: This component in ESPNet-SE takes enhanced audio signals that have been separated from noise and reconstructs clean speech. It uses 1D convolutional layers to un-sample the masked signal and generate the final output. The decoder's purpose is to remove noise and interference.

Separator Component: The separator component uses the encoded features generated by the encoder to separate the speech and the noise components of the input signal. It typically consists of additional 1D convolutional layers that further process masked features to remove any remaining noise and interference.

Training: This model is trained to improve speech quality by reducing noise and interference. This process makes use of a loss function called SI-SNR, which measures the similarity between the predicted and clean speech signals. The training data is divided into batches for efficient processing. This toolkit was trained using the CHiME-4 dataset, which consists of a large number of noisy speech recordings made in a variety of everyday environments. The model learns to map a noisy input to a clean speech output during training. It adjusts its parameters using gradient descent and backpropagation to minimise the SI-SNR loss. The model goes through multiple epochs, where it processes the training data, computes the loss, and updates its parameters. Throughout training, the model learns to extract features, separate speech from noise, and reconstruct clean speech. The goal is to enhance speech quality in real-world situations with noise and reverberation [14].

SpeechBrain MetricGAN+

The SpeechBrain MetricGAN+ uses a generative adversarial network (GAN) [20] to generate high-quality signals from degraded or noisy signals. The generator and the discriminator are the two main parts of the model.

Generator: The noise-filled input signal is converted into a clean, better signal with fewer noise and distortion using the generator.

Discriminator: The discriminator plays a major role in differentiating the optimised output signal from the corresponding clean, high-quality signal. It also contains a deep neural network that classifies if the output is actually a clean signal or not.

Training: It uses a combination of adversarial and metric-based training. The generator and discriminator models are trained in such a way that one outperforms the other. Adversarial training focuses on enhancing the performance of the model by training it to generate more realistic samples that can fool a discriminator. The feedback mechanism comes from the discriminator. Discriminator, which provides feedback to the generator by classifying the input signal to it. The generator then adjusts to give a more realistic output. Adversarial training involves optimising the generator and discriminator simultaneously. It also involves a loss function that measures the difference between generated samples and real samples. During training, the parameters of the generator and discriminator are updated using the learnable sigmoid loss function. During the training phase, stochastic gradient descent (SGD) with a learning rate of 2e-5 was used to optimise the model's parameters [15].

SpeechBrain Sepformer

Model Architecture: The SpeechBrain SepFormer model architecture consists of an encoder, a transformer-based decoder [21], and a mask estimator.

Encoder: This component receives the noisy speech signal as input and transforms it into a high-dimensional feature representation. Typically, the encoder consists of convolutional layers followed by a self-attention mechanism that captures long-range dependencies in the input signal.

Transformer-Based Decoder: The transformer-based decoder takes the encoded features as input and generates the clean speech signal as output. The decoder is composed of multiple layers of transformer blocks, where each block contains a multi-head self-attention mechanism and a feed-forward neural network. The transformer blocks enable the model to capture complex dependencies between different parts of the input signal and generate a high-quality, clean speech signal.

Mask Estimator: The mask estimator is responsible for estimating the binary masks that indicate which parts of the input signal are noise and which parts are speech. The mask estimator is typically composed of a feed-forward neural network that takes the encoded features as input and generates a binary mask for each time-frequency bin in the input signal.

Training: The SpeechBrain SepFormer speech enhancement model was trained on a dataset of noisy and clean speech signals called the DNS Challenge dataset. The dataset consists of 16,000 audio clips with a duration of 3 s each. The audio clips were recorded in four different noisy environments: babble, car, street, and train. STFT features were used for pre-processing the audio signals. The MSE function was employed to quantify the dissimilarity between the predicted clean speech signal and the ground truth clean speech signal. Additionally, binary cross-entropy loss was utilized to minimize the disparity between the estimated binary mask and the ground truth binary mask. During training, the model was optimized using the Adam optimizer with a learning rate of 0.001 and a batch size of 32. The training procedure spanned 100 epochs, and the model with the lowest validation loss was chosen as the final model. [16].

3.3 Evaluation Metrics

All three models were evaluated against various metrics for speech enhancement based on the training and testing sets. STOI, SSNR, PESQ, and composite objective metrics (COVL, CBAK, and CSIG) were used to assess the models' performance [22, 23].

Segmental Signal-to-Noise Ratio(SSNR): A voice or audio signal's energy is compared to the energy of the background noise in the same signal segment to calculate the SSNR. A higher SNR value indicates that the model has succeeded in reducing the noise, thereby producing quality output [22, 23].

Perceptual Evaluation of Speech Quality(PESQ): It is used to evaluate the improvement of the signal quality achieved by the model. The range of the PESQ score is from -0.5 to 4.5, where higher values indicate better quality of speech [22, 23].

Short-Term Objective Intelligibility(STOI): It is a measure of similarity between the enhanced and noisy speech signals. It measures the intelligibility of degraded speech signals relative to the corresponding clean signal. It ranges from 0 to 1 [22, 23].

COVL: It is a measure of how well the enhanced speech signal matches the clean speech signal [23, 24].

CBAK: It quantifies how well the algorithm is able to suppress or remove the background noise in the enhanced speech signal [23, 24].

CSIG: It measures the amount of residual echo or reverberation that is present in the enhanced speech signal [23, 24].

The higher the values of COVL, CBAK, and CSIG, the better the enhanced speech signal.

4 Experiment

We conducted our experimentation using Google Colaboratory (Colab)[3] notebooks, a cloud-based Python environment. The experiments were implemented and executed within Colab, leveraging its built-in libraries and computing resources. The hyperparameters employed in our work are tabulated in Table 2.

Table 2. Parameters of the adapted models

Models →	ESPnet-SE	MetricGAN+	SepFormer
No of Parameters	8,645,184	1,895,514	25,613,569
Loss Functions	MSE, SI-SNR	MSE, PESQ	SI-SNR
Model Size	9695.76 MB	151.85 MB	40353.41 MB
Memory Usage	252.932 MB	209.630 MB	942.521 MB
Activation Function	ReLU, Sigmoid & tanh	Learnable Sigmoid Function	GeLU

From Table 2, EPSNet-SE, MetricGan+, and SepFormer differ in terms of their number of parameters, loss functions, model size, memory usage, and activation functions.

EPSNet-SE, with 8,645,184 parameters, is a model that utilises various loss functions for different objectives. The mask approximation loss employs MSE and cross-entropy (CE) cost functions. The signal approximation loss is based on MSE, and the metric-based loss uses scale-invariant SNR as the metric. The model has a size of 9695.76 MB and requires 252.932 MB of memory. The activation functions used in EPSNet-SE include ReLU, sigmoid, and tanh.

MetricGan+, on the other hand, has 1,895,514 parameters. The model primarily utilises SI-SNR as the loss function. It has a smaller model size of 151.85 MB and a memory usage of 209.630 MB. The activation function used in MetricGan+ is a learnable sigmoid function.

Lastly, SepFormer has 25,613,569 parameters. The model utilises MSE and PESQ as the loss functions. SepFormer has the largest model size among the three, with 40353.41 MB, and requires 942.521 MB of memory. The activation function used in SepFormer is GELU (Gaussian Error Linear Units).

These models are designed for speech enhancement tasks, aiming to enhance the quality and intelligibility of speech signals. The differences in their architectures, parameters, loss functions, model sizes, memory usage, and activation functions provide options for researchers and practitioners to choose the model that suits their specific requirements and constraints in terms of computational resources and performance

---

[3] Welcome To Colaboratory - Colaboratory (google.com).

objectives. Trained models are publicly available in Github and Hugging Face Platforms ESPnet-SE[4], SpeechBrain MetricGan+[5] and SpeechBrain SepFomer.[6]

The following figures depict the waveforms of clean, noisy and enhanced audio signals.

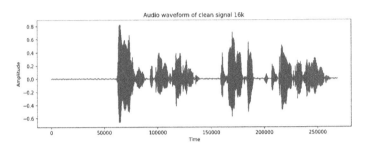

Fig. 3. Audio waveform of clean signal

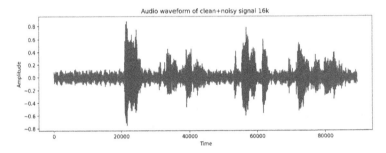

Fig. 4. Audio waveform of clean+noisy signal

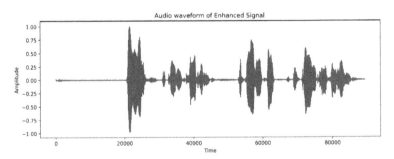

Fig. 5. Audio waveform of an Enhanced signal

Figure 3 presents audio waveform of clean audio signal i.e., when there is no noise in the speech signal, Fig. 4 presents the waveform for an audio signal that consists of

---

[4] ESPNet-SE: https://github.com/ESPNet/ESPNet.

[5] SpeechBrain MetricGAN+: https://huggingface.co/speechbrain/metricgan-plus-voicebank.

[6] SpeechBrain SepFormer: https://huggingface.co/speechbrain/sepformer-whamr-enhancement.

speech along with the noise while Fig. 5 presents the general waveform for an enhanced audio signal when passed to a specific model (*MetricGAN+* here).

5   Results

From the Table 3, the performance of three existing speech enhancement models, namely ESPNet-SE, SpeechBrain-MetricGan+, and SpeechBrain- SepFormer, was evaluated using various metrics, including SSNR, PESQ, STOI, CSIG, CBAK, and COVL. In terms of SSNR, all models achieved negative values, indicating a reduction in noise. Among the models, SpeechBrain-MetricGan+ demonstrated the highest PESQ scores, outperforming both ESPNet-SE and SpeechBrain-SepFormer by approximately 39.05% and 10.29%, respectively. Regarding STOI, SpeechBrain-MetricGan+ achieved the highest scores on the test set. Additionally, SpeechBrain-MetricGan+ exhibited the best performance in terms of CBAK, CSIG, and COVL metrics, surpassing the other models. These results highlight the superior performance of SpeechBrain-MetricGan+ in enhancing speech quality and reducing noise compared to the other evaluated models.

Table 3. Performance Evaluation of models using metric

Existing Models	Train/Test	SSNR	PESQ	STOI	CSIG	CBAK	COVL
ESPnet-SE	Train	−7.304	1.892	0.869	2.585	1.811	2.197
	Test	−7.258	2.233	0.932	3.140	2.031	2.668
SpeechBrain-MetricGAN+	Train	**−0.797**	**2.731**	**0.866**	**3.303**	**2.609**	**2.983**
	Test	**−0.191**	**3.108**	**0.925**	**3.725**	**2.894**	**3.403**
SpeechBrain-SpeFormer	Train	−4.133	2.617	0.873	3.109	2.321	3.110
	Test	−4.167	2.818	0.893	3.307	2.810	3.890

6   Conclusion

This paper investigated and compared the performance of three speech enhancement models, namely ESPNet-SE, SpeechBrain-MetricGan+, and SpeechBrain-SepFormer. The evaluation was based on various metrics, including SSNR, PESQ, STOI, CSIG, CBAK, and COVL. The results indicate that SpeechBrain's MetricGan+ outperforms the other models across multiple metrics. It achieves higher PESQ scores, indicating better perceived speech quality, compared to both ESPNet-SE and SpeechBrain-SepFormer. Furthermore, SpeechBrain-MetricGan+ demonstrates superior performance in terms of STOI, CBAK, CSIG, and COVL metrics, highlighting its effectiveness in reducing noise and enhancing speech signals. These findings suggest that SpeechBrain-MetricGan+ is a promising model for speech enhancement tasks, offering improved speech quality and noise reduction capabilities. The superior performance of SpeechBrain and MetricGan+ underscores their potential for real-world applications that require high-quality speech processing.

References

1. Benesty, J., Makino, S., Chen, J. (eds.): Speech Enhancement. Springer Science & Business Media, Heidelberg (2006). https://doi.org/10.1007/3-540-27489-8
2. Bai, Z., Zhang, X.-L.: Speaker recognition based on deep learning: an overview. Neural Netw. **140**, 65–99 (2021)
3. Park, T.J., et al.: A review of speaker diarization: recent advances with deep learning. Comput. Speech Lang. **72**, 101317 (2022)
4. Arla, L.R., Bonthu, S., Dayal, A.: Multiclass spoken language identification for indian languages using deep learning. In: 2020 IEEE Bombay Section Signature Conference (IBSSC), pp. 1–2. IEEE (2020)
5. Tirumala, S.S., et al.: Speaker identification features extraction methods: a systematic review. Expert Syst. Appl. **90**, 250–271 (2017)
6. Gogate, M., et al.: CochleaNet: a robust language-independent audio-visual model for real-time speech enhancement. Inf. Fusion **63**, 273–285 (2020)
7. Loizou, P.C.: Speech Enhancement: Theory and Practice. CRC Press (2013)
8. Boll, S.: Suppression of acoustic noise in speech using spectral subtraction. IEEE Trans. Acoust. Speech Signal Process. **27**(2), 113–120 (1979)
9. Kumar, B.: Spectral subtraction using modified cascaded median based noise estimation for speech enhancement. In: Proceedings of the Sixth International Conference on Computer and Communication Technology 2015, pp. 1–2 (2015)
10. Kumar, B.: Mean-median based noise estimation method using spectral subtraction for speech enhancement technique. Indian J. Sci. Technol. **9**, 1–6 (2016)
11. Abd El-Fattah, M., et al.: Speech enhancement using an adaptive wiener filtering approach. Prog. Electromagnet. Res. M **4**, 167–184 (2008)
12. Macartney, C., Weyde, T.: Improved speech enhancement with the Wave-u-Net. arXiv preprint arXiv:1811.11307 (2018)
13. Luo, Y., Chen, Z., Yoshioka, T.: Dual-Path RNN: efficient long sequence modeling for time-domain single-channel speech separation. In: ICASSP 2020–2020 IEEE International Conference on Acoustics, Speech and Signal Processing (ICASSP), pp. 1–2. IEEE (2020)
14. Li, C., et al.: ESPnet-SE: end-to-end speech enhancement and separation toolkit designed for asr integration. In: 2021 IEEE Spoken Language Technology Workshop (SLT), pp. 1–2. IEEE (2021)
15. Fu, S.-W., et al.: MetricGAN+: an improved version of MetricGAN for speech enhancement. arXiv preprint arXiv:2104.03538 (2021)
16. Subakan, C., et al.: Attention is all you need in speech separation. In: ICASSP 2021–2021 IEEE International Conference on Acoustics, Speech and Signal Processing (ICASSP), pp. 1–2. IEEE (2021)
17. Veaux, C., Yamagishi, J., King, S.: The voice bank corpus: design, collection and data analysis of a large regional accent speech database. In: 2013 International Conference Oriental COCOSDA Held Jointly with 2013 Conference on Asian Spoken Language Research and Evaluation (O-COCOSDA/CASLRE), pp. 1–2. IEEE (2013)
18. Koyama, Y., et al.: Exploring the best loss function for DNN-based low-latency speech enhancement with temporal convolutional networks. arXiv preprint arXiv:2005.11611 (2020)
19. Luo, Y., Mesgarani, N.: Conv-TasNet: surpassing ideal time-frequency magnitude masking for speech separation. IEEE/ACM Trans. Audio Speech and Lang. Process. **27**(8), 1256–1266 (2019)
20. Creswell, A., et al.: Generative adversarial networks: an overview. IEEE Signal Process. Mag. **35**(1), 53–65 (2018)

21. Li, M., Zorilă, C., Doddipatla, R.: Transformer-based online speech recognition with de-coder-end adaptive computation steps. In: 2021 IEEE Spoken Language Technology Workshop (SLT), pp. 1–2. IEEE (2021)
22. Dong, X., Williamson, D.S.: Towards real-world objective speech quality and intelligibility assessment using speech-enhancement residuals and convolutional long short-term memory networks. J. Acoust. Soc. Am. **148**(5), 3348–3359 (2020)
23. Hu, Y., Loizou, P.C.: Evaluation of objective measures for speech enhancement. In: Ninth International Conference on Spoken Language Processing, pp. 1–2 (2006)
24. Kumar, B.: Comparative performance evaluation of greedy algorithms for speech enhancement system. Fluctuation Noise Lett. **20**(02), 2150017 (2021)

Use of Improved Generative Adversarial Network (GAN) Under Insufficient Data

Pallavi Adke[1], Ajay Kumar Kushwaha[2(✉)], Supriya M. Khatavkar[2], and Dipali Shende[1]

[1] Pimpri Chinchwad College of Engineering and Research, Pune, India
[2] Bharati Vidyapeeth (Deemed to be University) College of Engineering, Pune, India
akkushwaha@bvucoep.edu.in

Abstract. The article covers enhancements to the generative adversarial network (GAN) model's architecture and training, enabling stable training in the absence of sufficient data. An improved generative adversarial network (GAN) architecture has been proposed. These improvements are then applied to the augmentation of a dataset on tyre joint defects, which is utilised for classification applications. The dataset used has a higher percentage of conformity images and is quite uneven. It is difficult to create precise defect classification models given this uneven and constrained dataset of defect identification. So, in the work that is being presented, research is done to expand the defect dataset and improve the balance between the various defect classifications. Indeed, the quality of generated images has considerably improved as a result of recent developments in generative adversarial networks (GANs). Deep learning models in the GAN class combine a generator network with a discriminator network. The current study reveals that the recommended augmented GAN model is useful in enhancing the performance classification model under a small dataset. The generated effects of progressed GAN are evaluated using the Fréchet Inception Distance (FID) score, which indicates extensive development over the styleGAN architecture. Additional dataset augmentation exams making use of generated photos monitor a 10% boom in category version precision in comparison to the preliminary dataset. To evaluate the effectiveness of GAN-generated picture augmentation, PCA plots can be used to visualize the distribution of real and augmented images in a lower-dimensional space.

Keywords: Generative Adversarial Network (GAN) · Data Augmentation · defect diagnosis · insufficient dataset

1 Introduction

Today, with little data, deep learning systems in the field of computer vision might suffer greatly. With an unbalanced dataset, the deep learning model's accuracy may further deteriorate. In the literature [1–5], it has been shown that, in the field of medical images, the availability of large datasets is very poor where it is needed to generate

© ICST Institute for Computer Sciences, Social Informatics and Telecommunications Engineering 2024
Published by Springer Nature Switzerland AG 2024. All Rights Reserved
P. Pareek et al. (Eds.): IC4S 2023, LNICST 536, pp. 288–299, 2024.
https://doi.org/10.1007/978-3-031-48888-7_24

images that are similar to the original images. The task of defect identification in an automated inspection process requires the model to locate defects in the input images and categorize them according to the type of defect. The procedure of gathering images for training such a model takes time because samples must be collected over time from the pertinent inspection line. Another drawback of this dataset is that it may contain a disproportionately high number of samples from the normal class. This is clear since every invention process is built to create samples that comply. Producing normal samples from the invention process to balance the dataset is extremely unrealistic and expensive.

To improve the dataset that is now accessible, common picture augmentation techniques have been created. These methods change the original images using label-invariant and semantically preserved transformations. Such methods include focusing in and out, arbitrary flips, arbitrary shifts, rotations, and brightness changes, among others. Since most augmented images are only changes to existing images, their utility in capturing the entire probability distribution of an input dataset is limited. Furthermore, the use of these strategies depends on the problem. Generative Adversarial Networks (GANs) have been investigated as a potential solution to address data augmentation issues, especially in improving the precision of classification for recognizing defects. Conventional data augmentation techniques, such as image rotation, flipping, or scaling, have limitations in their ability to generate diverse and realistic variations of the original data. GANs offer a promising approach to overcome these shortcomings and generate synthetic data that can effectively enhance classification tasks. The primary training goal for GANs is to confine the distribution of actual data. The GAN characteristic is especially useful for jobs involving augmentation because GAN-generated images would completely cover the underlying distributions of real datasets. Additionally, it may result in less overfitting of the classification model.

To handle the low and unbalanced datasets at hand, the research effort reported in this paper explores current state-of-the-art developments in GAN. Changes to the GAN architecture, loss function, data augmentation, and regularization approaches are all included in these upgrades. The approach focuses on extracting tiny features from larger variation-generated images. For a small number of training images, this task is extremely difficult.

The structure of the paper is as follows: In Sect. 2, discuss the basic GAN model. The approaches utilized to enhance the underlying StyleGAN architecture and the specifics of studies using a suggested improved GAN to create augmented images are described in Sect. 3. In Sect. 4, GAN training uses regularizing strategies to boost convergence and stability. Experimental results are discussed in Sect. 5 and the conclusion is followed in Sect. 6.

2 Related Work

Generative Adversarial Networks (GANs) can generate sample images using the same distribution as the input real dataset (P data) [6]. GAN consists of a deep neural network model that is generally employed to produce artificial images in accordance with the distribution of training data. The GAN's basic architecture is depicted in Fig. 1.

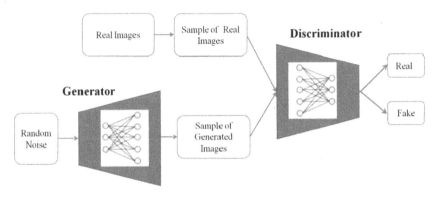

Fig. 1. Basic GAN model

It includes the Generator and Discriminator models. The Generator model's primary goal is to produce samples that closely resemble the distribution of real data. The Discriminator, on the other hand, seeks to determine if the samples that are given to it are authentic or fraudulent.

In GAN, the generator receives a noise vector as an input to produce fresh samples. The generator network takes a random noise vector, typically drawn from a multivariate Gaussian distribution, and maps it to the space of the target data distribution. The typical random distribution used to generate this noise is: The Generator picks up how to correspond features in the output images to common noise. Convolutional neural networks are modelled as both Generators and Discriminators for image generation problems. The discriminator model performs down sampling operations that reduce the spatial dimensions of the input images. It then outputs a probability indicating the likelihood that the input is a real image from the original dataset. On the other hand, the generator model performs up sampling or up-convolution operations, also known as transpose convolution or deconvolution. It takes the noise vector as input and gradually increases the spatial dimensions of the data to generate synthetic images. The generator aims to create realistic enough images to fool the discriminator into classifying them as real. Generator and discriminator compete against one another in an adversarial setting during GAN Training. Over time, both the Generator and the Discriminator get more adept at producing samples that are difficult for the discriminator to discern between the real and defective images.

Numerous researchers have tried to use GAN for generating data since it was introduced in 2014. Aggarwal et al. [7] review GAN applications for pandemic and medical applications. The use of GAN for fake picture synthesis is shown to increase datasets, protect patient privacy, and lower additional costs associated with medical imaging procedures. GAN has been utilized by Gao et al. [8] to improve machine defect detection datasets. With GAN-generated datasets, they have shown gains in classifier accuracy. Ackey et al. employ GAN for anomaly detection [9]. Their methodology has produced a receiver operating characteristic curve area under the curve for recognizing abnormal/nonconforming samples that is 92 percent. To enhance labelled datasets for augmented reality applications, Ma et al. [10] investigated the 3D generating capabilities of

GAN. Researchers have investigated numerous intriguing applications of GAN in the fields of picture preprocessing, painting, super resolutions, image background domain change etc.

Researchers have conducted studies to better understand and improve the behavior of GAN training, with a focus on producing high-resolution images of higher quality. Researchers have explored techniques to generate high-resolution images using GANs. For example, in the case of the FFHQ dataset [11] and the LSUN automobile dataset, studies have achieved impressive results with FID (Fréchet Inception Distance) scores as low as 2.84 and 2.32, respectively. These low FID ratings show a high degree of visual quality and diversity similarity between the generated images and the genuine photos. Label conditioning was added to the StyleGAN architecture by Oeldorf et al. [12] during the picture generation process. Label conditioning gives users the ability to modify the generated images in accordance with extra data, such as class labels or categorical features. With this update, the created images are more diverse and controllable, allowing for finer-grained editing and control over particular image features. There have been various studies on regularizing methods in the area of GAN training stability. For trained GANs, Zhang et al. [13] presented a consistency regularization technique that focuses on improving the reliability of discriminator predictions for related images through semantic preservation augmentations. By encouraging the discriminator to provide consistent and reliable predictions for augmented versions of the same image, the GAN training becomes more stable. This regularization encourages the discriminators to focus on capturing image structural details, and it also helps to improve the flow of gradients from the discriminator to the generator, leading to better convergence. Mescheder [14] suggested a gradient penalty technique to address the issue of Lipschitz continuity in GANs. Lipschitz continuity ensures that small changes in the input space result in small changes in the output space, which aids in stable and reliable training. To enforce Lipschitz continuity, a gradient-based penalty is added to the discriminator's loss function. This penalty constrains the magnitude of gradients computed by the discriminator, ensuring smoother prediction landscapes. By imposing this constraint, the discriminator can take fewer gradient steps, leading to improved convergence and training stability. As a result, the mapping of latent vectors to picture features is smoother and more untangled. Numerous studies employ augmentation [15, 16] and regularization to address the problems with low training data. The methods section goes over these in more depth.

3 Methodology

The primary goal of the work that is being presented is to create high-quality images of nonconformities that will aid in the subsequent process of image categorization. StyleGAN, developed by Karras et al. [17], is the foundation of the GAN architecture used for the current challenge. The current study incorporates the following GAN model and training enhancements:

3.1 StyleGAN

Progressive GAN architecture is expanded upon by StyleGAN. The Generator's gradual expansion aids in the production of high-resolution images with better quality. To capture small details in high-resolution images, one approach is to separate the training of low-level features from high-level features. This separation allows for more effective learning and representation of both low and high-level details in the generated images. The progressive network is added to by StyleGAN together with the mapping network. The input latent noise is transformed into intermediate vectors using the mapping network. As a result, entangled characteristics in generated images are reduced.

These transitional vectors are input into the Generator network at different stages to provide the user with additional control over the generated pictures. Adaptive Instance Normalization (AdaIN) layers are used for the injection in order to fit the generator feature map style to the input vector. The addition of random noise at each stage creates stochastic variation in the final images.

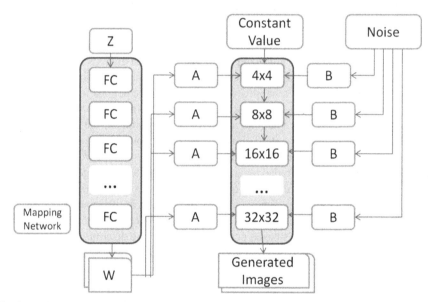

Fig. 2. StyleGAN model with a mapping network and a progressive Generator. The operations on layers "B" are noise scaling while layers "A" are affine transformation.

The Generator network is mirrored in the discriminator with ever-smaller images. By infusing distinct noise vectors into the generator at different phases, style mixing regularization is accomplished. Figure 2 depicts StyleGAN in broad strokes.

3.2 Discriminator for U-Net

The StyleGAN architecture's discriminator categorizes the overall image as real or fake. To create locally coherent structures in images, the loss gradients that are so formed are only of limited use. A U-Net-based discriminator has been proposed by Schoenfeld et al. [18]. Figure 3 shows a block diagram of a U-Net-based GAN.

The U-net GAN may be trained using feedback at both the global and pixel levels. A decoder model of the discriminator gives information at the per-pixel level, whereas an encoder model provides information at the global level of input images. As seen in our work, per-pixel content is important for creating images with conceptual relatedness based on true organization and capturing superior, nuanced information in images. Both complex and minute details of an image are transferred via skip links between the encoder and decoder models.

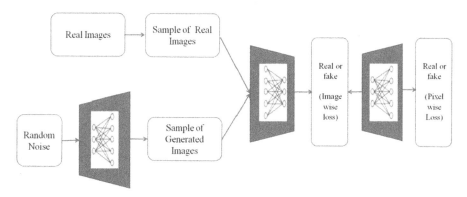

Fig. 3. Model U-net GAN

The U-net structure is added to the StyleGAN architecture created for the training. According to U-net GAN, StyleGAN's discriminator and loss algorithms were adjusted. The architecture's generator doesn't alter.

3.3   Data Augmentation in GAN Training

With insufficient training data, the image quality of GAN-generated images might substantially degrade. The discriminator may quickly overfit and stop delivering usable gradients to train the generator if it memorizes the key features from the training dataset. This results in bad-quality images and approaches collapse. Numerous studies have been conducted to use augmentation for training GAN, according to the literature [12]. The generator may create images that are both false and comparable to real photographs when conventional data augmentation is only used on real images. Unwanted distributions result in the samples that are created as a result. Instead, both real and fake images can be enhanced. This would provide a discriminator that is better at categorizing solely augmented images. Consequently, unconnected gradient flows following manipulations may prevent it from correctly identifying non-augmented generated images.

The application of differential augmentation offers a remedy for this [12]. Since all modifications made to both true and false images are differentiable, as the name suggests, gradients can proceed continuously from the discriminator to the generator. As a result, the generator continues to get precise training and the discriminator learns to recognize undamaged images from the appropriate target distribution. It is possible to differentiate between augmentations by utilizing the standard primary operations offered by deep learning frameworks.

The work by Karras et al. on various transformations that don't cause leakage in generated images can be found in [17]. They discovered that invertible transformations, such as pixel blitting, geometric, and colour transformations, have a better effect on produced images in terms of measuring metrics. These modifications are made with a nonzero probability (preferably lower than 0.8) during training to include non-augmented images as well.

3.4 Loss Functions

The current study's choice of loss function is mostly influenced by the presence of mode collapse in generated images. When the discriminator is overfitted to a small number of features in real picture distributions, mode collapse occurs. As a result, the generator often creates images that are solely appropriate for deceiving the discriminator based on those features. As a result, the generator is no longer able to make unique images. Mode collapse is more likely when there is less available data. Wasserstein loss with a gradient penalty is mostly used to address this issue (WGAN-GP) [19]. It teaches the discriminator to decrease the Wasserstein distance between the real distribution of the real samples (P_r) and the generated distribution of the produced samples (P_g). To ensure Lipschitz continuity close to the real data manifold, a consistency term is additionally attached to the WGAN-GP loss term. As described below, Wasserstein loss is applied in a non-saturating form.

Loss of the critic (discriminator):

$$L_{i \sim Pr}[D(i)] - L_{j \sim Pg}\big[D(G(j))\big] \tag{1}$$

where, P_r is distribution of real samples,

P_g is distribution of produced samples,

L_i is expected value of real inputs,

$D(i)$ is output of discriminator for real input (i),

$G(j)$ is output of generator for random noise (j),

$D(G(j))$ is discriminator output for input $G(j)$.

Generator loss:

$$L_{j \sim Pg}\big[D(G(j))\big] \tag{2}$$

where, L_j is expected value of generated data,

$D(G(j))$ is discriminator output for input $G(j)$.

The discriminator in WGAN-GP is known as the "critic" since it does not distinguish between real and fake images. A critic rates whether an image is authentic or fake. Here, the critic must adhere to 1-Lipschitz continuity to ensure that a loss assessed based on the output of the critic adheres to the Wasserstein distance metric. Lipschitz continuity is imposed by applying the gradient penalty, which is defined by the equation below, to reduce the norm of the gradients of the critic's output relative to an input to zero.

Gradient Penalty term:

$$GPT = L_{i \sim Pr, Pg}\Big[(\|\nabla_x D(i)\|_2 - 1)^2\Big] \tag{3}$$

where, L_j is expected value of generated data,
$\nabla_x D(i)$ is gradient of discriminator output of real data.
Term for consistency:

$$TFC = L_{i\sim \text{Pr}}\left[(\|\nabla_x D(x)\|_2 - 1)^2\right] \tag{4}$$

The formula for total critic loss is as follows:

$$L_{i\sim \text{Pr}}[D(i)] - L_{j\sim \text{Pg}}\left[D(G(j))\right] + \lambda * GPT + \lambda_1 * TFC \tag{5}$$

Scaling factors for the gradient penalty term and the consistency term, respectively, are in this case and 1. In the computation of critic loss, writers advise scaling the GPT term by a factor of 10 and the TFC term by 2.

4 Regularization

GAN training uses regularizing strategies to boost convergence and stability. These techniques can be classified according to how they implement network weights, gradients, and layer outputs. On the discriminator, most regularizing approaches are used. On generator weights, very few approaches, such as perceptual path length regularization, are used. The current research focuses on regularizing the discriminator to improve mode collapse and training stability. The discriminator subjected to consistency regularization to impose equivariant behaviour for differential augmentation. CutMix-augmented pictures are used to put it into effect. Crops of actual and synthetic photos are mixed to supply those graphics. Consistency loss guarantees that a discriminator prediction for a CutMix photo and a group of predictions from its impartial vegetation fluctuate little or nothing from one another. Gradient penalty terms, mentioned within the preceding phase and as utilized in loss evaluations, even have a regularizing impact with the aid of maintaining gradient beneath cohesion and using Lipschitz continuity. The generator weights' exponential weight averaging tune is saved at some point in training. These averaging weights are implemented while developing photos for augmentation. Since common weights are much less sensitive to noise and outlier iterations at some stage in training, they offer photos of better quality.

The enhancements outlined above are used in the current work on picture augmentation utilizing GAN generation to create higher-quality images. To capture pixel-level information, a discriminator from a styleGAN model is converted to a U-NET architecture. To solve the lack of training dataset availability, differential augmentation is used. In order to lessen the problem of mode collapse and produce images with more variations, a better WGAN-GP loss term is utilized. By including a consistent loss term and a gradient penalty term in loss evaluations, a regularization effect is produced. Finally, images for augmentation are produced using the generator with exponentially moving average weights. The remainder of the essay will refer to this adapted GAN architecture as improved GAN.

5   Discussion of Experiments and Results

The tyre texture image recognition dataset [20] is used for training and testing. The dataset consists of 1028 images of cracked (oxidized) and normal images tyres. The dataset is divided into training and testing images. Using a dataset on tyre joint compliance, the proposed improved GAN's applicability is assessed. Images are produced by combining cracked (oxidized) and normal images from the dataset in a number of tests. Three methods are used to supplement the data. Table 1 provides a summary of all methods used for image production. Each defect's GAN model is trained using the single noise vector approach. The style merging method is used to train both distinct GAN models for each defect and normal image as well as a single GAN model for all defective photos. Whereas the latent interpolation technique is used for training all GAN models.

For each defect category, a separate GAN model is trained in the first method. Then, independent augmented images of each defect are produced using these trained models. In the second approach, a GAN model is trained using photos from all categories. Style merging on the trained generator creates enhanced visuals. Two separate nonconforming images' latent vectors are input into the styleGAN generator at various resolutions. This method of injecting style creates visuals that shift from one defect to another. As a result, we can create a dataset that allows us to change an image's defect category. The third technique develops a new GAN model using a collection of conforming and nonconforming photos from a single category. This demonstration can be utilized to present the imperfections with which it was prepared into a standard picture by utilizing fashion combining. The momentary and third strategies of information increase, which also include inactive vector addition to move pictures from one category to another, were created using Python 3.6 and the TensorFlow 2.1.0 system. Table 1 shows the pictures made using the proposed GAN.

The Microsoft Azure Machine Learning Services are used to train all models. One NVIDIA Tesla K80 GPU is used for computation. The final image resolution is 256 x 256 pixels. The quality of the generated images is evaluated using the Fréchet inception distance (FID) [21, 22]. A classification model that has been trained to categorize images into each defect or conformity (OK) category is used to assess the efficacy of augmentation. Convolutional neural networks are the foundation of the categorization model.

The basic StyleGAN model architecture is contrasted with the improved GAN model that is proposed. Utilizing FID, their performance is assessed. It has been noticed that improved image quality and variety are correlated with lower FID scores. The identical tyre joint defect datasets are used to train both designs and the results are compared. Their difference is shown in Table 2.

The FID score for improved GAN has significantly increased when compared to the styleGAN model, according to these data. The outcomes further demonstrate the value of improved GAN in enhancing generation quality with a constrained quantity of training images. Changes made to Improved GAN's architecture and training have improved the outcomes. In order to deal with limited dataset regimes, differential augmentation and consistency regularization have been implemented. Additionally, it maintains training for improved convergence. The UNET discriminator offers pixel-by-pixel feedback that

Table 1. GAN generated images for different defects in tyres.

Clusters	GAN Image_1	GAN Image_2	GAN Image_3	GAN Image_4	GAN Image_5
Cluster_1					
Cluster_2					
Cluster_3					
Cluster_4					

Table 2. Comparison of StyleGAN's performance with that of the planned Improved GAN FID Scores GAN Architecture

	FID Scores		
GAN Architecture	Defects 1	Defects 2	Defects 3
StyleGAN	162.9	159	158.1
Proposed Improved GAN	95.6	91.9	94.8

enhances the generated image quality and lowers the FID score. By reducing outlier noisy iterations and averaging the generator weights' exponential weights, the training oscillations and FID score are further decreased.

To investigate the impact of augmentation, the classifier model is initially trained on just real photos without incorporating any GAN-generated images. For classifier model training, all tests use the same conventional augmentations, such as horizontal flip, crop, and translate. To test the classifier model, only genuine photographs that were arbitrarily chosen from the original dataset are used. 10% of the real photos are used for testing, with the remaining 90% being used for training and validation. The accuracy of the trained classifier model is compared with other augmentation studies. Precision on a test dataset is assessed and appears to be the cruellest of all test tests across all classes. Table 2 gives an outline of the results of all tests utilizing both created and genuine photographs. The results displayed here are the result of numerous classification models prepared on the same dataset to decrease instability. Accuracy on a test dataset is surveyed, and it shows up to be the cruellest of all test scores across all classes. Table 2 gives a diagram of the results of all tests utilizing both conveyed and verifiable photos. The results shown here are the result of various classification models arranged on the same dataset to decrease insecurity.

The dataset for this study was created using data from two stages of a manufacturing line. In the initial phase, 1183 samples were collected altogether. An additional 1108 samples were collected in the second stage, bringing the total to 2291. GAN models are trained using real data from the first stage, and reinforcement comes from generated images. Afterward, all genuine photographs from both stages are utilized for GAN preparation. The proficiency of the increase is surveyed freely for each set of created pictures from the two stages.

6 Conclusion

Incorporating recent innovations into GAN models for improved generated image quality is covered in the paper. Compared to styleGAN, the proposed improved GAN architecture yields significantly lower FID scores, which suggests improved image quality and generation variation. The paper discusses some architectural and training enhancements that can help GAN training converge more smoothly. Therefore, the improved GAN that has been proposed may produce a variety of images with fine features. Improved GAN is very helpful when supplementing small, unbalanced datasets. The accuracy of the picture classification tasks that follow has significantly improved thanks to an upgraded, balanced dataset. Experimental evidence from the principal components of the enhanced dataset supports the claim that generated images from the improved GAN can aid in improving the differentiation between various classification classes.

Due to limitations in computing power and processing speed, the experiments presented in this paper were only able to produce images that were 256 x 256 Pixels in size. In the case of smaller datasets, the effectiveness of picture augmentation by GAN produced images is high. Its applicability to huge datasets requires additional research. The present work's future scope includes using style-merged images for augmentation and adding GAN model enhancements with styleGAN2 architecture. For classes with lower classification recall, class wise augmentation can be explored.

References

1. Kulkarni, P., Madathil, D.: A review of echocardiographic image segmentation techniques for left ventricular study. ARPN J. Eng. Appl. Sci. **13**(10), 3536–3541 (2018)
2. Kulkarni, P., Madathil, D.: A review on echocardiographic image speckle reduction filters. Biomed. Res. **29**, 12 (2018)
3. Kulkarni, P., Madathil, D.: Adaptive thresholding method for speckle reduction of echocardiographic images. IETE J. Res. **68**(2), 1034–1042 (2022)

4. Kulkarni, P., Madathil, D.: Fully automatic segmentation of LV from echocardiography images and calculation of ejection fraction using deep learning. IJBET **40**(3), 241 (2022)
5. Kulkarni, P., Madathil, D.: Echocardiography image segmentation using semi-automatic numerical optimisation method based on wavelet decomposition thresholding. Int. J. Imaging Syst. Tech. **31**(4), 2295–2304 (2021)
6. Goodfellow, I.J., et al.: Generative adversarial networks (2014)
7. Aggarwal, A., Mittal, M., Battineni, G.: Generative adversarial network: an overview of theory and applications. Int. J. Inform. Manage. Data Insights **1**, 100004 (2021)
8. Gao, X., Deng, F., Yue, X.: Data augmentation in fault diagnosis based on the Wasserstein generative adversarial network with gradient penalty. Neurocomputing **396**, 487–494 (2020)
9. Akcay, S., Atapour-Abarghouei, A., Breckon, T.P.: GANomaly: semi-supervised anomaly detection via adversarial training. In: Jawahar, C.V., Li, H., Mori, G., Schindler, K. (eds.) ACCV 2018. LNCS, vol. 11363, pp. 622–637. Springer, Cham (2019). https://doi.org/10.1007/978-3-030-20893-6_39
10. Ma, Q., et al.: Learning to dress 3D people in generative clothing. In: IEEE/CVF Conference on Computer Vision and Pattern Recognition (CVPR), Seattle, WA, USA, pp. 6468–6477. IEEE (2020)
11. Kramberger, T., Potočnik, B.: LSUN-Stanford car dataset: enhancing large-scale car image datasets using deep learning for usage in GAN training. Appl. Sci. **10**(14), 4913 (2020)
12. Oeldorf, C., Spanakis, G.: LoGANv2: Conditional style-based logo generation with generative adversarial networks. In: 2019 18th IEEE International Conference on Machine Learning and Applications (ICMLA), Boca Raton, FL, USA, pp. 462–468. IEEE (2019)
13. Zhang, H., Zhang, Z., Odena, A., Lee, H.: Consistency regularization for generative adversarial networks. arXiv:1910.12027 (2020)
14. Mescheder, L, Geiger, A., Nowozin, S.: Which training methods for GANs do actually converge. arXiv:1801.04406 (2018)
15. Zhao, S., Liu, Z., Lin, J., Zhu, J., Han, S.: Differentiable augmentation for data-efficient GAN training. arXiv:2006.10738 (2020)
16. Sinha, A., Ayush, K., Song, J., Uzkent, B., Jin, H., Ermon, S.: Negative data augmentation. arXiv:2102.05113 (2021)
17. Karras, T., Laine, S., Aila, T.: A style-based generator architecture for generative adversarial networks. In: IEEE/CVF Conference on Computer Vision and Pattern Recognition (CVPR), pp. 4396–4405 (2019)
18. Schonfeld, E., Schiele, B., Khoreva, A.: A U-net based discriminator for generative adversarial networks. In: IEEE/CVF Conference on Computer Vision and Pattern Recognition (CVPR), pp. 8204–8213 (2020)
19. Wei, X., Gong, B., Liu, Z., Lu, W., Wang, L.: Improving the improved training of Wasserstein GANs: a consistency term and its dual effect (2018)
20. Siegel, J.: Oxidized and non-oxidized tire sidewall and tread images, Harvard Dataverse (2021)
21. Heusel, M., Ramsauer, H., Unterthiner, T., Nessler, B., Hochreiter, S.: GANs trained by a two-timescale update rule converge to a local Nash Equilibrium. In: NeurIPS Proceedings, vol. 30 (2017)
22. Kushwaha, A.K., Khatavkar, S.M., Biradar, D.M., Chougule, P.A.: Depth estimation and navigation route planning for mobile robots based on stereo camera. Lecture Notes of the Institute for Computer Sciences, Social Informatics and Telecommunications Engineering **472**, 180–191 (2023). https://doi.org/10.1007/978-3-031-28975-0_15

Unraveling the Techniques for Speaker Diarization

Ganesh Pechetti⬚, Anakapalli Rohini Durga Bhavani⬚, Abhinav Dayal$^{(\boxtimes)}$⬚, and Sreenu Ponnada⬚

Computer Science and Engineeering Department, Vishnu Institute of Technology, Bhimavaram, Andhra Pradesh, India
{21pa1a05d1,21pa1a0508,abhinav.dayal}@vishnu.edu.in

Abstract. This research paper aims to contribute to the field of speaker diarization by providing an in-depth analysis of existing audio datasets and evaluating prominent models. The study focuses on the suitability of these datasets for studying speaker diarization tasks and examines the performance of models such as pyannote-speaker diarization and NVIDIA NeMo speaker diarization. For aspiring researchers in the field, this paper serves as a solid foundation, offering valuable guidance and resources for experimentation in speaker diarization. The evaluation of the models reveals important insights. While each model has its advantages, their limitations must be considered. Overall, this research paper provides valuable insights into audio dataset analysis, model evaluation, and selection considerations for speaker diarization tasks. It equips researchers with essential knowledge to make informed decisions and lays the groundwork for further advancements in the field.

Keywords: Speaker Diarization · Segmentation · Voice Activity Detection · Pyannote · Kaldi · NeMo

1 Introduction

Speaker Diarization is the task of dividing an audio sample, which contains multiple speakers, into segments that belong to individual speakers based on their homogeneous characteristics [1]. Throughout the years, numerous speaker diarization models have been proposed, each with its distinctive approach and underlying techniques. As the demand for accurate and efficient speaker diarization systems continues to grow, it becomes essential to compare and evaluate the existing models.

The main steps involved in the speaker diarization are VAD(Voice Activity Detection), segmentation, feature extraction, clustering, and labeling. VAD identifies voice activity regions in an audio sample, while segmentation splits the large audio into smaller samples. Feature extraction techniques are applied to these smaller chunks to extract features and convert them into embeddings. Clustering techniques group the embeddings into clusters based on the extracted features. Finally, the audio sample is annotated,

P. Pareek et al. (Eds.): IC4S 2023, LNICST 536, pp. 300–308, 2024.
https://doi.org/10.1007/978-3-031-48888-7_25

assigning labels to the clusters [1]. To have a deeper understanding of these steps go through the blog at[1].

Speaker diarization is a very important step in speaker identification. Because it allows for the accurate identification by separating the speakers within audio recordings. It has a lot of applications some of them are transcription service [2] which refers to the process of converting audio recording into written text, used in forensic investigations [3], call centers [4, 5], speaker identification [6, 7] etc.

Initially, traditional clustering methods like Gaussian Mixture Models (GMMs) [8] and Hidden Markov Models (HMMs) [9] were used for audio sample diarization. Subsequently, the Bayesian Information Criterion(BIC) [10]. Later, an i-vector [11] based method was proposed. The field experienced a breakthrough with the application of AI and Deep Learning networks, leading to the proposal of revolutionary architectures for diarization. Now-a-days, the majority of models with lower Diarization Error Rate(DER) make use of Deep Neural Network (DNN) based architectures.

In this study, we focused on exploring various models for speaker diarization and we selected some models which have less error rate namely DER. With these models we try to compare their performance. Later, we evaluated and compared the selected models based on their ability to accurately separate speakers.

The key contributions of this research work are as follows. Firstly, an in-depth analysis of existing audio datasets is provided, specifically focusing on their suitability for studying the speaker diarization task. Secondly, a comprehensive examination and evaluation of prominent models, including pyannote-speaker diarization and NVIDIA NeMo speaker diarization, are conducted. Moreover, for aspiring researchers embarking on speaker diarization research, this paper serves as a solid foundation, offering guidance and resources for experimentation in this domain. By consolidating knowledge and highlighting relevant tools and methodologies, this work facilitates a smoother initiation and exploration of speaker diarization research endeavors.

2 Related Work

This section introduces the existing mechanisms to perform speaker diarization. Speaker diarization can be performed with the help of i-vectors [11], x-vectors [12] and through deep neural networks [13]. With the advent of deep learning technology, significant advancements have been made in the field of speaker diarization. These advancements have propelled the development of more advanced techniques and methodologies.

There have been various methods developed over time for different steps in diarization which includes VAD, segmentation, feature extraction, and clustering etc. Traditional approaches like GMMs [8] and HMMs [9] which are basically clustering models struggled with handling overlapped speech, leading to inaccurate segmentation and speaker assignment. They also have limitations in capturing the full range of speaker characteristics. Landini et al., [10] proposed BIC clustering was introduced to determine the optimal number of clusters, but it also faced challenges with overlapping speech. Subsequently, Wang et al., [11] proposed i-vectors and Kim et al., [12] x-vectors

---

[1] https://medium.com/@21pa1a05d1/speaker-diarization-fec87f839f52.

which are used for extracting the features from an audio sample, were introduced as low-dimensional representations using neural networks. Garcia-Romero et al., [13] proposed deep neural networks embeddings. However, extracting i-vectors and x-vectors requires significant computational power. Nowadays, advanced methods such as neural speaker segmentation, multi-modal approaches, and deep neural networks techniques for VAD and Multi-Scale Diarization Decoder (MSDD) [14] are being used.

3   Methodology

This section provides the methods adopted to carry out the study. The architecture used to perform this study is presented through Fig. 1. It provides insights on the adopted datasets, models, pre-processing approaches used and performance evaluation.

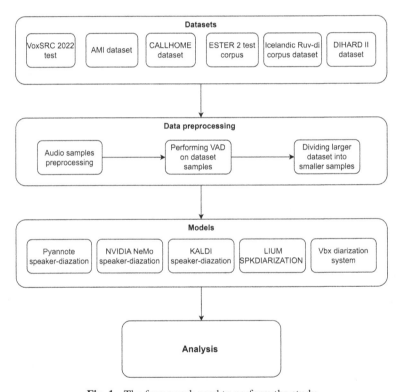

Fig. 1. The framework used to perform the study

The first block of the Fig. 1 showcases the datasets that are used by most of the researchers. The second block includes the preprocessing steps we performed on the datasets. Next block includes the various models used to evaluate the performance of the speaker diarization task on the datasets and the final block includes the analysis section where we compare different models.

In our study, we conducted preprocessing steps to enhance the quality of our data. Firstly, we performed noise removal to eliminate unwanted noise from the dataset [14]. Additionally, we applied VAD [16] to identify and remove silent portions of the audio samples. Subsequently, we segmented the audio samples to create segments that are suitable for the diarization process.

To evaluate the performance of a model the commonly used metrics are DER [1] and JER(Jaccard Error Rate) [1]. In our study, we have chosen to focus on DER as it is the preferred metric used by the majority of researchers in the diarization field. DER is the sum of speaker error, false alarm speech, missed speech.

$$DER = \frac{SER + FA + Miss}{Total\_speech} \tag{1}$$

3.1 Corpora

VoxSRC-2022 Test Dataset: The VoxSRC-2022 dataset[2], created for the VoxCeleb Speaker Recognition Challenge in 2022, is a valuable collection of speech obtained 'in the wild.' It features audio from diverse sources like celebrity interviews, news shows, talk shows, and debates, representing real-world scenarios. The dataset includes professionally edited videos and casual conversational audio, offering a wide range of speech styles and acoustic conditions.

With 5,994 speakers and 1,092,009 utterances, it provides a substantial amount of data for analysis and model training. The dataset's inclusion of background noise, laughter, and other natural artifacts adds realism to the evaluation of speaker recognition methods. Researchers can leverage this dataset to enhance the performance and reliability of speaker recognition systems, making them more applicable in practical applications.

Icelandic Ruv-di Corpus Dataset: The Icelandic Ruv-di corpus dataset[3] consists of speech data sourced primarily from the Icelandic national broadcasting service, RÚV. It offers authentic recordings from various programs, interviews, and news broadcasts, making it valuable for speech and language processing tasks in Icelandic. The dataset captures natural speech patterns, accents, and styles and represents variation in these characteristics.

Depending on the release, it may include annotations like transcriptions or speaker identities. The dataset's focus on Icelandic ensures targeted solutions for language-specific challenges and its availability facilitates research and enables the development of speech processing models tailored to Icelandic. In summary, the Icelandic Ruv-di corpus dataset offers authentic and diverse speech data focused on Icelandic with the potential for annotations.

CALLHOME Dataset: The CALLHOME American English Speech dataset[4] was developed by the Linguistic Data Consortium (LDC). It consists of 120 unscripted 30-min telephone conversations between native speakers of English language. This means

---

[2] http://mm.kaist.ac.kr/datasets/voxceleb/voxsrc/competition2022.html.

[3] https://clarin.is/en/resources/j_ruv/.

[4] https://catalog.ldc.upenn.edu/LDC97S42.

that there are a total of 240 speakers in the dataset, with two speakers in each audio recording. The total duration of the data in the dataset is 60 h. However, it only contains telephone conversations between native speakers of English and was collected in the 1990s, which may limit its applicability to research on other languages or current speech patterns. Despite these limitations, the availability of this dataset for research purposes can facilitate the development and evaluation of speech processing technologies.

AMI Dataset: The AMI Meeting Corpus is a dataset[5] that combines multiple modes of data and comprises 100 h of recorded meetings. It includes both elicited and naturally occurring meetings and provides a rich source of data for research. The dataset consists of 171 meetings recorded at 4 locations, with each meeting having 4–5 speakers. However, the dataset primarily consists of recordings from controlled meeting environments and may not fully capture the diversity of real-world multi-party interactions. Additionally, the scale of the dataset may be relatively small compared to other speech or audio datasets and the annotation process can be resource-intensive. Despite these limitations, the AMI Meeting Corpus is a valuable resource for research on multi-party interactions.

DIHARD II Dataset: The DIHARD II dataset[6] is designed for evaluating speaker diarization systems. It includes diverse audio recordings from various sources and provides manual annotations for ground truth evaluation. With approximately 144 unique speakers and around 44 h of audio data, it serves as a valuable benchmark for advancing speaker diarization technology. The dataset offers advantages such as its evaluation focus, varied acoustic conditions, diverse data sources, and established evaluation metrics. However, using the dataset may require significant computational resources, updates may be infrequent, and there could be copyright or licensing restrictions.

ESTER 2 Test Corpus: The ESTER 2 test corpus dataset[7], designed for evaluating French speech transcription systems, consists of approximately 150 h of audio data. This substantial amount of data enhances the dataset's utility for comprehensive training and evaluation of automatic speech recognition (ASR) models. With its diverse audio sources, manual transcriptions, evaluation metrics, and language-specific focus, the ESTER 2 test corpus serves as a valuable resource for advancing ASR technology in the context of the French language.

3.2 Models

Pyannote Diarization Model: The pyannote.audio provides a neural speaker diarization pipeline, which is available through Hugging Face. The pipeline contains neural speaker segmentation which is a method for automatically detecting speaker changes in an audio recording using a neural network. The neural network is trained to analyze the acoustic characteristics of the speech signal and to identify points in time where the speaker changes. This can be done by sliding a fixed-length window over the speech signal and predicting, for each window, whether it contains a speaker change or not. The

---

[5] https://groups.inf.ed.ac.uk/ami/download/

[6] https://dihardchallenge.github.io/dihard2/.

[7] https://catalogue.elra.info/en-us/repository/browse/ELRA-S0338/.

output of the neural network can then be post-processed to obtain a final segmentation of the audio recording into speaker-homogeneous segments, the SpeechBrain implementation of the ECAPA-TDNN model for extracting feature embeddings. Agglomerative hierarchical clustering is used for clustering embeddings.

It reaches a DER = 5.6% on VoxSRC 2022 test dataset [17].

Nvidia NeMo: NVIDIA NeMo's speaker diarization system consists of several modules: a Voice Activity Detector (VAD) model namely MarbleNet model which is a deep 1D neural network, detects the presence or absence of speech to generate timestamps for speech activity from the given audio recording; a Speaker Embeddings model namely TitaNet, which extracts speaker embeddings on speech segments obtained from VAD time stamps; and a Multi-Scale Diarization Decoder (MSDD), which is a speaker diarization model based on initializing clustering and multi-scale segmentation input. This model has two significant improvements to enhance diarization importance of each scale at each step.

It reaches a DER = 3.92% dataset and DER = 1.05% on CALLHOME and AMI datasets [14].

Kaldi: Kaldi is an open-source toolkit for speech recognition that includes support for speaker diarization and also consists of models for x-vectors. Kaldi's speaker diarization system uses x-vectors, a type of speaker embedding, to represent speech segments. The x-vector extractor is a Time Delay Neural Network (TDNN) that is trained on a large amount of labeled speech data to learn a mapping from speech segments to a fixed-dimensional embedding space. The extracted x-vectors are then used in combination with clustering algorithms such as Agglomerative Hierarchical Clustering (AHC) to group speech segments by speaker. The kaldi model used x-vectors, MFCCS, PLDA trained on Althingi Parliamentary Speech corpus.

It reaches a DER = 26.27% on Icelandic Ruv-di corpus dataset [18].

LIUM SPKDIARIZATION: LIUM SpkDiarization is an open-source toolkit for speaker diarization developed by the LIUM (Le Mans University). It inlcudes a fullset of tools that facilitates the creation of an entire speaker diarization system, starting from the audio signal and progressing towards speaker clustering using CLR/NCLR metrics. These toolset encompasses MFCC computation, speech/non-speech detection, and various speaker diarization methods. The LIUM_SPKDIARIZATION model uses hierarchical agglomerative clustering methods using measures such as BIC and CLR(Classification Likelihood Ration).

It reaches a DER = 10.01% on ESTER 2 test corpus [19].

VBX: VBx is a recently proposed speaker diarization method that uses a Bayesian Hidden Markov Model (BHMM) to cluster x-vectors, which are fixed-dimensional representations of variable-length speech segments. The VBx method utilizes a BHMM to group x-vectors and identify speaker clusters within a sequence of x-vectors. The VBx model applies the same BHMM approach to detect speaker clusters in a sequence of x-vectors.

It reaches a diarization error rate of DER = 21.77% on CALLHOME dataset and DER = 18.55 on DIHARD II dataset [10].

4   Results and Discussion

Our study adopted six different corpora and compared them based on several parameters namely the number of hours of available data, the number of speakers, language and pricing type. The results of this comparison are tabulated in Table 1. Among the corpora, the AMI and ESTER 2 datasets contained the largest amount of audio data in terms of hours. For scenarios requiring a high number of speakers, the Icelandic Ruv-di corpus emerged as a suitable choice.

Fortunately, there are also freely available datasets that can be utilized to initiate diarization research. These include the VoxSRC-2022 test dataset, the Icelandic Ruv-di corpus, and the AMI dataset. These datasets were sourced from various multimedia platforms such as YouTube, radio broadcasts, news shows, debates, and celebrity interviews. Additionally, some of these datasets encompass telephonic recordings, such as the CALLHOME datasets, while others focus on meeting recordings, such as the AMI dataset.

Table 1. Comparison of datasets

S. No	Name of the dataset	No of hours	No of speakers	Style	Language	Pricing type
1	VoxSrc-2022 test	50	4–6	Celebrity, interviews, news, shows, talk, debates	English	Free
2	Icelandic Ruv-di corpus	46 (min)	20	Programs, Interviews, News broadcasts	Icelandic	Free
3	CALLHOME	60	2	Telephone conversations	English	Paid
4	AMI	100	4–5	Meeting recordings	English	Free
5	DIHARD II	44	1–10	YouTube, court rooms, meetings	English	Paid
6	ESTER 2 test corpus	150	1–3	Radio broadcast	French	Paid

Our study adopted five different speaker diarization models, and their performance was evaluated on various datasets. The results, including the DER, are presented in Table 2. The models selected for our study were Pyannote-speaker diarization, NVIDIA NeMo speaker diarization, Kaldi speaker diarization, LIUM speaker diarization, and VBX speaker diarization.

Table 2. Comparison of different models.

S.No	Name of the model	Dataset	DER%
1	Pyannote-speaker diarization	VoxSRC 2022 test	5.6
2	Nvidia Nemo speaker diarization	CALLHOME, AMI	3.92, 1.05
3	Kaldi speaker diarization	Icelandic Ruv-di corpus	26.27
4	LIUM speaker diarization	ESTER 2 test corpus	10.01
5	Vbx speaker diarization	CALLHOME, DIHARD II	21.77, 18.55

Among these models, Pyannote-speaker diarization demonstrated better performance with a DER of 5.6, while NVIDIA NeMo achieved a DER of approximately 3.92. In some of the rows two dataset names are provided in the dataset column, along with two corresponding DER values in the DER column. This signifies that the first dataset name corresponds to the first DER value, while the second dataset name corresponds to the second DER value.

While each model has its unique advantages, it is important to consider their limitations as well. If the primary goal is to minimize the DER and time is not a limiting factor, the pyannote diarization model can be a suitable choice. However, it should be noted that this model might require more time to generate the output. The NVIDIA NeMo speaker diarization model is designed to work with specific audio requirements, including bitrate, duration, and other properties. It is crucial to ensure that the original audio sample meets these specifications in order to effectively utilize the NVIDIA NeMo model. When the properties are matched, selecting the NVIDIA NeMo model can lead to lower DER compared to the pyannote-diarization model. Additionally, the NVIDIA model exhibits faster processing times compared to pyannote-diarization, offering a more efficient solution for speaker diarization tasks. We also mentioned some other models which are good to use but have more DER.

5   Conclusion

In addition to the models mentioned earlier, there are other models available for speaker diarization, although they may have a higher DER compared to the previously discussed options. It is important to consider that the DER changes depending on the dataset used. Therefore, it is crucial to select a model that best fits the dataset is very important. The choice of model should be based on evaluation of its performance for the given dataset, considering factors such as accuracy, efficiency, and compatibility. This ensures that the selected model aligns well with the dataset and maximizes the effectiveness of speaker diarization outcomes.

References

1. Tae Jin, P., et al.: A review of speaker diarization: recent advances with deep learning. Comput. Speech Lang. **72**, 101317 (2022)

2. Claude, B., et al.: Multistage speaker diarization of broadcast news. IEEE Trans. Audio Speech Lang. Process. **14**(5), 1505–1512 (2006)
3. Joyanta, B., et al.: An overview of speaker diarization: approaches, resources, and challenges. In: 2016 Conference of The Oriental Chapter of International Committee for Coordination and Standardization of Speech Databases and Assessment Techniques (O-COCOSDA). IEEE (2016)
4. Zajíc, Z., Kunešová, M., Müller, L.: Applying EEND diarization to telephone recordings from a call center. In: Karpov, A., Potapova, R. (eds.) SPECOM 2021. LNCS (LNAI), vol. 12997, pp. 807–817. Springer, Cham (2021). https://doi.org/10.1007/978-3-030-87802-3_72
5. Rosenberg, Aaron E., et al.: Unsupervised speaker segmentation of telephone conversations. In: INTERSPEECH (2002)
6. Aleksandar, M., Gerazov, B., Ivanovski, Z.: Delay based optimization of an integrated online call recording speaker diarisation and identification system. In: IEEE EUROCON 2017–17th International Conference on Smart Technologies. IEEE (2017)
7. Lakshmana Rao, A., Bonthu, S., Dayal, A.: Multiclass spoken language identification for Indian Languages using deep learning. In: 2020 IEEE Bombay Section Signature Conference (IBSSC). IEEE (2020)
8. Tantan, L., Liu, X., Yan,Y.: Speaker Diarization System Based on GMM and BIC. In: International Symposium on Chinese Spoken Language Processing, Singapore (2006)
9. Jeremy, H.M.W., Xiao, X., Gong, Y.: Hidden markov model diarisation with speaker location information. In: ICASSP 2021–2021 IEEE International Conference on Acoustics, Speech and Signal Processing (ICASSP). IEEE (2021)
10. Federico, L., et al.: Bayesian hmm clustering of x-vector sequences (VBx) in speaker diarization: theory, implementation, and analysis on standard tasks. Comput. Speech Lang. **71**, 101254 (2022)
11. Wei, W., et al.: I-vector features and deep neural network modeling for language recognition. Procedia Comput. Sci. **147**, 36–43 (2019)
12. Myungjong, K., Apsingekar, V.R., Neelagiri, D.: X-Vectors with Multi-Scale Aggregation for Speaker Diarization. arXiv preprint arXiv:2105.07367 (2021)
13. Garcia-Romero, D., et al.: Speaker diarization using deep neural network embeddings. In: 2017 IEEE International Conference on Acoustics, Speech, and Signal Processing (ICASSP). IEEE (2017)
14. Park, T.J., et al.: Multi-scale speaker diarization with dynamic scale weighting. arXiv preprint arXiv:2203.15974 (2022)
15. Fu, S.-W., et al.: MetricGAN+: an improved version of metricgan for speech enhancement. arXiv preprint arXiv:2104.03538 (2021)
16. Hao, Z., Deming, L.: Research of voice activity detection algorithm. In: 2011 International Conference on Computational and Information Sciences. IEEE (2011)
17. Bredin, Hervé: pyannote. audio speaker diarization pipeline at VoxSRC 2022
18. Fong, J.Y., Gudnason, J.: RÚV-DI Speaker Diarization (20.09) (2020)
19. Sylvain, M., Merlin, T.: LIUM SpkDiarization: an open source toolkit for diarization. In: CMU SPUD Workshop (2010)

LUT-Based Area-Optimized Accurate Multiplier Design for Signal Processing Applications

B. V. V. Satyanarayana⬥, B. Kanaka Sri Lakshmi, G. Prasanna Kumar$^{(\boxtimes)}$⬥, and K. Srinivas

Department of ECE, Vishnu Institute of Technology, Bhimavaram 534202, India
godiprasanna@gmail.com

Abstract. Multipliers play a role in various aspects of smart cities, which can be used in many applications like Traffic management, energy management and environmental management etc. The wide variety of applications of multipliers are in the field of signal processing and image processing. FPGA design of multiplier is one of the complex tasks in Digital electronics. Most of the designs uses DSP blocks, these multipliers are complex and occupies much area in FPGA. Accurate multiplier design with low area on FPGA is the challenging task. The proposed method is accurate multiplier design, which is designed only using lookup table (LUT). The proposed design has low power and reduced area because of using simple LUT's for generating partial product. The proposed accurate multipliers reduce 10% less Hardware on vertex 7 FPGA compared to existing designs.

Keywords: Multiplier · FPGA · LUT · Power · vertex 7

1 Introduction

In arithmetic operators' multiplier is one basic operator, which is very important and crucial. In the digital era multiplier design has challenges in accuracy and complexity. FPGA implementation of multipliers has two major challenges, one is hardware complexity and another is accuracy. In some applications in the area of image processing and signal processing has that much significance of accuracy. The approximate multipliers are best suited for such applications. This work majorly focuses on the accuracy, no approximation techniques are used.

Many applications in the field of signal processing [1] and image processing requires high speed and efficient sub components. Those are mainly adders and multipliers. In [2] Fast Fourier transform requires high speed and accurate twiddle factor multiplier for generation of higher order FFT. Many applications that's example presented in [3] gives the counter on FPGA, which gives the detecting and counting in industries.

In [4] authors proposed a fast multiplier, in which parallel counter algorithm is introduced to compress the columns. The design is efficient with respect to LUT utilization and DSP blocks, but utilization of high-power resources is the one issue found in this work. In [5] efficient soft-core multiplier on FPGA vertex-6, in which compressed tree

© ICST Institute for Computer Sciences, Social Informatics and Telecommunications Engineering 2024
Published by Springer Nature Switzerland AG 2024. All Rights Reserved
P. Pareek et al. (Eds.): IC4S 2023, LNICST 536, pp. 309–317, 2024.
https://doi.org/10.1007/978-3-031-48888-7_26

algorithm is used instead of partial products. This structure has complex and more routing delay compared to existing structures. The main advantage of this design is fast carry chain and simultaneous multiplication and accumulation. In [6] authors present 16-bi multiplier on FPGA Spartan 3 which is consumes more LUTs and dsp blocks. In [7] large multipliers was designed on FPGA which uses less resources, the main components in this design are DSP blocks and small multipliers. The integer linear programming problem was solved using this method. The main findings in this method are usage more DSP blocks.

In multiplier design Power and area are major parameters, high level components like DSP blocks in FPGA uses more power and area. In [8] area and power optimized multiplier is designed proposed which is approximate multipliers based on wallac tree algorithm. In the process of reducing the area, in this work compromising the accuracy. This work suitable for applications where accuracy is not a big issue, example in image processing applications. In [9] efficient multiplier approximation is proposed in which soft logic is introduced to implement the multiplier. The high-level performing resources has limitation in size and quantity. This work majorly focused on performance analysis on embedded multiplier and soft-core multiplier and clearly address how to overcome the problem using soft core techniques. Hybrid multiplier combining dadda algorithm and booth [10], which is superior than the traditional booth multiplier. There is no significant reduction in high level hardware, but speed is the main advantage in this method.

The performance analysis of Baugh Wooley multiplier is presented in [11], in which parallel architecture is used to produce high speed output. Efficiency in the design of adders is also one significant modification in this method. The improved speed and low power are main advantage in this method. This multiplier is very popular and widely used in various fields of applications.

In [12] proposed a new multiplier design for signed and unsigned data, which is very optimizes power and delay, in [13,14] multiplier design techniques are based on LUT only. In [15] default multiplier in Xilinx vivado version.

The proposed multiplier is a LUT based multiplier which optimizes the area because of no DSP blocks use and efficient carry chain generation using LUT.

The paper organized as follows Sect. 1 gives the detailed introduction and literature review, Sect. 2 describes the proposed method, Sect. 3 gives the simulation results and comparisons of proposed method with existing designs. Section 4 concludes the project with future scope.

2   Proposed Method

The literature review of existing methods gives the following identifications, most of the accurate multipliers uses the high-level hardware components on FPGA, approximate multiplier is the solution for reducing the hardware complexity and utilizing the resources efficiently. The proposed method is accurate multiplier design using efficient resource utilization. The general structure for multiplication is shown in Fig. 1. In which M is the multiplicand and N is the multiplier.

The proposed method uses two types of LUT's for partial product generation, one is Type-1 and another is Type-2 these LUT's generates the partial products and accumulation of two stages. The Type-1 LUT is LUT6_2 in Xilinx FPGA vertex-7 series.

			M_3N_0	M_2N_0	M_1N_0	M_0N_0
		M_3N_1	M_2N_1	M_1N_1	M_0N_1	
	M_3N_2	M_2N_2	M_1N_2	M_0N_2		
M_3N_3	M_2N_3	M_1N_3	M_0N_3			

Fig. 1. Binary multiplier of size 4 bit

It is a 6 input and 2 output LUT. The Type-1 LUT is shown in Fig. 2, carry (C) and sum (S) are the output ports of this LUT, which are O5 and O6 ports in LUT6_2. The detailed configuration is shown in Table 1, the logical and gates performs the partial products and the resultant outputs C and S outputs are generated by O5 and O6 ports of LUT. The INIT attribute defines the functioning of LUT, it is 64-bit hexadecimal value in which least 32 bits defines the carry O5 and the remaining sum O6. The Verilog INIT value of Type-1 LUT 64'h7888788880008000 will produce our desired output of O6 = 7888 and O5 = 8000. The functionality of Type-2 LUT is shown in Fig. 3 and the configuration is shown Table 2. The accurate multiplier was designed using Type-1 and Type-2 multiplier.

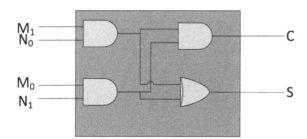

Fig. 2. Type-1 LUT structure

The main blocks in accurate multiplier design is partial product generation (PPG). The PPG for 4x4 accurate multiplier is shown in Fig. 4, generally for a 4x4 multiplier there are 4 stages of products are generated as shown in Fig. 1. Each Type-1 LUT generates two products and accumulation of stage-2 output. So that the proposed design requires two layers of four LUT's to generate all the partial products. Initially accumulation not required so Type -2 LUT used. The output of each stage is given to carry chain module to propagate the carry. The accumulated result of partial products of first two stages is p00 to p04 and for last two stages is p01 to p14. The final result is obtained from PPG unit using full adders and half adders.

The final product generation of the proposed multiplier is done by using LUT based slice adder which was shown in Fig. 5. The final output bits of multiplier is

Table 1. Configuration of Type-1 LUT

M1	N0	M0	N1	M1 N0	M0 N1	S(O6)	C(O5)
0	0	0	0	0	0	0	0
0	0	0	1	0	0	0	0
0	0	1	0	0	0	0	0
0	0	1	1	0	1	1	0
0	1	0	0	0	0	0	0
0	1	0	1	0	0	0	0
0	1	1	0	0	0	0	0
0	1	1	1	0	1	1	0
1	0	0	0	0	0	0	0
1	0	0	1	0	0	0	0
1	0	1	0	0	0	0	0
1	0	1	1	0	1	1	0
1	1	0	0	1	0	1	0
1	1	0	1	1	0	1	0
1	1	1	0	1	0	1	0
1	1	1	1	1	1	0	1

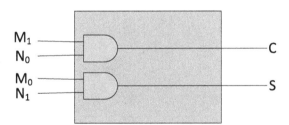

Fig. 3. Type-2 LUT structure

Table 2. Configuration of Type-2 LUT

M_1	N_0	M_0	N_1	$M_1 N_0$	$M_0 N_1$	S	C
0	0	0	0	0	0	0	0
0	0	0	1	0	0	0	0
0	0	1	0	0	0	0	0

(*continued*)

Table 2. (*continued*)

M_1	N_0	M_0	N_1	$M_1 N_0$	$M_0 N_1$	S	C
0	0	1	1	0	1	0	1
0	1	0	0	0	0	0	0
0	1	0	1	0	0	0	0
0	1	1	0	0	0	0	0
0	1	1	1	0	1	0	1
1	0	0	0	0	0	0	0
1	0	0	1	0	0	0	0
1	0	1	0	0	0	0	0
1	0	1	1	0	1	0	1
1	1	0	0	1	0	1	0
1	1	0	1	1	0	1	0
1	1	1	0	1	0	1	0
1	1	1	1	1	1	1	1

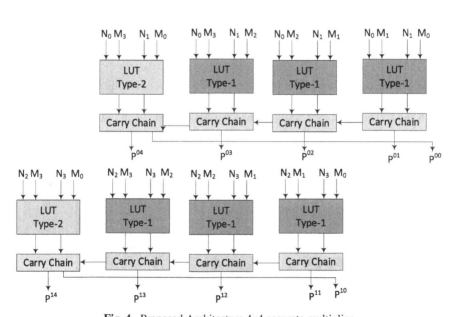

Fig. 4. Proposed Architecture 4x4 accurate multiplier

$P^6, P^5, P^4, P^3, P^2, P^1$ and P^0 in which P^6, P^5, P^1 and P^0 are directly obtained from the outputs of two stage adders but P^4, P^3 and P^2 requires additional circuit that is slice adder. The LUT takes the inputs from two stage adders and gives the final output based on the carry. The carry inclusion can be done by using exor gate and multiplexer.

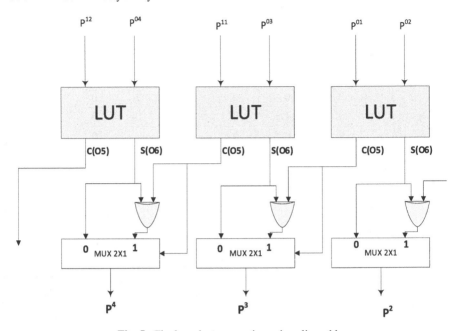

Fig. 5. Final product generation using slice adder

3   Results and Discussions

The proposed accurate multiplier is designed in Verilog HDL coding and synthesized in vertex-7 7v300T device. The proposed design was implemented in various bit lengths of 4, 6 and 8. The simulation results of 6x6 multiplier is shown in Fig. 6 and Fig. 7. The RTL schematic is shown in Fig. 8 and Fig. 9, where A and B are the inputs and C is the output.

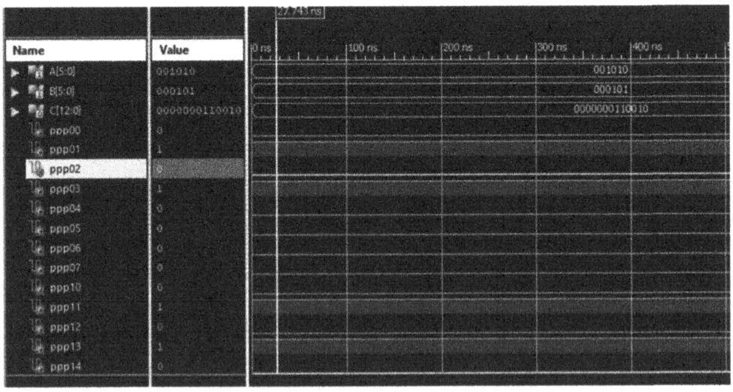

Fig. 6. Simulation result of Accurate multiplier Result in binary

The RTL schematic of Type-1 and Type-2 LUT is shown in Fig. 9.

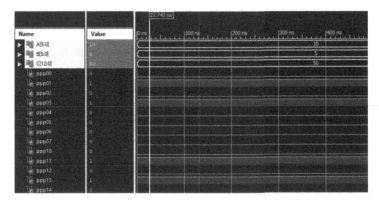

Fig. 7. Simulation result of Accurate multiplier Result in decimal

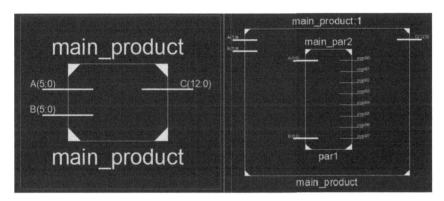

Fig. 8. RTL Schematic of main module with PPG unit

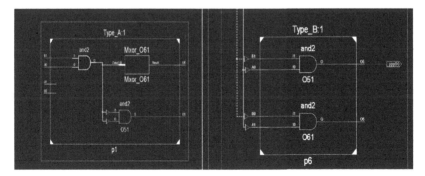

Fig. 9. RTL of Type-1 and Type-2 LUT

The performance of proposed method is evaluated by taking the parameters as number of LUT's, Delay and Power. The detailed comparison of 4x4 and 8x8 multiplier is presented in Table 3 and Table 4 respectively. The proposed method has significant reduction in LUT compared to [13] and [14]. When power is concerned which is lower than the [15] that is predefined multiplier in xilinx.

The following are the significant advantages of proposed method.

- Number of DSP blocks are zero
- Number of LUTs are reduced than predefined multiplier
- Low power than predefined multiplier
- Delay is high

Table 3. Performance analysis of 4x4 accurate multiplier

Design	LUT	Delay (ns)	Power(pJ)
[13]	18	1.65	1.13
[14]	14	2.59	1.31
[15]	18	2.91	2.25
Proposed	14	2.71	1.14

Table 4. Performance analysis of 8x8 accurate multiplier

Design	LUT	Delay (ns)	Power(pJ)
[13]	66	2.8	6.06
[14]	54	4.5	7.26
[15]	88	3.48	9.07
Proposed	48	4.7	6.82

4 Conclusion

This paper mainly focuses on the accurate multiplier design with low area. The following conclusions were made after detailed analyses of proposed accurate multiplier. The proposed accurate multiplier was designed with only LUT's without using DSP blocks on FPGA. The area and power of the proposed design is reduced compared to existing designs. This accurate multiplier best fit for applications in signal and image processing. In future this work can be extended to application level in image processing for convolution operation.

References

1. Bhagyasri, G., Prasannakumar, G., Murthy, P.S.N.: Underwater image enhancement using SWT based image fusion and colour correction. In: 2019 International Conference on Intelligent Computing and Control Systems (ICCS), IEEE, pp. 749–754 (2019)

2. Godi, P.K., Krishna, B.T., Kotipalli, P.: Design optimisation of multiplier-free parallel pipelined FFT on field programmable gate array. IET Circ. Devices Syst. **14**, 995–1000 (2020)
3. Arvapally, S., Khan, M.A., Vemula, P.C., Sonnaila, P. Chary, U.G.: FPGA implementation of industry automated bottle counter. In: 2017 International conference of Electronics, Communication and Aerospace Technology (ICECA), vol. 2, IEEE, pp. 461–465 (2017)
4. Kakacak, A., Guzel, A.E., Cihangir, O., Gören, S., Ugurdag, H.F.: Fast multiplier generator for FPGAs with LUT based partial product generation and column/row compression. Integration **57**, 147–157 (2017)
5. Kumm, M., Abbas, S., Zipf, P.: An efficient softcore multiplier architecture for Xilinx FPGAs. In: 2015 IEEE 22nd Symposium on Computer Arithmetic, IEEE, pp. 18–25 (2015)
6. Ram, G.C., Subbarao, M.V., Kumar, D.G., Terlapu, S.K.: FPGA implementation of 16-Bit wallace multiplier using HCA. In: Advances in Micro-Electronics, Embedded Systems and IoT: Proceedings of Sixth International Conference on Microelectronics, Electromagnetics and Telecommunications (ICMEET 2021), Vol. 1, pp. 419–427 (2022)
7. Kumm, M., Kappauf, J., Istoan, M., Zipf, P.: Resource optimal design of large multipliers for FPGAs. In: 2017 IEEE 24th Symposium on Computer Arithmetic (ARITH), IEEE, pp. 131–138 (2017)
8. Bhardwaj, K., Mane, P.S., Henkel, J.: Power-and area-efficient approximate wallace tree multiplier for error-resilient systems. In: Fifteenth international symposium on quality electronic design, IEEE, pp. 263–269 (2014)
9. Parandeh-Afshar, H., Ienne, P.: Measuring and reducing the performance gap between embedded and soft multipliers on FPGAs. In: 2011 21st International Conference on Field Programmable Logic and Applications, IEEE, pp. 225–231 (2011)
10. Millar, B., Madrid, P.E., Swartzlander, E.E.: A fast hybrid multiplier combining Booth and Wallace/Dadda algorithms. In: [1992] Proceedings of the 35th Midwest Symposium on Circuits and Systems, IEEE, pp.158–165 (1992)
11. Biradar, V.B., Vishwas, P.G., Chetan, C.S., Premananda, B.S.: Design and performance analysis of modified unsigned braun and signed Baugh-Wooley multiplier. In: 2017 International Conference on Electrical, Electronics, Communication, Computer, and Optimization Techniques (ICEECCOT), IEEE, pp. 1–6 (2017)
12. Ullah, S., Schmidl, H., Sahoo, S.S., Rehman, S., Kumar, A.: Area-optimized accurate and approximate softcore signed multiplier architectures. IEEE Trans. Comput. **70**, 384–392 (2020)
13. Ullah, S., Nguyen, T.D.A., Kumar, A.: Energy-efficient low-latency signed multiplier for FPGA-based hardware accelerators. IEEE Embed. Syst. Lett. **13**, 41–44 (2020)
14. Ullah, S., Rehman, S., Shafique, M., Kumar, A.: High-performance accurate and approximate multipliers for FPGA-based hardware accelerators. IEEE Trans. Comput. Des. Integr. Circ. Syst. **41**, 211–224 (2021)
15. LogiCORE, I.P.: LogiCORE IP V12. 0. San Jose, CA, USA Xilinx (2015)

Speaker Recognition Using Convolutional Autoencoder in Mismatch Condition with Small Dataset in Noisy Background

Arundhati Niwatkar[1], Yuvraj Kanse[2], and Ajay Kumar Kushwaha[3]([⊠])

[1] Sivaji University, Kolhapur, Maharashtra, India
[2] Karmaveer Bhaurao Patil College of Engineering, Satara, Maharashtra, India
[3] Bharati Vidyapeeth (Deemed to be University) College of Engineering, Pune, India
akkushwaha@bvucoep.edu.in

Abstract. The objective of this paper is to increase the success rate and accuracy of speaker recognition and identification systems through the proposal of a novel approach. Data augmentation techniques have been employed to enhance a small dataset comprising audio recordings from five speakers, encompassing both male and female voices. The Python programming language is used for data processing. The chosen model is a convolutional autoencoder. In order to convert the speech signal into an image, their respective spectrograms have been used. Consequently, a set of images serves as the input for training the autoencoder. A speaker recognition and identification system are developed using the convolutional autoencoder, a deep learning technique. A comparative analysis is conducted of the results against traditional systems reliant on the MFCC feature extraction technique. The proposed system exhibits a high success rate, indicating its efficacy in accurately recognising and identifying speakers. To account for a "mismatch condition," different time durations of the audio signal are utilised during both the training and testing phases. Through a series of experiments involving various activation and loss functions in permutation and combination, the optimal pair for the small dataset is successfully identified, yielding favorable outcomes. In matched conditions, this system has achieved 92.4% accuracy rate.

Keywords: convolutional autoencoder · deep-learning · speaker recognition · MFCC · mismatch condition

1 Introduction

Over the past few years, research scholars have been making significant progress in the field of speaker recognition and identification [1]. This system aims to identify speakers based on their voices and may be divided into two groups: text-dependent and text-independent [2]. Speaker recognition involves determining which trained speech sample best matches the voice of a speaker, and it serves as a means of verifying or denying a speaker's claimed identity. Several traditional systems have been employed

P. Pareek et al. (Eds.): IC4S 2023, LNICST 536, pp. 318–330, 2024.
https://doi.org/10.1007/978-3-031-48888-7_27

for speaker recognition, such as GMM (Gaussian Mixture Model), i-vectors, and HMM (Hidden Markov Models) [3]. A Gaussian Mixture Model is a probabilistic model that represents the weighted sum of Gaussian mixtures. Historically, the Gaussian Mixture Model has demonstrated great success in creating accurate speaker recognition models. However, modern speaker recognition systems increasingly recommend the use of deep learning approaches. Many researchers are actively exploring this approach to develop precise speaker recognition systems. The convolutional neural network (CNN) has been increasing in prominence among deep learning approaches [4]. Designing a speaker recognition system can involve several challenges and hurdles, such as variability in speech signals, limited training data, computational complexity, and adverse recording conditions. Overcoming these hurdles requires a combination of robust algorithms, large and diverse datasets, careful system design, and continuous improvement based on feedback and evaluation. A new algorithm utilising convolutional autoencoders is proposed in this paper to address the challenge of achieving higher accuracy in speaker recognition. Despite attempting various traditional methods previously, the desired level of accuracy has not been achieved. Recognising the limitations of existing approaches, a novel solution based on a convolutional autoencoder architecture has been proposed in this paper. By leveraging the power of convolutional neural networks and autoencoders, the proposed algorithm aims to overcome the hurdles faced by traditional speaker recognition systems.

This paper begins with a comprehensive literature survey, which provides an overview of existing research and advancements in the field of speaker recognition. This section establishes the background and contextualizes the proposed methodology. Following the literature survey, the paper delves into the methodology, presenting the details of the proposed algorithm based on a convolutional autoencoder architecture for speaker recognition. The algorithm's components, such as the convolutional neural network and autoencoder, are described, along with the specific techniques and approaches employed. Subsequently, the paper moves on to the results and discussion section, where the outcomes of the experiments and evaluations conducted on the dataset are presented. The performance of the proposed algorithm is analyzed, compared to existing methods, and discussed in detail. Any significant findings, limitations, or interesting observations are also explored and discussed. Next, the paper concludes with a concise conclusion section summarizing the key contributions of the research, highlighting the strengths of the proposed algorithm, and discussing potential areas for future improvement and exploration. Finally, the references section lists all the cited sources throughout the paper, ensuring proper attribution and facilitating further reading and research for interested readers.

2   Literature Survey

Speaker recognition systems rely on accurate feature extraction from speech signals to distinguish between different speakers. This paper [5] addresses the need for more robust feature extraction methods that can handle different types of speech signals and noise conditions. Another challenge is to achieve domain robustness in speaker recognition systems. Domain robustness refers to the ability of a system to perform well in different domains or environments, such as different acoustic conditions or speaking styles [6]. A small neural network architecture could also limit the accuracy of the speaker identification system [7]. Speaker verification systems may perform poorly when faced with speakers or acoustic conditions that are not well represented in the training data. This could be due to limitations in the size or diversity of the available datasets. Even with large amounts of training data, it can be challenging to accurately model the wide range of acoustic variations that can occur between speakers, such as differences in accent, age, or gender. Speaker verification systems may perform poorly in real-world scenarios due to the presence of noise, reverberation, or other environmental factors that can degrade the quality of the speech signal [8]. So, using different CNN architectures, feature extraction techniques, and training methods, one can enhance the model's performance [9]. In this paper, several research gaps are addressed, including the small dataset issue, the search for a good loss function, considerations for different acoustic conditions, and domain robustness, through the utilization of this methodology. An additional step is taken in this study by creating a speaker recognition system using a convolutional autoencoder. Different types of autoencoders, such as vanilla autoencoders and denoising autoencoders, are available. In this experiment, a convolutional autoencoder is employed with a very small dataset collected from five different speakers. Furthermore, all recorded samples are kept without any preprocessing. Each utterance has a duration of 3 s. For training and testing, different texts are used, as this system is text independent. It is important to note that the focus of this paper is on the speaker's speech features, so the language of the training and test data does not matter in this case. The proposed methodology is initially explained in this paper, followed by a discussion of the experimental setup, the results of all the experiments, the conclusion, and the future scope of the proposed model.

The work presented has been focused on developing a new automatic speech recognition (ASR) system based on a sparse auto-encoder neural network architecture inspired by the hunting behavior of Harris hawks [10]. The authors propose a new ASR system that uses a sparse auto-encoder network to learn features from speech signals and recognize speech using a deep neural network (DNN) classifier. The proposed system is assessed using the TIMIT dataset and contrasted with other ASR systems that are already in use. The experimental findings demonstrate that the proposed Harris Hawks Sparse Auto-Encoder Networks (HHSAEN) approach outperforms other traditional and deep learning-based ASR systems in terms of recognition accuracy, achieving state-of-the-art performance. In-depth analysis of the suggested system's learned properties and information on the effectiveness of the suggested approach to voice recognition are also included in the study. This research claims that the challenge lies in expanding the dataset used for training and testing in order to assess its efficacy in more diverse scenarios. This study [11] provides an in-depth examination of deep learning algorithms for voice emotion recognition. The authors explore numerous publicly available databases that contain emotional expressions in speech recordings, as well as the challenges and limits connected with these databases. Several deep learning techniques, including convolutional neural networks (CNNs), recurrent neural networks (RNNs), and long short-term memory (LSTM) networks, are then discussed in relation to voice emotion recognition. The authors describe the architectures and training methods for these models, and they compare their performance on different databases. The paper also includes a detailed discussion of the pre-processing steps that are necessary to prepare speech data for deep learning models, such as feature extraction and normalization. The authors discuss the potential for utilizing larger and more diverse datasets, improving model accuracy, and developing more robust models that can adapt to different languages and cultures. They also mention the possibility of integrating other types of data, such as physiological signals, to further enhance the accuracy of emotion recognition systems. Overall, the authors suggest that there is still much to be explored and improved upon in this field. This [12] describes a method to modify the accent of non-native speakers to improve their speech recognition accuracy. The authors propose a technique that utilizes neural style transfer to modify the accent of non-native speakers' speech by transferring the style of a reference speaker's speech to the non-native speaker's speech. The authors trained a deep neural network to learn the mapping between the spectrograms of the non-native speaker's speech and the reference speaker's speech. They then applied this network to transform the non-native speaker's speech spectrogram to match the reference speaker's spectrogram while preserving the content of the speech. The resulting modified speech was then used as input to a speech recognition system, and the authors found that this approach improved the recognition accuracy of non-native speakers' speech. The authors evaluated their method on two datasets, and the results showed that their approach out-performed several baseline methods for accent modification. In a paper [13], the authors propose a text-independent speaker identification system based on a deep learning model of a convolutional neural network (CNN). The system aims to identify the speaker of an input speech signal without relying on any specific text or speech content. To achieve this, the authors pre-process the speech signal using Mel-frequency cepstral coefficients (MFCCs) and use them as input to the CNN model. The CNN model is trained using a large dataset of speech signals from multiple speakers, and it learns to extract relevant features from the input speech signals that are specific to each speaker. The authors evaluate the performance of their system using two standard datasets, and they report high accuracy rates, demonstrating the effectiveness of their proposed approach. The proposed system has potential applications in various domains, including security, surveillance, and forensics. It is suggested that modifications be made to the deep learning model to increase the accuracy rate. This paper [14] presents the development and evaluation of a deep learning-based Arabic autoencoder speech recognition system for an electro-larynx device, which is a communication aid used by individuals who have lost their natural voice due to laryngectomy. The proposed system aims to improve the recog-nition accuracy and usability of the device by addressing the challenges of noisy and limited data and the specific characteristics of electro-larynx speech. Several deep learn-ing models, including convolutional neural networks (CNNs), long short-term memory (LSTM) networks, and autoencoder-based models, are trained and assessed on a dataset of electro-larynx speech recordings. In order to increase the amount of training data, they also used data augmentation techniques. They evaluated the models' performance using a range of metrics, such as accuracy, precision, recall, and F1 score. The outcomes of the experiment demonstrated that the autoencoder-based models performed better than the other models in terms of recognition accuracy and robustness to noise and limited data. On the test set, the suggested system's accuracy of 88.37% outperformed the baseline system's accuracy of 67.86% by a significant margin. Overall, the application of deep learning-based voice recognition systems in this work offers a promising method for enhancing the usability and efficacy of electro-larynx devices. A laryngectomy patient's quality of life and communication skills may be improved by the suggested system. Further studies suggest exploring more features of the speech signal for better accuracy. In their paper [15], the authors present a comprehensive review of speaker identification techniques using artificial intelligence (AI) and machine learning (ML) methods. The paper discusses various AI techniques, such as neural networks, support vector machines, and deep learning, and how they can be used for speaker identification. The authors also review the challenges and limitations of current speaker identification techniques, including issues related to data pre-processing, feature extraction, and model selection. The paper concludes by discussing potential research directions to address these challenges and improve the accuracy of speaker identification systems. Overall, the work done in this paper provides a valuable overview of the current state of the art in speaker identification using AI and highlights the research challenges that need to be addressed to improve the accuracy and reliability of these systems.

According to previous work, speaker recognition systems have several challenges that can affect their accuracy and reliability. One of the primary challenges is the presence of background noise, which can significantly affect the quality of the audio signal and make it difficult to distinguish between different speakers. Another challenge is speaker variability, which can be caused by differences in speech patterns, accents, and language fluency. The use of voice disguising techniques, such as pitch shifting or speaking in different accents, can also pose a significant challenge to speaker recognition systems. Additionally, speaker recognition systems may encounter challenges in handling large datasets, dealing with impostor attacks, and ensuring the privacy and security of the stored voiceprints. To overcome these challenges, this paper proposes a new system that can enhance the accuracy and robustness of speaker recognition systems.

3   Methodology

Figure 1 shows the proposed model for this work. In this study, a dataset comprising voice samples from five speakers was collected in.wav format. The duration of the samples varied from 3 s to 10 s while maintaining a consistent sampling rate of 16 kHz. Since the system under investigation is a text-independent system, different texts were used for training and testing. This means that the speech samples used for training the model contained diverse content, allowing the system to learn speaker-specific characteristics independent of the spoken text. Similarly, during testing, separate texts were employed to evaluate the system's ability to accurately recognize the speakers without relying on specific textual content. Some studies have proven that even with a small dataset, one can build a successful model [16]. Since the dataset is very small, a data augmentation technique has been employed. Techniques such as time stretching, pitch shifting, noise injection, and speed perturbation can introduce variations and increase the effective size of the dataset. In this work, the time stretching technique is employed to augment the small dataset of speech signals. Time stretching involves altering the duration of the speech signal without changing its pitch. By compressing or expanding the time axis, variations in the temporal characteristics of the speech are introduced. By applying time stretching to the existing speech samples, new instances of the same speech content are generated, but with different durations. This effectively increases the size of the dataset and provides additional training examples for the speaker recognition system. The data augmentation technique can provide a good amount of data for training the model. Hence, the modified dataset is now ready for further experimental analysis. No preprocessing has been done on the voice samples collected from the speakers. After modifying the database, the next step is to convert all voice samples into spectrograms. Convolutional autoencoders work really well when the inputs are images. Creating spectrograms is a way to represent voice samples in the form of an image. Therefore, all the voice samples are converted into spectrograms.

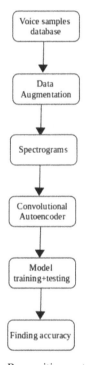

Fig. 1. Speaker Recognition system Framework

Convolutional Autoencoders (CAEs) leverage the power of convolutional operators to capture spatial information effectively. Compared to conventional methods, where convolutional filters are manually designed, CAEs allow the model to discover the ideal filters by minimizing the reconstruction error. Because of their capacity to learn filters, CAEs are the most advanced method for convolutional filter unsupervised learning. CAEs excel at learning concise and useful representations of input data when performing computer vision tasks. CAEs can extract pertinent features from any input data by utilizing the learned filters. These extracted features can then be utilized for various tasks, including classification or any other task that requires a concise representation of the input. While CAEs are a type of Convolutional Neural network (CNN), there is a fundamental distinction between them. CNNs are typically trained end-to-end, aiming to learn filters and combine features to classify input data. On the other hand, CAEs focus solely on learning filters that can extract features used to reconstruct the input. This differentiation underscores the unique purpose and objective of CAEs in comparison to traditional CNNs. The advantages of using convolutional autoencoders are that they can extract high-level features from raw audio signals, which can result in more accurate speaker recognition compared to traditional feature extraction techniques. They can also effectively filter out noise and other distortions from audio signals, making speaker recognition systems more robust in noisy environments. Speaker recognition systems using CAEs do not require physical contact with the user, making them non-intrusive and convenient to use. The important benefit is that convolutional autoencoders can learn from new data, which makes them adaptable to new speakers and dialects [17, 18]. This adaptability also means that the system can continuously improve its accuracy over time. Hence, by utilizing the convolutional autoencoder in this proposed methodology, the research gaps are overcome.

3.1 Representation of the Database

As shown in the block diagram, all voice samples are converted into spectrograms to be used as inputs for training purposes. Figure 2 depicts the representation of the voice sample as a spectrogram.

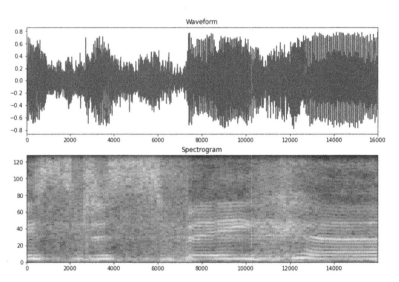

Fig. 2. Representation of voice sample into Spectrogram

3.2 Selection of Activation and Loss Function

Activation functions play a very important role in the system's performance. Its function is to trigger the non-linearity in the cells. The participation of neuron cells will be determined by the activation function. So, in making decisions, it plays a very important role. There are many activation functions, for example, sigmoid. In the proposed model, after experimenting with various combinations, the rectified linear unit (ReLU) is implemented as the activation function, and the mean squared error is chosen as the loss function.

3.3 Details of the Model Used in the Experiment

In this experiment, a convolutional autoencoder is used, which has two parts: an encoder and a decoder. Figure 3 indicates the encoder part, and Fig. 4 indicates the decoder part. The function of pooling layers is the minimization of features. Once the features are minimized, it becomes easy to compute. Here, normalization is also used along with the activation function.

Layer (type)	Output Shape	Param #
resizing_11 (Resizing)	(None, 32, 32, 1)	0
normalization_11 (Normaliza tion)	(None, 32, 32, 1)	3
conv2d_23 (Conv2D)	(None, 30, 30, 32)	320
conv2d_24 (Conv2D)	(None, 28, 28, 64)	18496
conv2d_25 (Conv2D)	(None, 26, 26, 128)	73856

Fig. 3. Convolutional Autoencoder (encoder part)

```
conv2d_transpose_23 (Conv2D    (None, 28, 28, 128)    147584
Transpose)

conv2d_transpose_24 (Conv2D    (None, 30, 30, 64)     73792
Transpose)

conv2d_transpose_25 (Conv2D    (None, 32, 32, 32)     18464
Transpose)

conv2d_transpose_26 (Conv2D    (None, 34, 34, 16)     4624
Transpose)

max_pooling2d_11 (MaxPoolin    (None, 17, 17, 16)     0
g2D)

dropout_22 (Dropout)           (None, 17, 17, 16)     0

flatten_11 (Flatten)           (None, 4624)           0

dense_22 (Dense)               (None, 128)            592000

dropout_23 (Dropout)           (None, 128)            0

dense_23 (Dense)               (None, 5)              645
```

Fig. 4. Convolutional Autoencoder (decoder part)

4   Result and Discussion

The Python programming platform is used for implementing this work. A dataset consisting of 50 voice samples from five distinct speakers has been gathered. The dataset contains utterances ranging from 3 to 10 s, and different texts are used for training and testing purposes. The dataset includes a mixture of languages, but since the expectation is to develop a speaker recognition model based on the speaker's voice features, the language is unlikely to impact the results. Each speaker has a unique characteristic, enabling classification based on the speaker and achieving speaker recognition.

An augmented dataset has been used for training and testing purposes, with the aim of increasing the size of the dataset. The entire dataset has been divided into three sets, namely the training dataset, testing dataset, and validation dataset. In both training and testing, experiments have been conducted under two conditions: the matching condition and the mismatching condition. The matching condition is met when the duration of the training and testing utterances is the same, while in the mismatched condition, the durations are different. The accuracy curve for the matched condition is shown in Fig. 5.

4.1   Result Analysis

In matched conditions, this system has achieved 92.4% accuracy rate. Figure 5 shows the accuracy curve against the number of epochs. It is observed that for a small dataset, the epoch rate should be small. Figure 6 indicates the training loss and validation loss. Every system should be perfectly fitted. Here there is no problem of over- or under-fitting of the model. Figure 6 shows the training loss and validation loss curves in the matched condition (Table 1).

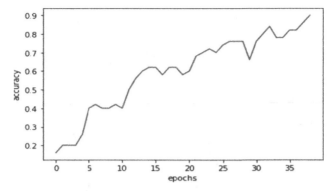

Fig. 5. Accuracy curve for the matched condition.

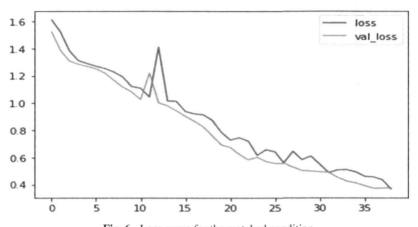

Fig. 6. Loss curve for the matched condition.

Table 1. % Accuracy comparison table of system accuracy in matched and in mismatched condition.

Training sample duration (sec)	Testing sample duration (sec)	
	3	10
3	**92.4**	85.3
10	87.1	**92**

The results indicate that the system has achieved better accuracy in the matched condition, meaning when the test conditions were similar to the training conditions. However, in the mismatched condition, where the test conditions differed significantly from the training conditions, the system's performance decreased. Figure 7 displays the confusion matrix for the matched condition.

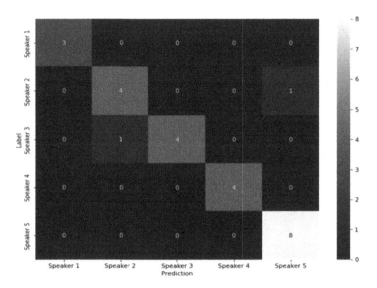

Fig. 7. Confusion matrix for the matched condition.

Table 2 shows the comparison of the proposed model with other methods. As shown in Fig. 7, the confusion matrix shows the performance of the system. It shows the rate of prediction for various labels. Here, five labels have been used. Speaker1, Speaker2, Speaker3, Speaker4, and Speaker5. In the case of matched conditions, the system is performing very well. But in the case of mismatched conditions, its accuracy rate is low. Hence, during training and testing, one can use utterances of the same length.

Table 2. Comparison table of proposed model with other existing methods with same dataset

Method	AUC	CA	F1	Precision
SVM	0.785	0.877	0.839	0.815
Random Forest	0.933	0.853	0.811	0.796
proposed Model	**0.969**	**0.953**	**0.974**	**0.960**

5 Conclusion

In this research paper, a new speaker recognition system is proposed that utilises a convolutional autoencoder. The system has achieved a good success rate under matched conditions for utterances. However, its performance was not satisfactory under mismatched conditions. Various activation functions were experimented with, and it was observed that the ReLU activation function produced better results. The system used raw voice samples without any pre-processing, which made it somewhat resilient to background noise. A comparison was conducted between the proposed system's results and existing techniques such as SVM and Random Forest using the same dataset. According to Table 2, the system achieved a good accuracy rate. Additionally, parameters like Area under Curve, F1 score, CA, and precision were compared. Previous studies in the related section revealed the use of MFCC features and readily available, clean datasets. However, the novelty of this paper lies in feeding speech signals in the form of images by converting them into spectrograms. Thus, instead of MFCC, the system used spectrograms as a feature of speech signals. Moreover, a specifically collected dataset was used for this research purpose. Based on the experimental results, it has been identified that the main issue lies in the mismatched conditions between the training and testing utterances. Therefore, future researchers should focus on addressing this problem to improve the system's performance. Additionally, since the system was tested with voices containing background noises, it would be beneficial to find a solution to remove these noises, thereby enhancing its performance. Furthermore, it is recommended that researchers explore the use of different types of autoencoders, such as denoising autoencoders and vanilla autoencoders, as well as other feature extraction techniques like pitch, jitter, or LPCC-based feature extraction, to further improve the system's performance.

References

1. La Mura, M., Lamberti, P.: Human-machine interaction personalization: a review on gender and emotion recognition through speech analysis. In: IEEE International Workshop on Metrology for Industry 4.0 & IoT, pp. 319–323 (2020)
2. Shelke, P.P., Wagh, K.P.: Review on aspect based sentiment analysis on social data. In: International Conference on Computing for Sustainable Global Development, pp. 331–336 (2021)
3. Ishak, Z., Rajendran, N., Al-Sanjary, O.I., Razali, N.A.M.: secure biometric lock system for files and applications: a review. In: IEEE International Colloquium on Signal Processing & Its Applications, pp. 23–28 (2020)
4. Hourri, S., Nikolov, N.S., Kharroubi, J.: Convolutional neural network vectors for speaker recognition. Int. J. Speech Technol. **24**, 389–400 (2021)
5. Vaessen, N., Van Leeuwen, D.A.: Fine-Tuning Wav2Vec2 for speaker recognition. In: IEEE International Conference on Acoustics, Speech, and Signal Processing, pp. 7967–7971 (2022)
6. Hu, H.R., Song, Y., Liu, Y., Dai, L.R., McLoughlin, I., Liu, L.: Domain robust deep embedding learning for speaker recognition. In: IEEE International Conference on Acoustics, Speech and Signal Processing, pp. 7182–7186 (2022)
7. Loina, L.: Speaker identification using small artificial neural network on small dataset. In: International Conference on Smart Systems and Technologies, pp. 141–145 (2022)
8. Lin, W., Mak, M.W.: Robust speaker verification using population-based data augmentation. In: IEEE International Conference on Acoustics, Speech and Signal Processing, pp. 7642–7646 (2022)
9. Abdulqader, H.A., Rahman Al-Haddad, S.A., Abdo, S., Abdulghani, A., Natarajan, S.: Hybrid feature extraction MFCC and feature selection CNN for speaker identification using CNN: a comparative study. In: International Conference on Emerging Smart Technologies and Applications, pp. 1–6 (2022)
10. Ali, M.H., et al.: Harris hawks sparse auto-encoder networks for automatic speech recognition system. Appl. Sci. **12**(3), 1091–1095 (2022)
11. Abbaschian, B.J., Sierra-Sosa, D., Elmaghraby, A.: Deep learning techniques for speech emotion recognition, from databases to models. Sensors **21**(4), 1249–1255 (2021)

12. Radzikowski, K., Wang, L., Yoshie, O.: Accent modification for speech recognition of non-native speakers using neural style transfer. EURASIP J. Audio Speech Music Process. **2021**, 11 (2021)
13. Bunrit, S., Inkian, T., Kerdprasop, N., Kerdprasop, K.: Text-independent speaker identification using deep learning model of convolution neural network. Int. J. Mach. Learn. Comput. **9**(2), 143–148 (2019)
14. Zinah J. Mohammed Ameen, Abdul kareem Abdulrahman Kadhim.: Deep learning methods for arabic autoencoder speech recognition system for electro-larynx device. Adv. Hum. - Comput. Interact. **2023**(5), 1–11, (2023)
15. Jahangir, R., Teh, Y.W., Nweke, H.F., Mujtaba, G., Ali Al-Garadi, M., Ali, I.: Speaker identification through artificial intelligence techniques: a comprehensive review and research challenges. Expert Syst. Appl. **171**, 114591 (2021)
16. Jagiasi, R., Ghosalkar, S., Kulal, P., Bharambe, A.: CNN based speaker recognition in language and text-independent small-scale system. In: International conference on IoT in Social, Mobile, Analytics and Cloud, pp. 176–179 (2019)
17. Tirumala, S.S., Shahamiri, S.R.: A deep autoencoder approach for speaker identification. In: International Conference on Signal Processing Systems (2017)
18. Kushwaha, A.K., Khatavkar, S.M., Biradar, D.M., Chougule, P.A.: Depth estimation and navigation route planning for mobile robots based on stereo camera. In: LNICS, Social Informatics and Telecommunications Engineering, vol. 472, pp. 180–191 (2023)

Face Emotion Recognition Based on Images Using the Haar-Cascade Front End Approach

G. Gowri Pushpa[1]([⊠]), Jayasri Kotti[2], and Ch. Bindumadhuri[3]

[1] Department of Computer Science and Engineering, Anil Neeru Konda Institute of Technology and Sciences, Visakhapatnam, Andhra Pradesh, India
gowripushpa11@gmail.com
[2] Department of Computer Science and Engineering, Vignan's Institute of Engineering for Women, Visakhapatnam, Andhra Pradesh, India
[3] Department of Information and Technology, University College of Engineering Vizianagaram, JNTUK, Vizianagaram, Andhra Pradesh, India
chbmadhuri.it@jntukucev.ac.in

Abstract. Facial expression recognition (FER) has emerged as a major research topic, with human-computer interactions. Nonverbal messages which are face expressions are crucial in our day-to-day life, which is an example of non-verbal communication. As a human, detecting facial expressions and understanding human emotions is a simple process, but doing it with the assistance of a machine is more challenging. With the remarkable success of deep learning, the different types of architectures of this technique are exploited to achieve a better performance with an accuracy of 87%. In this research, our proposed model shows how to identify and recognize facial emotions from images using neural networks with help of preprocessing techniques and the whole process comprises various stages of classifying the detected features, involving human face detection and classifying them then into any of the seven basic emotion classes using the convolutional neural networks (CNN). Haar-cascade frontal face algorithm was utilized in order to detect human faces from the images. Our model was trained and tested on the FER-2013 dataset.

Keywords: Haar-cascade frontal face algorithm · FER-2013 · Facial expression recognition · Face detection · Convolution neural network (CNN)

1 Introduction

Facial Emotions are mostly communicated through these facial expressions, gestures, and physiological signals. With better deep learning techniques, we can solve these types of complex problems using neural networks. Numerous uses exist in the field of human-computer interface for recognizing these human facial expressions [9]. All the non-verbal communications basically fall under facial emotion recognition. A person's facial gestures can be used to assess both his or her emotional state and cerebral outlook. The basic facial emotions include Angry, Disgust, Sad, Happy, Fear, Neutral, Surprise.

© ICST Institute for Computer Sciences, Social Informatics and Telecommunications Engineering 2024
Published by Springer Nature Switzerland AG 2024. All Rights Reserved
P. Pareek et al. (Eds.): IC4S 2023, LNICST 536, pp. 331–339, 2024.
https://doi.org/10.1007/978-3-031-48888-7_28

Convolutional Neural Networks are widely applied in the field of image recognition [1]. As it can have a huge number of network layers and it also extracts more high-level features [2]. Facial recognition requires several phases which are detection of faces from images, pre-processing, retrieval of those facial features, and identification of emotions in the facial images. For detection of human faces there are widely used other algorithms like Local binary Pattern (LBP) and MTCNN and haar cascade which are advantageous in their own way based on the requirements. The purpose of this study is to characterise an emotional face input image using efficient deep learning techniques.

2 Literature Review

Face emotions play a significant role in our everyday lives, as is universally acknowledged. We need a system that can recognize our facial expressions of emotion and react accordingly. For recognizing face expressions, a compact convolutional neural network is created. The completely linked layer from the conventional convolutional neural network is replaced with a global average pooling layer. The completely connected layer's [17] "black box" characteristics are somewhat eliminated by this technique, which can also be connected to global data to learn more in-depth and complete aspects of facial expressions.

Additionally, the pooling layer is devoid of any parameters that would enable parameter reduction and prevent overfitting. This model is a MTCNN [10] that accomplishes the task of detecting frontal faces which has more complexity of time and space. Many of the models were built using RESNET, ALEXNET, MOBILENETV2, INCEPTION architectures which are pre trained with better weights and resulting in an accuracy of around 70%-75%. In order to avoid overfitting, norms were added to the weight coefficient that not only prevents overfitting but makes the model stable and fast. At last, the final model is made combining a neural network with four convolutions and an MTCNN detection [10] to accomplish the aim of human emotion recognition. Batches and a RELU activation function [16] follow each of these four convolutions, while a soft-max activation function and global average pooling are used in the final layer to classify data. This system has a total of 58423 characteristics, 56951 of which are trainable. As a consequence, when attempting to categorize the facial gestures into one of the seven basic feelings, the prediction accuracy is 67%.

3 Dataset

FER-2013 has been utilized for the purpose of training and testing this model. This FER-2013 data set is available in both images and pixel values. We have implemented the model using the pixel values which are in csv format. These pixel values comprise almost every sort of the 7 recognized facial emotions (Fig. 1).

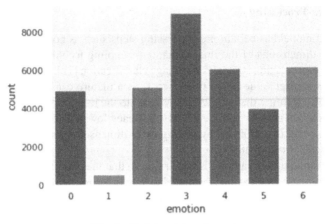

Fig. 1. Training data set count

4 Proposed Methodology

We proposed an approach to recognize the human from the images using traditional CNN [2] which is capable of classifying seven different types of emotions. The proposed traditional CNN is composed of 5 convolutional layers which have conv2d, batch normalization [4], maxpooling2d and dropout. A dense layer of 200 neurons attached to the completely connected layer and a SoftMax activation function are used to categorize human emotions (Fig. 2).

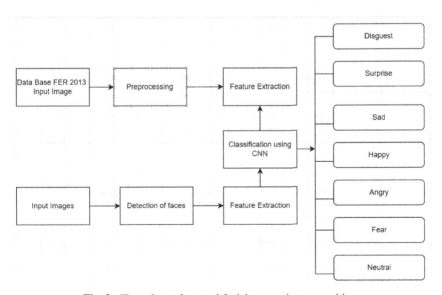

Fig. 2. Flow chart of general facial expression recognition

4.1 Data Pre-Processing

The proposed model has certain pre-processing steps such as normalizing the images, adjusting the dimensions of the image, random sampling techniques all of which are considered pre-processing strategies for image classification [12]. After normalizing the picture, it has a pixel value range from 0 to 1, as a picture can have a pixel value of max 255 and by splitting them by 255, they fall into the range of 0 and 1, making math computations much simpler and the model's intricacy lower. Then the reshaping of the dimensions of data is performed by adding extra dimensions for pixel values and we generate a shape of (-1,48,48,1).

Then the random over sampler is performed so that the majority of the class will be shuffled randomly so that it avoids the model to memorize and helps in generalizing. This sampler can also be used for minority classes based on the requirement of the model. Once these pre-processing methods have been successfully completed, we will divide the data set into both testing and training with the aid of the test size parameter and random state parameter, which shuffles the training data. We had introduced a shuffle size of 45 i.e., for every iteration a 45 new image pixel values will be generated replacing the old pixel values.

4.2 Model Training

This Traditional CNN [14] design consists of 5 convolutional layers and a flatten and finally three fully connected layers. Each has a Conv2D, Activation, Batch normalization, dropout and Maxpooling2D. The first layer is the input layer which takes input in with padding value 1, stride value 1 and kernel size of 3X3. The Conv2D extracts all possible feature maps and identifies the hidden internal representations. This was managed by the filters of a kernel of conv2d. This is also known as a feature extraction layer (Fig. 3).

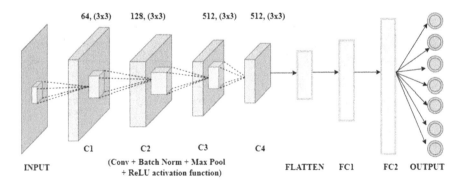

Fig. 3. Proposed Architecture of CNN model

Batch normalization [5] functions on extracted data by conv2d which takes as input batches. This makes the model learn much faster. We had introduced a batch size of 128 for our particular model. Maxpooling2D helps in finding all prominent places of the feature maps and thereby reducing the dimensions by extracting these prominent features. The RELU Activation function [7] brings the nonlinearity for the feature maps generated by convolutional layers. A dropout layer [3] is used to avoid the over-fitting of the model. We had dropped 20% of neurons for every new iteration. The flattened layer then converts into the 1d array and then the fully connected layer and makes its predictions according to the input.

4.3 Model Testing

During the testing phase, validation accuracy and loss of the proposed model is saved in their best state having higher accuracy. We have implemented certain call backs like early stopping, lr scheduler and model checkpoint in order to stop the model from over-fitting and save it in the best possible condition with a higher learning rate. We ran the model for 80 epochs using the Adam algorithm and category cross entropy [6] as the loss function. For prediction purposes, test data will be fed to the trained model to find final accuracy. This model implemented a haar-cascade frontal face feature [12] which detects the frontal faces present in the images. Once detecting the faces by this frontal face algorithm [8] they are reshaped and pixel values are divided by 255 and sent to the model to make predictions by recognizing facial emotions. The prediction is in the following way (Fig. 4).

Fig. 4. Predictions of model for test data set

5   Experimental Results and Analysis

The model was evaluated, and our research's findings indicate that it has a training accuracy of 91%, a testing accuracy of 87%, a training loss of 0.2469, and a testing loss of 0.5012. By The values we can state that this model has predicted effectively with higher accuracy and loss values. With the help of loss curve, it is possible to detect if the model has got overfit or underfit or optimal fit [11].

From the below loss curves we can say that the proposed model is perfectly designed with no overfitting or underfitting of the data with the better loss values and accuracy (Figs. 5 and 6).

The confusion matrix, which displays both the number of accurate forecasts and the number of inaccurate guesses produced by the proposed CNN algorithm, is shown

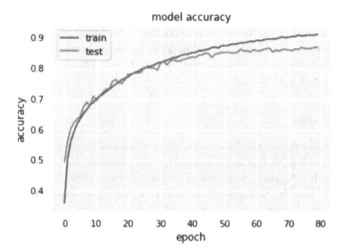

Fig. 5. Loss curve model.

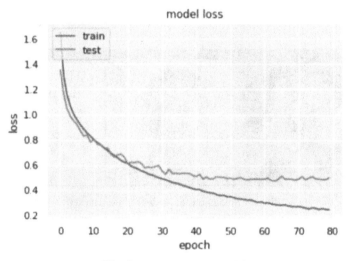

Fig. 6. Accuracy curve model

below, can be used to analyze how well this model predicted nearly every type of human emotion. Each model has a total of 4,478,278 params where 4,475,759 are trainable params and 3,968 are non-trainable (Fig. 7).

Model is shown in the confusion matrix above, along with the model's total number of incorrect predictions. The model is evaluated using the accuracy, recall, precision and f-1 score which are shown in the Fig. 8 i.e., classification report [13].

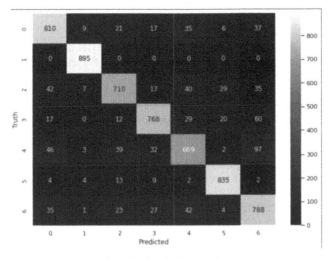

Fig. 7. Confusion matrix

	precision	recall	f1-score	support
0	0.86	0.87	0.87	935
1	0.99	1.00	0.99	895
2	0.83	0.82	0.82	880
3	0.87	0.86	0.87	906
4	0.77	0.76	0.76	888
5	0.92	0.96	0.94	869
6	0.81	0.80	0.80	920
accuracy			0.87	6293
macro avg	0.87	0.87	0.87	6293
weighted avg	0.86	0.87	0.87	6293

Fig. 8. Classification report of Proposed model

6 Conclusion and Feature Work

This article discussed current FER study and informed us of the most recent advancements in this field. The FER system in this article offers a method for identifying emotions from images that is more effective than all other conventional methodologies, with an accuracy of 87% and a prediction rate that is higher. We also have a study that showcases the high rate at which researchers were able to figure out that machines will become better at interpreting feelings in the future, suggesting that interactions between humans and machines will become more natural over time. This method's design does not include any pre-trained models. The suggested work can be improved as a video summarization and broadened to recognize human face emotions based on speech and video.

References

1. Sariyanidi, E., Gunes, H., Cavallaro, A.: Automatic analysis of facial affect: a survey of registration, representation, and recognition. IEEE Trans. Pattern Anal. Mach. Intell. **37**(6), 1113–1133 (2015)
2. O'Shea, K., Nash, R.: An Introduction to Convolutional Neural Networks. arXiv:1511.084 58v1 [cs.NE], 26 Nov 2015
3. Srivastava, N., Hinton, G., Krizhevsky, A., Sutskever, I., Salakhutdinov, R.: Dropout: a simple way to prevent neural networks from overfitting. J. Mach. Learn. Res. **15**(56), 1929–1958 (2014)
4. Daneshmand, H., Kohler, J., Bach, F., Hofmann, T., Lucchi, A.: Batch normalization provably avoids rank collapse for randomly initialised deep networks. arXiv:2003.01652v3 [stat.ML11], June 2020
5. Ioffe, S., Szegedy, C.: Batch normalization: accelerating deep network training by reducing internal covariate shift. arXiv:1502.03167v3 [cs.LG], 2 March 2015
6. Valeria, A., Nadiia, S.: Generalization of cross-entropy loss function for image classification. Mohyla Math. J. **3**, 3–10 (2021)
7. Agarap, A.F.M.: Deep Learning using Rectified Linear Units (ReLU). arXiv:1803.08375v2 [cs.NE], 7 February 2019
8. Padilla, R., Costa Filho, C.F.F., Costa, M.G.F.: Costa evaluation of Haar cascade classifiers designed for face detection international science index. Comput. Inf. Eng. **6** (2012)
9. Bansal, H., Khan, R.: A review paper on human computer interaction. Int. J. Adv. Res. Comput. Sci. Softw. Eng. **8**, 53–56 (2018)
10. Zhang, K., Zhang, Z., Li, Z., Qiao, Y.: Joint face detection and alignment using multitask cascaded convolutional networks. IEEE Sig. Process. Lett.**23**(10), 1499–1503 (2016)
11. Ekundayo, O., Viriri, S.: Facial Expression recognition: a review of methods, performances and limitations. In: Conference on Information Communications Technology and Society (ICTAS). IEEE, Durban, South Africa 2019
12. Pitaloka, D.A., Wulandari, A., Basaruddin, T., Liliana, D.Y.: Enhancing CNN with preprocessing stage in automatic emotion recognition. In: 2nd International Conference on Computer Science and Computational Intelligence (ICCSCI), pp. 523–529, Bali, Indonesia (2017)
13. Jumani, S.Z., Ali, F., Guriro, S., Kandhro, I.A., Khan, A., Zaidi, A.: Facial expression recognition with histogram of oriented gradients using CNN. Ind. J. Sci. Technol. **12**, 1–9 (2019)
14. Mehendale, N.: Facial emotion recognition using convolutional neural networks (FERC). SN Appl. Sci. **2**, 446 (2020)
15. Lin, G., Shen, W.: Research on convolutional neural network based on improved Relu piecewise activation function. In: 8th International Congress of Information and Communication Technology (ICICT), pp. 977–984 (2018)
16. Wang, Y., Wu, J., Hoashi, K.: Lightweight deep convolutional neural networks for facial expression recognition. IEEE, Kuala Lumpur, Malaysia (2019)
17. Chang, T., Wen, G., Hu, Y., Ma, J.: Facial expression recognition based on complexity perception classification algorithm, arXiv:1803.00185 [cs.CV] (2018)
18. Abdulsalam, W.H., Alhamdani, R.S., Abdullah, M.N.: Facial emotion recognition from videos using deep convolutional neural networks. Int. J. Mach. Learn. Comput. **9**, 14–19 (2019)
19. Pantic, M., Rothkrantz. L.J.M.: Towards an affect – sensitive multimodal human-computer interaction. In: Proceedings of the IEEE, vol. 91, pp. 1370–1390, September 2003
20. Debnath, T., Reza, M.M., Rahman, A., Beheshti, A., Band, S.S., Alinejad-Rokny, H.: Four-layer ConvNet to facial emotion recognition with minimal epochs and the significance of data diversity. Sci. Rep. (2022)

21. Mellouka, W., Handouzia, W.: Facial emotion recognition using deep learning: review and insights. In: The 2nd International Workshop on the Future of Internet of Everything (FIoE), Leuven, Belgium, pp. 9–12, August 2020
22. Bie, M., Xu, H., Gao, Y., Che, X.: Facial expression recognition from a single face image based on deep learning and broad learning. Wirel. Commun. Mob. Comput. 1–10 (2022)

TextRank – Based Keyword Extraction for Constructing a Domain-Specific Dictionary

Sridevi Bonthu[1]([⊠]) [iD], Hema Sankar Sai Ganesh Babu Muddam[2] [iD],
Koushik Varma Mudunuri[1] [iD], Abhinav Dayal[1] [iD], V. V. R. Maheswara Rao[3] [iD],
and Bharat Kumar Bolla[4] [iD]

[1] Computer Science and Engineering Department, Vishnu Institute of Technology, Bhimavaram, Andhra Pradesh, India
sridevi.b@vishnu.edu.in
[2] Tata Consultancy Services, Synergy Park, Hyderabad, India
[3] Computer Science and Engineering Department, Shri Vishnu Engineering College for Women, Bhimavaram, Andhra Pradesh, India
[4] University of Arizona, Tucson, AZ 85721, USA

Abstract. Extracting domain-related keywords from text documents is a crucial task in both Information Retrieval and Natural Language Processing (NLP). This paper presents an approach that combines the TextRank algorithm with various NLP techniques to effectively identify domain-specific keywords. Our method utilizes the power of unsupervised graph-based ranking algorithms and the semantic understanding of NLP models to extract key terms that are highly relevant to a specific domain. The work is carried out on an arXiv research abstract dataset. This work preprocesses the input text to capture linguistic features, extracts the keywords using TextRank and POS filtering approaches, extracts the definitions and finally evaluates the performance. The performance of the extracted keywords is done with the help of manually annotated labels. The proposed method has obtained 83% accuracy. The proposed approach is flexible and adaptable to different domains, as it can be trained on domain-specific data to further improve its performance.

Keywords: Extraction · TextRank · POS tagging · Text mining · domain-specific dictionary · Natural Language Processing

1 Introduction

The proliferation of digital data and the imperative to analyze it efficiently have spurred the emergence of numerous methodologies for data analysis [1]. Among these methodologies is text mining, a process that entails extracting valuable insights from unstructured or semi-structured data. Text mining finds utility across a diverse array of applications, including sentiment analysis, recommendation systems, and content analysis [2]. Within the realm of text mining, a crucial undertaking involves the extraction of keywords along with their corresponding definitions [3]. Keywords represent terms or

© ICST Institute for Computer Sciences, Social Informatics and Telecommunications Engineering 2024
Published by Springer Nature Switzerland AG 2024. All Rights Reserved
P. Pareek et al. (Eds.): IC4S 2023, LNICST 536, pp. 340–349, 2024.
https://doi.org/10.1007/978-3-031-48888-7_29

phrases that hold significance within a specific subject area or field, while their definitions offer a succinct and accurate explanation of their significance. Extracting keywords serves the purpose of identifying the most pertinent topics addressed in a single document or a group of documents [4]. Moreover, the definitions associated with these keywords enhance comprehension and foster a deeper understanding of the concepts under examination.

The internet has witnessed an unprecedented surge in digital content and data, presenting a formidable challenge in efficiently accessing the most pertinent information amidst this vast volume of data. Keyword extraction serves as a foundational method in NLP, enabling the identification of the crucial words or set of words (phrases) within a text corpus [5]. This process of identifying and extracting relevant keywords holds immense value in applications like information retrieval, text classification, and summarization. In our work, we delve into the utilization of the TextRank algorithm for keyword extraction, as well as the identification of prevailing defining patterns to facilitate definition extraction [6]. This paper presents a novel approach to extract keywords and their corresponding definitions from textual data. Our proposed methodology harnesses the power of the TextRank algorithm, an unsupervised graph-based ranking algorithm renowned for identifying significant terms within a document. By integrating regular expressions, we leverage common patterns in keyword definitions. To assess the effectiveness of our methodology, we conduct evaluations on a dataset comprising arXiv paper abstracts, comparing its performance against other cutting-edge keyword extraction techniques.

This study holds great importance as it has the potential to enhance the effectiveness and precision of keyword extraction and definition extraction tasks. The outcomes of this research offer valuable insights that can significantly benefit various natural language processing applications, including text classification, summarization, and sentiment analysis. Additionally, the proposed approach demonstrates its versatility and broad applicability by being adaptable to diverse domains and languages.

The study is driven by the subsequent research inquiries.

1. To what amount can the TextRank algorithm be utilized for effective keyword extraction?
2. What are the prevailing patterns for definition extraction achieved through regular expressions?
3. How does the proposed approach compare to other advanced keyword extraction methods?
4. What are the potential implications of the findings?

The rest of this paper is organized as follows: Sect. 2 provides an extensive literature review on keyword extraction. Section 3 outlines the details of our proposed methodology. Following that, Sect. 4 presents the experimental results. Finally, in Sect. 5, the paper concludes by summarizing the findings and suggesting potential directions for future research.

2   Related Work

Keyword extraction plays a pivotal role in natural language processing by identifying the most crucial words or phrases within a given text [7]. Multiple methodologies have been devised for this purpose, encompassing statistical, linguistic, and graph-based approaches [4]. Keyword extraction techniques can be categorized into supervised, semi-supervised, or unsupervised methods [8]. One commonly used unsupervised technique is based on TF-IDF, which establishes a baseline by scoring and selecting key-phrases according to their TF-IDF values [9]. Another approach for topic modeling is Latent Dirichlet Allocation (LDA), which performs unsupervised learning to identify the main topics present in a document [10]. In a study conducted by Gu Yijun et al., LDA was utilized to extract document keywords, and their association with keywords within the document itself was found to enhance the results of keyword extraction [11]. Additionally, Rapid Automatic Keyword Extraction (RAKE) is a widely employed algorithm for domain-independent keyword extraction [12]. Among the graph-based techniques, the TextRank algorithm has gained considerable popularity. It operates by analyzing word co-occurrences to determine the significance of individual words within the text. TextRank is an influential ranking algorithm that operates on the principles of graph-based analysis, leveraging the PageRank algorithm [6]. By constructing a graph representation of a given text, where individual keywords or keyphrases serve as nodes of the graph, and establishing connections between nodes that co-occur within the text, TextRank calculates the importance score of each node. This score is determined by applying the PageRank algorithm, which assesses the significance of a node based on the quantity and quality of incoming edges it possesses [13].

In the field of NLP, a notable obstacle entails the identification and extraction of definitions for words or phrases within a given text. To address this challenge, one approach involves identifying prevalent defining patterns commonly employed to introduce definitions, such as the format "A [word] is a [definition]." By utilizing regular expressions, these patterns can be automatically recognized, enabling the extraction of definitions from the text.

Numerous studies have delved into the realm of keyword extraction and definition extraction techniques. However, many of these studies suffer from limitations in their scope, failing to adequately address the challenges encountered in real-world applications. For instance, certain studies rely on small datasets that fail to capture the diverse range of texts and writing styles found in real-world scenarios [14]. Additionally, some studies focus exclusively on specific text types, such as scientific papers, which restricts their applicability to other text genres [15]. Furthermore, certain studies employ outdated or less effective algorithms for keyword extraction and definition extraction, thereby diminishing their overall efficacy [16].

3   Methodology

The research design adopted for this study encompasses a comprehensive and systematic approach to address the objectives of extracting domain-specific keywords and their corresponding definitions from a large corpus of research papers. We have employed a data-driven methodology, leveraging advanced NLP techniques and machine learning algorithms to systematize the extraction process and ensure scalability. The methodology followed is present the Fig. 1.

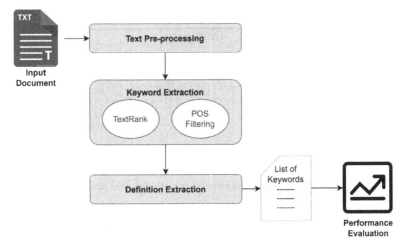

Fig. 1. A framework to identify domain-specific keywords within a given document.

3.1 Data Acquisition and Pre-Processing

To gather the necessary data for our research, we have curated a robust and diverse dataset sourced from the ArXiv database[1]. The ArXiv database hosts a vast collection of research papers spanning multiple disciplines, thereby providing a rich and extensive source of technical content. The dataset was meticulously selected to ensure its relevance and representativeness, enabling us to perform a comprehensive extraction of technical terms which are domain-specific.

Data preprocessing plays a crucial role in any NLP task as it aims to convert raw text data into a suitable format for analysis [17, 18]. This essential stage involves several key steps. Firstly, non-relevant characters, stopwords, and punctuation marks are eliminated. Additionally, tokenization is performed, breaking the text into individual words or phrases. Following this, part-of-speech tagging is applied to determine the grammatical role of each token. This valuable information is utilized to construct a co-occurrence matrix, which captures the frequency of term appearances within the same sentence as other terms. Ultimately, the generated co-occurrence matrix is utilized as the input for the TextRank algorithm. Figure 2 illustrates the sequential steps undertaken during text preprocessing in this work.

Text normalization is a crucial step in achieving consistency within textual data. It encompasses various operations such as eliminating special characters, converting text to lowercase, and expanding contractions [19]. The process involves segmenting the text

---

[1] https://www.kaggle.com/datasets/Cornell-University/arxiv.

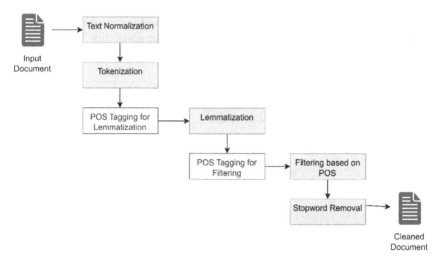

Fig. 2. The pre-processing techniques adopted in transforming the raw text to a clean format.

into individual units or tokens, which can range from words and phrases to individual characters. These tokens serve as the foundation for subsequent processing steps, such as part-of-speech tagging or constructing a co-occurrence matrix. In this particular study, the input text (referred to as "Abstract data") was normalized through the removal of non-printable characters, converting all text to lowercase, eliminating special characters, and removing excessive spaces.

POS tagging and lemmatization are performed in tandem, in the preprocessing of text data for NLP tasks. POS tagging entails assigning a part-of-speech label, such as noun, verb, or adjective, to each word in the text [20]. This labeling information plays a crucial role in lemmatization, which involves transforming each word into its base or dictionary form. For instance, through the process of POS tagging and lemmatization, the word "running" would be converted to "run". This combined approach aids in simplifying the complexity of the text data and enhancing the accuracy of subsequent processing tasks. In this study, the WordNetLemmatizer was utilized to perform lemmatization on the tokens generated in the preceding step.

Filtering the words, by identifying and retaining only certain parts of speech, such as nouns, adjectives and gerunds to improve the relevance and accuracy. Any word in the lemmatized text that does not fall into the categories of noun, adjective, gerund, or a foreign word is classified as a stopword (non-content). Based on the specified conditions, a filter is added to the lemmatized and POS tagged tokens to filter out the non-content (stopwords). This filter includes the POS tags such as *NN, NNS, NNP, NNPS, JJ, JJR, JJS, VBG* and *FW*.

3.2 Keyword Extraction Using TextRank Algorithm and POS Filtering

To identify the most significant technical and domain-specific keywords from research papers, we employed a combined approach of the TextRank algorithm and POS filtering. TextRank, a graph-based ranking algorithm, enables the extraction of important terms by analyzing their co-occurrence patterns within the document [6]. Leveraging the inherent structure and word relationships, TextRank effectively identifies the essential concepts and ideas discussed in the research papers. In conjunction with TextRank, we implemented POS filtering [21] to further refine the extracted keywords. By considering specific parts of speech commonly associated with technical terms, such as nouns, adjectives, and verb forms, we filtered out irrelevant words while retaining those more likely to hold technical significance. This additional filtering step significantly enhances the precision and accuracy of the extracted keywords, ensuring their close alignment with the technical domain being investigated.

3.3 Definition Extraction Using Regular Expression

After obtaining the extracted keywords, the subsequent task was to retrieve their corresponding definitions from the research papers. To achieve this, we employed the use of regular expressions, a powerful tool for pattern matching, to identify common textual patterns that indicate the presence of definitions. Regular expressions enable us to capture specific structures and linguistic cues within the text that are typically associated with definitions. Through thorough analysis and domain expertise, we meticulously designed a set of well-crafted regular expressions tailored to the specific context of technical literature. These patterns encompass diverse sentence structures, syntactic cues, and linguistic patterns commonly employed in technical definitions. By matching these predefined patterns with the surrounding text, we successfully extracted the relevant definitions for the identified keywords.

3.4 Performance Assessment

To assess the effectiveness and reliability of our methodology, we conducted a comprehensive assessment using a variety of evaluation metrics and statistical analyses. The accuracy and coverage of the extracted keywords were evaluated by comparing them to manually created reference glossaries. Precision, recall, and F1-score were calculated as quantitative measures to assess the performance of the keyword extraction process.

For evaluating the extracted definitions, we adopted a multi-faceted approach. Firstly, a sample of extracted definitions underwent meticulous manual evaluation by domain experts who assessed their accuracy and relevance. Their expert insights provided valuable feedback and ensured the quality of the extracted definitions. Additionally, we utilized semantic similarity measures, such as cosine similarity, to compare the extracted definitions with reference definitions available in external resources. This analysis allowed us to estimate the degree of alignment between our extracted definitions and established definitions, further enhancing the evaluation process.

4 Experimentation and Results

In Glossary Term Extraction, training data is commonly provided as a substantial corpus of text documents, such as books or articles, that encompass the domain-specific terminology of interest. In our experimentation, we utilized straightforward abstracts that encompassed multiple keywords. By leveraging this training data, the Textrank algorithm was employed to identify and rank the most significant terms within a new text document, considering their frequency and co-occurrence with other crucial terms. This automated approach facilitates the extraction of relevant terminology and streamlines the creation of a comprehensive glossary.

The computation of graph-based ranking, which considers edge weights when determining the score allied with each vertex in the graph, is performed using the formula outlined in the provided Eq. 1. Where, E and V are set of edges and vertices, $In(v_i)$ set of vertices that point to it, $Out(v_i)$ set of vertices that vertex v_i points to. In our approach, we establish a co-occurrence relationship, wherein lexical units are connected as vertices in the graph if they co-occur within a specific word space, with the maximum number of words allowed. This window size can be adjusted, typically ranging from 2 to 10 words.

$$WS(V_i) = (1 - d) + d * \sum_{V_j \in In(V_i)} \frac{w_{ji}}{\sum_{V_k \in Out(V_j)} w_{jk}} WS(V_j) \tag{1}$$

Once the graph is constructed, each vertex is initially assigned a score of 1. The ranking algorithm mentioned earlier is then executed on the graph for multiple iterations until convergence, typically around 20 to 30 iterations, with a threshold of *1e-4*. Once scores are assigned to all the keywords, we consider the top one-third of keywords for further analysis and consideration.

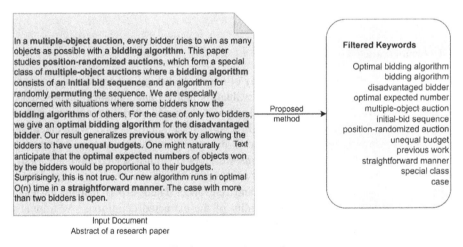

Input Document
Abstract of a research paper

Fig. 3. Keyword extraction

To evaluate the performance of our model, it is essential to have either a predefined list of keywords or a list of annotated keywords. In our experimentation, we manually annotated a list of keywords for this purpose. We then compare these annotated keywords with the filtered keywords generated by the model. Figure 3 presents both the annotated keywords and filtered keywords for a sample abstract. By comparing both lists of keywords, we calculate the confusion matrix, which provides valuable insights. From the

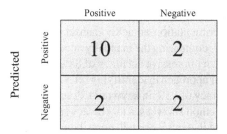

Fig. 4. Confusion matrix

values obtained in the confusion matrix, we can derive metrics such as Precision, Recall, and F1-scores. These metrics allow us to assess the accuracy and effectiveness of our model. The confusion matrix is shown in the Fig. 4. The obtained accuracy is 83%. A sample input, annotated keywords and the filtered keywords are shown in Fig. 3. The words and phrases highlighted in blue are the annotated keywords by the domain expert and the filtered words are present in the right of the figure. Most of the keywords are matching the annotated keywords as conveyed in table. Figure 5 presents the outcome of the definition extraction. For the supplied input document, the word and the definitions will come as *key: value* pairs in the form of a json file.

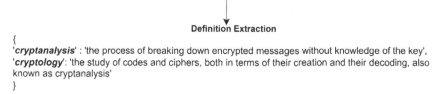

Fig. 5. Domain-specific definition extraction

The significance of our research lies in its practical implications. The extracted domain-related keywords can be utilized in various applications, including information retrieval, content analysis, and document categorization within specific domains. This can greatly improve the efficiency of these tasks and provide valuable insights for domain experts.

5 Conclusion

In this work, we presented a comprehensive framework for domain-related keyword extraction, incorporating the TextRank algorithm and several NLP approaches. Our approach not only focused on keyword extraction but also extended to definition extraction, aiming to provide a more comprehensive understanding of domain-specific content. Through extensive experimentation on arXiv dataset, we evaluated the performance of our proposed method. By comparing the manually annotated keywords with the extracted keywords, we achieved an impressive accuracy of 83%. This demonstrates the effectiveness and reliability of our approach in identifying relevant terms within a specific domain. With a remarkable accuracy of 83% in keyword extraction and the ability to extract definitions, our proposed method showcases its efficacy in capturing domain-specific terms accurately. We anticipate that our work will contribute to advancing the field of keyword extraction and provide valuable insights for domain experts in diverse industries.

References

1. Daniel, B.K.: Big Data and data science: a critical review of issues for educational research. Br. J. Edu. Technol. **50**(1), 101–113 (2019)
2. Allahyari, M., et al.: A brief survey of text mining: classification, clustering and extraction techniques. arXiv preprint arXiv:1707.02919 (2017)
3. Campos, R., Mangaravite, V., Pasquali, A., Jorge, A., Nunes, C., Jatowt, A.: YAKE! Keyword extraction from single documents using multiple local features. Inf. Sci. **509**, 257–289 (2020)
4. Bharti, S.K., Babu, K.S.: Automatic keyword extraction for text summarization: a survey. arXiv preprint arXiv:1704.03242 (2017)
5. Liu, D., Li, Y., Thomas, M.A.: A roadmap for natural language processing research in information systems (2017)
6. Pan, S., Li, Z., Dai, J.: An improved TextRank keywords extraction algorithm. In: Proceedings of the ACM Turing Celebration Conference-China, pp. 1–7 (2019)
7. Firoozeh, N., Nazarenko, A., Alizon, F., Daille, B.: Keyword extraction: issues and methods. Nat. Lang. Eng. **26**(3), 259–291 (2020)
8. Thushara, M.G., Mownika, T., Mangamuru, R.: A comparative study on different keyword extraction algorithms. In: 2019 3rd International Conference on Computing Methodologies and Communication (ICCMC). IEEE (2019)
9. Grineva, M., Grinev, M., Lizorkin, D.: Extracting key terms from noisy and multitheme documents. In: Proceedings of the 18th International Conference on World Wide Web (2009)
10. Mulukutla, V. K., et al.: Sentiment analysis of Twitter data on 'The Agnipath Yojana'. In: Morusupalli, R., Dandibhotla, T.S., Atluri, V.V., Windridge, D., Lingras, P., Komati, V.R. (eds.) Multi-disciplinary Trends in Artificial Intelligence. MIWAI 2023. Lecture Notes in Computer Science, vol. 14078. Springer, Cham (2023). https://doi.org/10.1007/978-3-031-36402-0_50
11. Yijun, G., Tian, X.: Study on keyword extraction with LDA and TextRank combination. Data Anal. Knowl. Discov. **30**(7), 41–47 (2014)
12. Rose, S., et al.: Automatic keyword extraction from individual documents. In: Text Mining: Applications and Theory, pp. 1–20 (2010)
13. Florescu, C., Caragea, C.: A position-biased pagerank algorithm for keyphrase extraction. In: Proceedings of the AAAI Conference on Artificial Intelligence, vol. 31, no. 1. (2017)
14. Yang, F., Zhu, J., Lun, J., Zheng, Z., Tang, Y., Wu, J.: A keyword-based scholar recommendation framework for biomedical literature. In: 2018 IEEE 22nd International Conference on Computer Supported Cooperative Work in Design (CSCWD), pp. 247–252. IEEE (2018)
15. Li, S., et al.: DuIE: a large-scale Chinese dataset for information extraction. In: Tang, J., Kan, MY., Zhao, D., Li, S., Zan, H. (eds.) Natural Language Processing and Chinese Computing: 8th CCF International Conference, NLPCC 2019, Dunhuang, China, 9–14 October 2019, Proceedings, Part II, vol. 8, pp. 791–800. Springer, Heidelberg (2019). https://doi.org/10.1007/978-3-030-32236-6_72

16. Liang, H., Sun, X., Sun, Y., Gao, Y.: Text feature extraction based on deep learning: a review. EURASIP J. Wirel. Commun. Netw. **2017**(1), 1–12 (2017)
17. Anandarajan, M., et al.: Text preprocessing. Practical text analytics: Maximizing the value of text data, pp. 45–59 (2019)
18. Silpa, N., Rao, V.M.M.: Machine learning-based optimal segmentation system for web data using genetic approach. J. Theor. Appl. Inf. Technol. **100**(11) (2022)
19. Millstein, F.: Natural language processing with python: natural language processing using NLTK. Frank Millstein (2020)
20. Kumawat, D., and Jain, V.: POS tagging approaches: a comparison. Int. J. Comput. Appl. **118**(6) (2015)
21. Liu, F., et al.: Unsupervised approaches for automatic keyword extraction using meeting transcripts. In: Proceedings of Human Language Technologies: The 2009 Annual Conference of the North American Chapter of the Association for Computational Linguistics (2009)

Creating a Protected Virtual Learning Space: A Comprehensive Strategy for Security and User Experience in Online Education

Mohan Sai Dinesh Boddapati[1] [ID], Sri Aravind Desamsetti[1] [ID], Karunasri Adina[1(✉)] [ID], Padma Jyothi Uppalapati[1] [ID], P T Satyanarayana Murty[2] [ID], and RajaRao P. B. V[2] [ID]

[1] Department of Computer Science and Engineering, Vishnu Institute of Technology, Bhimavaram 534202, Andhra Pradesh, India
karunasri.adina@gmail.com
[2] Department of Computer Science and Engineering, Shri Vishnu Engineering College for Women, Bhimavaram 534202, Andhra Pradesh, India

Abstract. The pandemic has a significant impact on how people conduct meetings, both in corporations and in schools. Online meetings have become a popular way to connect people from all over the world, lowering the expenses and time associated with travel. Various video conferencing systems and communication tools have aided in this trend towards online meetings. Many countries have moved to online classrooms as an alternative to traditional face-to-face instruction in the educational sector. It has also enabled educational institutions to adapt to changing circumstances and continue to educate students. One of the major concerns is security. As online platforms become more popular, the potential of infiltration activities such as hacking or unauthorized access increases. This study proposes a comprehensive strategy for improving security and user experience in online education. The framework focuses on detecting existing participants, detecting intruders, restricting intruders, and restricting abusive messages. It employs authentication mechanisms, user behaviour analysis, network monitoring, and machine learning algorithms to validate participant identities, differentiate legitimate users from prospective invaders, restrict unauthorized access, and promote courteous conversation. The framework proves its usefulness in minimizing security concerns and promoting a secure online learning environment through simulations and case studies.

Keywords: Intrusion detection · Abusive messages · BERT model

1 Introduction

1.1 Content Creation

Covid-19 is the most lethal virus in ages. It is spreading like wildfire, and the only way to stop it is through social isolation. Schools began offering online classes to comply with this rule [1]. It may have been challenging at first, but everyone finally adapted to

P. Pareek et al. (Eds.): IC4S 2023, LNICST 536, pp. 350–361, 2024.
https://doi.org/10.1007/978-3-031-48888-7_30

the concept. You cannot deny that online classes are far more convenient than traditional classes. You can dress as you choose, access the lesson from anywhere in the world, and record the class for future reference. The most significant advantage of having classes online is that you can record all of them and refer to them later when studying as shown in Fig. 1 [2].

Online programmers are less expensive because they eliminate the need to maintain a physical site. Institutes began investing in online tools that were far less expensive than maintaining big parts of their physical properties. We understand that many teachers and kids struggled to acclimatize to technology. Eventually, everyone learned and is now aware of the different functions of a laptop or computer. The best part about online education is that you be-come Tech Savvy, and there is always something new to learn.

Fig. 1. Online Meeting

Eventually, everyone learned and is now aware of the different functions of a laptop or computer. The parent largest disadvantage was the high cost of purchasing laptop computers. Many low income parents had to use their savings to purchase laptops, as it became vital for pupils to be able to attend classes with ease. Problems with the internet, computers not working, and a lack of electricity are just a few examples. These are some of the issues that kids and teachers frequently face and are powerless to address. There are no such options available. You cannot deny that school children are capable of exploiting the circumstance and being less attentive in class.

1.2 Related Work

Several studies have looked into the use of data and technology analysis to improve security in online meeting procedures. Using facial detection or fingerprint technology, algorithmic models for machine learning have been utilized to detect the abusive messages and block such intruders in online meetings.

Karim [1] stated that the Google Meet app was the most secure against cyber-attacks, followed by Microsoft Teams and, finally the Zoom app but they didn't detect the intruders in the online meetings. Abudhagir [11] comprises one of the best face detection performances, as the images in the dataset are one shot learned it has the triplet loss which helps to avoid more unrelated example of pictures while passing through the convolutional sheets.

De la Cruz [2] presents a framework for raising awareness about the need for more robust security measures such as threat prevention, identification security, compliance with academic institution law, and ethical conduct to protect student personal information now and in the future.

Chen [6] uses a comprehensive evaluation technique to compare user experiences before and after the outbreak of COVID-19, and eventually determines how user's concerns about the online education platform have changed. This paper investigates the supporting abilities and response levels of online education platforms during COVID-19, and proposes corresponding measures to improve how these platforms function in terms of access speed, reliability, timely transmission technology of video information, course management, communication and interaction, and learning and technical support.

For deep face recognition, the Zulfiqar [5] pretrained CNN model and a set of hyper parameters are experimentally chosen. The usefulness of deep facial recognition in automated biometric identification systems is demonstrated by promising testing findings with an overall accuracy of 98.76%. With the use of lexicon-based encodings, Koufakou [9] investigates several applications for lexical characteristics and offers a thorough dataset evaluation that tack-les both in-domain and cross-domain abusive content identification.

Caselli [3] created a pre-trained BERT model for detecting abusive words in English. This model was trained using RALE, a huge dataset of Reddit comments in English from communities banned for being offensive, abusive, or hateful that we gathered and made public. Nobata [4] created a machine learning based solution that outperforms a state of the art deep learning strategy for detecting hate speech in online user comments from two do-mains[14].

On the basis of the current pre-trained language model, Huang [10] suggests a multi task framework (MFAE) combining abuse detection and emotion categorization to increase the algorithm's representational capacity. Founta [7] proposed a deep learning architecture that uses a wide range of available metadata and combines it with automatically derived hidden patterns inside tweet text to detect various abusive behavioral standards that are strongly interconnected.

1.3 Contributions

A few students exchange URLs to academic sessions/webinars/CREs with miscreants, who subsequently login to the meeting/sessions using the IDs or names of other recognized attendees. After entering the meeting, mischievous kids cause indiscipline, confusion, and use nasty abusive language to disrupt the entire conference.

The COVID Pandemic/Lockdown has caused students to be frustrated. It has also damaged their mental health, resulting in undesirable behaviors and actions that disrupt the entire class and the decorum of the session/academic activity. A few students exchange URLs to academic sessions/webinars/CREs with miscreants, who subsequently login to the meeting/sessions using the IDs or names of other recognized attendees. After entering the meeting, mischievous kids cause indiscipline, confusion, and use nasty abusive language to disrupt the entire conference (as shown in Fig. 2).

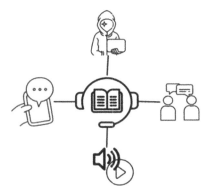

Fig. 2. Objectives of the work

Objective:

- Solutions need to be devised so that the intruders/miscreants are identified.
- They should not be able to use IDs and Names of the identified Participants/Students.
- It should get easy for the Host to block such intruders which does not happen usually.
- Messages (which are usually disturbing and offensive) need to be blocked in such a way that they are not shown in the chat box or displayed during the sessions.

2 Analysis

The effectiveness of teaching and learning, student involvement, technological utilization, and student success are all factors to consider while analyzing online classes. Here are some elements to consider when analyzing online classes:

- Learning outcomes: Providing excellent education and improving learning outcomes are two of the key goals of online classrooms. As a result, it is critical to assess the course's efficacy in delivering the desired learning outcomes. This can be accomplished by comparing student performance, like exam scores or assignment grades, to traditional in-person classes.
- Student engagement: Students must be self-motivated and proactive in their learning when taking online programmes. As a result, it is critical to assess the amount of student engagement in the course [16]. This can be accomplished by examining participation rates in class discussions, assignment and quiz completion rates, and overall attendance rates.
- Technology is frequently used in online classes to convey content and improve communication between professors and students. As a result, it is critical to assess the usefulness of the technology utilized in the course. This can include evaluating the online platform's dependability and functionality, as well as its usability for both instructors and students.

- Performance of the instructor: The role of the instructor is critical to the success of online classes. As a result, it is critical to evaluate the instructor's success in delivering course content, offering feedback and support, and building a collaborative learning environment.
- Student feedback: Finally, it is critical to collect feedback from students regarding their experiences in the course. Surveys or other forms of feedback mechanisms can be used to obtain insights on what worked well, what could be improved, and suggestions for future improvements.

Overall, online course analysis necessitates a thorough examination of all components of the course in order to establish its efficacy and identify opportunities for improvement. It is critical to collect data from many sources and stakeholders in order to acquire a comprehensive picture of the course and its impact on student learning.

3   Working Methodology

3.1   Identifying Intruders

Using facial detection or fingerprint technology [5] to identify intruders is a typical strategy used for access control and security. Here's a quick rundown of how it works:

- Facial Detection: Facial detection systems record and analyze the distinctive aspects of a person's face using cameras and specialized software. This technology can recognize certain facial characteristics such as the distance between the eyes, nose shape, and facial curves. Against find a match, these characteristics are matched against a database of authorized individuals.

  Face pattern recognition in machine learning is a multi-step process. First, a dataset of facial images is compiled, comprising both positive and negative examples of faces and non-face images. To maintain uniformity, preprocessing procedures such as scaling and normalization are used. To extract relevant facial features, feature extraction methods such as Haar cascades, HOG, or deep learning based approaches such as CNNs [11] are used. These features, together with the labels associated with them, are then used to train a machine learning model, such as SVM or KNN. A distinct dataset is used to evaluate the trained model's accuracy and performance. This procedure allows the ML model to recognize facial patterns and identify faces with a high degree of accuracy.
- Fingerprint identification: To identify people, fingerprint identification technology employs unique patterns and ridges on their fingertips. Fingerprint scanners take an image or sequence of images of the ridges on a person's finger and compare them to a database of authorized fingerprints.

  In machine learning, finger pattern recognition follows a similar procedure. A dataset of fingerprint pictures is compiled, including a wide range of samples from various individuals. To separate the fingerprint pattern, preprocessing techniques such as picture enhancement and segmentation are used. The fingerprint photos are used to extract distinguishing information such as ridge orientation and minutiae (ridge ends, bifurcations). These extracted features and labels are used to train machine learning models or statistical models such as HMMs or Neural Networks. The trained models are then assessed using different validation or test images and metrics such as FAR, FRR, or EER. This allows the machine learning model to recognize and match fingerprint patterns, allowing for more accurate fingerprint recognition.

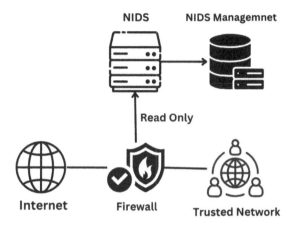

Fig. 3. Identifying Intruders in meetings

Both facial detection and fingerprint recognition systems give an extra degree of protection to the classroom by ensuring that only authorized personnel have access to it as shown in Fig. 3. These technologies are frequently utilized to improve security and prevent unauthorized entrance in a variety of industries, including education, government, and corporate environments.

3.2 Detecting Existing Participants

The following actions can be taken to recognize existing meeting attendees based on their email addresses and alert both the existing participant and the meeting organizer:

- Registration and Email Association: Each participant is needed to register for the meeting or class by entering their email address. Each participant's email address acts as a unique identification.
- Participant Database: Maintain a database that contains the email addresses of all registered attendees for the meeting or class.
- Participant Verification: When a participant enters a meeting, the meeting platform collects or extracts their email address.
- Email Address Comparison: The system compares the joining member's email address to the email addresses in the participant database.
- Detection and Notification: A pop-up notification is given to both the existing participant and the meeting organizer if a match is identified, indicating that the person already exists in the meeting or class.
- Existing Participant message: A pop-up message informs the existing participant that another participant with the same email address is attempting to join the meeting.

• Organizer Alert: A pop-up or alert notifies the meeting organizer that there are two participants in the meeting with the same email address. This advises the organizer to take appropriate action, such as validating the participants' identity or dealing with the problem.

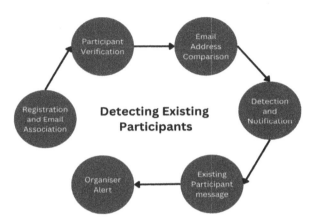

Fig. 4. Detecting Existing participants in online meetings

By following this procedure as shown in Fig. 4, both the present participant and the meeting organizer are notified of the potential duplicate participant as soon as possible. This allows the organizer to investigate and resolve any concerns linked to several participants using the same email address, maintaining the meetings or classes integrity and security.

3.3 Block Intruders

Various techniques can be used to prevent intruders and manage participant behavior during a lesson [12]. Here's an outline on how to deal with these issues:

Microphone use in class is critical for supporting good communication and engagement. Participants can express themselves, ask questions, and participate in discussions. However, in order to avoid disturbances and preserve a constructive learning environment, microphone usage must be managed. When participants should mute or un mute their microphones, instructors or meeting organizers can establish guidelines. Background noise and unexpected interruptions can be reduced by muting participants by default at admission and allowing them to un mute themselves when they want to contribute. Moderators can also actively monitor microphone usage and urge students to quiet themselves when not speaking, guaranteeing clear and focused audio throughout the session.

Video content in a class can boost engagement and promote a sense of community among students. Visual clues, nonverbal communication, and the sharing of pertinent materials are all possible. However, in order to create a focused and appropriate learning environment, video content must be managed. Video permissions can be controlled by the meeting organizer or designated moderator, ensuring that only authorized attendees have video access. Monitoring the video feeds of participants during the lesson also aids in identifying any instances of irrelevant or inappropriate content being broadcast. To protect the integrity and relevance of the class content while allowing participants to offer relevant visual contributions, prompt action, such as eliminating disruptive videos or disabling video sharing, can be taken.

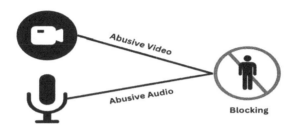

Fig. 5. Blocking intruders in meeting

Machine learning algorithms can be used in meetings to recognize irrelevant or off-topic spoken content and video, ensuring that discussions remain relevant and focused as shown in Fig. 5. The procedure entails assembling a collection of audio recordings and video feeds from meetings, which includes both linked and unconnected pieces. Noise reduction and frame extraction are two examples of preprocessing processes used on audio and video data. The audio and video streams are then analyzed to extract relevant information.

Classification algorithms or deep learning networks are taught using the retrieved characteristics and labels that indicate whether the segments are related or unrelated. These algorithms learn the properties and patterns of unrelated speech and video information. A different dataset is used to evaluate the trained models' performance in reliably detecting unrelated parts. Meeting systems or applications that use this ML-based method can automatically analyze both spoken content and video streams, indicating unconnected segments in real-time. This allows participants and organizers to stay focused and keep meetings on track, resulting in more productive and efficient talks.

3.4 Block Abusive Messages

A method for addressing abusive communications can be established to maintain a courteous and secure environment in the chat box during a meeting or class [14]. When an abusive message is discovered, the system warns the responsible person, reminding them of proper behavior. This initial warning provides the participant with an opportunity to correct their actions. If the abusive behavior continues, the system will take more serious action. The participant may be barred from further communications or perhaps removed outright from the meeting. This proactive strategy guarantees that abusive communications are addressed as soon as possible, while also promoting a good and inclusive learning environment for all participants.

It is critical to keep track of participant information and activities in order to retain responsibility and facilitate future action, if necessary. The system records pertinent information such as the participant's name, the time of the abusive communication [4], and any measures taken, such as warnings, blocks, or deletions. These log files provide a thorough record of participant behavior, which is useful evidence if additional inquiry or intervention is required. Meeting organizers and administrators can efficiently address instances of abusive messages, protect participants' well-being, and ensure a productive and courteous learning experience by logging participant information.

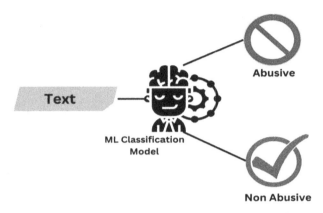

Fig. 6. Blocking abusive messages using BERT model

Using the BERT [8] model to determine whether a message is normal or abusive (as shown in Fig. 6) necessitates the use of natural language processing techniques. BERT (Bidirectional Encoder Representations from Transformers) is a sophisticated pre-trained language model that can be modified for a variety of NLP applications such as text categorization.

A labeled dataset is required to train a BERT [3] model for classifying normal and abusive texts. This dataset should include both typical and abusive message instances. To prepare the text data for entry into the BERT model, preprocessing processes such as tokenization and padding are used. BERT is then fine-tuned using labeled data, where the model learns to grasp the context and sentiment underlying the messages.

For BERT training we used Masked Language Model (MLM) approach Once fine-tuned, the BERT model that can be used to identify new messages as normal or abusive. The message is sent via the model (as shown in Fig. 7), which creates a forecast based on the training data's learnt patterns and context. To determine the classification, a threshold might be set, with a forecast above the threshold being abusive. The model's performance is assessed using metrics such as accuracy, precision, recall, and F1 score, which are compared to the ground truth labels. The model can be monitored and updated on a regular basis to increase its accuracy and adapt to changing language patterns.

i) Our text is tokenized. We begin with text tokenization, just as we would with transformers. We will obtain three different tensors as a result of tokenization:

- input_ids

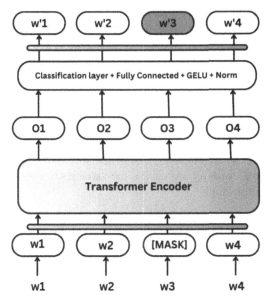

Fig. 7. Multiple layers of BERT model

- Token_type_ids
- attention_mask

ii) Make a tensor of labels. Because we're training our model here, we'll need a labels tensor to calculate loss — and optimize towards.
iii) Tokens in input_ids are masked. We can mask a random selection of tokens now that we've produced a duplicate of input_ids for labels.
iv) Determine your loss. We run the input_ids and labels tensors through our BERT model and compute the difference between them.

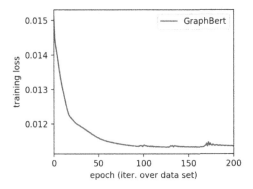

Fig. 8. Epoch vs Loss

A fine-tuned model may accurately identify communications as normal or abusive by exploiting BERT's (as shown in Fig 8) epochs vs training loss that has the ability to recognize contextual meaning. This can be implemented to a variety of applications, such as chat moderation, social media monitoring, or content filtering, to ensure that users are safe and polite.

4 Evaluation

The effectiveness of measures used to identify and prevent unauthorized access to the online class by those who are not authorized to attend is assessed during the evaluation of online class intruder tracking. Here are some variables to consider while evaluating online class intruder tracking:

- False positive: False positives occur when people are mistakenly detected as intruders. As a result, it is critical to assess the tracking system's false positive rate, which may be done by comparing the number of false positives to the total number of attempted logins.
- Response time: When an intruder is spotted, it is critical to act swiftly to prevent them from gaining access to the class. As a result, it is critical to assess the tracking system's response time, which may be accomplished by evaluating the time it takes for the system to recognize an intruder and take action to prevent them from entering the class.
- User experience: The tracking system should not interfere with legitimate users' experiences. As a result, it is critical to assess the tracking system's impact on legitimate users, such as whether it creates delays or necessitates additional authentication processes.
- Effectiveness of prevention measures: Aside from detecting and stopping intruders, it is critical to assess the effectiveness of prevention measures in place to detect possible intruders. This can include things like demanding strong passwords or two-factor authentication, which can be measured by the frequency of successful unauthorized attempts.

Overall, evaluating online class intruder tracking necessitates a thorough examination of numerous elements in order to determine its efficiency in preventing unauthorized access to the class. It is critical to regularly monitor and update the tracking system in order to resolve any vulnerabilities or weaknesses and ensure the online class environment's security.

5 Conclusion

Intruder tracking is a critical component of assuring the security and integrity of online learning environments. The tracking system is designed to detect and prevent unauthorized access to the class by people who are not authorized to be there. The efficiency of the procedures employed to identify and prevent unauthorized access to the class is assessed during the evaluation of online class intruder tracking. The detection rate, false positive rate, reaction time, user experience, and effectiveness of preventative measures are all factors that can be examined when evaluating online class intruder surveillance.

It is critical to monitor and update the tracking system on a regular basis in order to fix any vulnerabilities or weaknesses and ensure the security of the online learning environment for all genuine users. Finally, a good online class intruder tracking sys-tem can improve overall class effectiveness and promote a safer learning environment for all students and instructors.

References

1. Karim, N.A., Ali, A.H.: E-learning virtual meeting applications: a comparative study from a cybersecurity perspective. Indonesian J. Electr. Eng. Comput. Sci. **24**(2), 1121–1129 (2021)
2. De la Cruz, J.: Online Class: Student Data Privacy. Int. J. **10**(7) (2022)
3. Caselli, T., et al.: HateBERT: retraining BERT for abusive language detection in English. arXiv preprint arXiv:2010.12472 (2020)
4. Nobata, C., et al.: Abusive language detection in online user content. In: Proceedings of the 25th International Conference on World Wide Web (2016)
5. Zulfiqar, M., et al.: Deep face recognition for biometric authentication. In: 2019 International Conference on Electrical, Communication, and Computer Engineering (ICECCE). IEEE (2019)
6. Chen, T., et al.: The impact of the COVID-19 pandemic on user experience with online education platforms in China. Sustainability **12**(18), 7329 (2020)
7. Founta, A.M., et al.: A unified deep learning architecture for abuse detection. In: Proceedings of the 10th ACM Conference on Web Science (2019)
8. Hugging Face. BERT 101. Retrieved from http://www.springer.com/lncs
9. Koufakou, A., et al.: HurtBERT: incorporating lexical features with BERT for the detection of abusive language. In: Proceedings of the Fourth Workshop on Online Abuse and Harms. Association for Computational Linguistics (2020)
10. Huang, Y., et al.: A multitask learning framework for abuse detection and emotion classification. Algorithms **15**(4), 116 (2022)
11. Abudhagir, U.S., Anuja, K., Patel, J.: Faster RCNN for face detection on a FaceNet model. In: Vijayanand, R., Devaraj, D., Kannapiran, B. (eds.) Advances in Mechanical and Materials Technology: Select Proceedings of EMSME 2020, pp. 283–293. Springer Singapore (2022) https://doi.org/10.1007/978-981-16-2794-1_25
12. Vijayanand, R., Devaraj, D., Kannapiran, B.: Support vector machine-based intrusion detection system with reduced input features for advanced metering infrastructure of the smart grid. In: 2017 4th International Conference on Advanced Computing and Communication Systems (ICACCS). IEEE (2017)
13. Bushetty, S., et al.: Analysis of Online Comments Using Machine Learning Algorithms
14. Subbarao, M. V., Padavala, A. K., & Harika, K. D.: Performance Analysis of Speech Command Recognition Using Support Vector Machine Classifiers. In Communication and Control for Robotic Systems, pp. 313-325. Springer Singapore (2021).
15. Bonthu, S., Dayal, A.: Maximizing student engagement by integrating social media in assignments of an online course. J. Eng. Edu. Transformations, 35(Special Issue 1) (2022)

Lightweight Cryptography Model for Overhead and Delay Reduction in the Network

Rajesh Yamparala[✉] and T. Kamaleshwar

Department of CSE, Vel Tech Rangarajan Dr. Sagunthala R&D Institute of Science and Technology, Avadi, Chennai, Tamil Nadu 600062, India
`rajeshyamparala@gmail.com`

Abstract. Mobile Ad hoc Network (MANET) refers to a group of multi-hop wireless networks that can configure themselves. The success of MANET-based applications hinges on a number of criteria, with reliability being a key one. While many security procedures already exist, the new characteristics and vulnerabilities of this networking paradigm may make the old ones obsolete. The ability of a network to scale up in size without sacrificing performance is known as scalability. Proactive routing algorithms in mobile wireless networks face a significant scalability difficulty due to the frequent need to send and receive control signals. The key to fixing the scalability issue is cutting down on overhead. In order to lessen the burden of data transmission, a overhead and delay reduction model is designed in this research. The suggested technique is based on a weight function that integrates the node's connection and battery life. In this research, MANET network latency and traffic load reduction model is designed. The proposed technique is based on the idea of only transmitting certain, predetermined packets avoiding loss. Congestion in the network can be avoided by first ensuring that all neighbor nodes are valid before sending a packet. When it comes to the evolution of MANET, Quality of Service (QoS) is a crucial factor. This research proposes a Trusted Node Feedback based Node Authentication Model with Node Transmission Analysis (NAM-NTA) model for decreasing the network overhead and delay levels in the MANET. The emphasis of this research is on constraints based on latency and neighbor connectivity. This technique is able to detect superfluous connections and eliminate them from the network structure. The proposed achieved 98.3% accuracy in network delay reduction. The proposed model is compared with the traditional model and the results show that the proposed model performance in overhead and delay reduction is high.

Keywords: Mobile Ad hoc Network · Trusted Nodes · Node Authentication · Network Overhead · Network Delay · Node Feedback · Quality of Service

1 Introduction

A MANET, also known as a mobile communication network, is a network of mobile devices that connects to one another wirelessly and manages itself [1]. Multi-hop communication channels are used in a MANET [2]. Eventually, it will not be feasible or

P. Pareek et al. (Eds.): IC4S 2023, LNICST 536, pp. 362–375, 2024.
https://doi.org/10.1007/978-3-031-48888-7_31

even physically practicable to have a set architectural for this kind of network due to the rapidly evolving nature of wireless communication technology [3]. Due to the constantly shifting nature of the mobile structure, ad hoc wireless networks require the ability to self-organize and self-configure [4]. A MANET is a wireless network that establishes connectivity through multi-hop peer-to-peer routing rather than fixed network infrastructure [5]. MANETs find use in highly mobile and ever-changing military and civilian infrastructure. In a MANET, the structure of the network is dynamic [6].

MANETs are made up of a group of freely moving nodes. These nodes require no supporting infrastructure and can spontaneously form networks of any desired configuration [7]. Designing and implementing dynamic routing protocols with lower overhead and higher speed is a significant challenge in MANETs. For mobile ad hoc networks, researchers have developed a number of routing protocols, including Ad hoc On-demand Distance Vector Routing (AODV) and Dynamic Source Routing (DSR) [8]. Any network can be transformed into a MANET by connecting a group of mobile wireless devices together. Every moving node can act as a host as well as a router [9]. The wireless interface has a limited broadcast range because data packets must travel from their source to their destination via a series of intermediate nodes. In a MANET, packets are sent from one node to another without the use of a central hub or other infrastructure [10]. As a result, MANET can be used where a wired network would be dangerous or impractical. In addition, it is used in challenging environments where it would be impractical to replace the batteries in any of the nodes. As a result, the packet forwarding process relies heavily on the routing protocol [11]. The structure of a MANET is shown in Fig. 1.

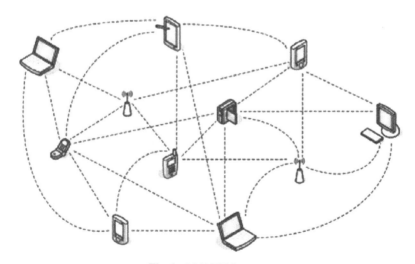

Fig. 1. MANET Structure

Ad hoc networks are becoming increasingly common in numerous contexts, including on military battlefields [12] to set up information networks between soldiers, weapon systems, command control centre, and vehicles due to their adaptability and ease of deployment [13]. Many real-time commercial applications are using the ad hoc network architecture to boost productivity and effectiveness and optimize profits for their respective corporations. These applications are used in a wide variety of situations and services [15], including tactical networks, disaster relief, [14], home the environment, education system, recreation, sensor network, and context aware services. Link failure among nodes, lack of centralized administration in the absence of infrastructure, susceptibility to attack, high power consumption, low bandwidth, scalability of the network [16], device homogeneity, multi-hop routing, self-creation, autonomy, and self-administration are just some of the problems that need to be studied in order to be solved [17]. Many researchers have developed numerous solutions to the problem of network overhead and delay by utilizing transmission requirements for energy [18], residual consumption of electricity, or both.

The nodes in a MANET are mobile computers that can move around and connect with other nodes in the network over wireless links that can go in both directions [19]. The current approach in MANET relies on cooperation between nodes to allow for long-distance communication by forwarding each other's data packets [20]. However, even in cooperative settings, some nodes can refuse to do so in order to conserve power or to purposefully disrupt regular communications, which in turn reduces the efficiency of the network [21]. This kind of malicious activity is known as a packet dropping attack or black hole attack [22], and it is one of the most serious types of attacks that can bring down a network. In order to reduce the routing cost, these strategies are used to identify selfish nodes, bypass them, and select an alternative path for data transmission. Massive routing overhead is produced after switching to an alternative method for data transmission [23]. As a result, the innovative trans-mission method is implemented in this research to reduce routing overhead and improve network stability avoiding delay [24].

The processing delay in a packet-switched network is the time it takes for routers to process a packet's header. Delays in processing time are a major cause of overall network slowness. While handling a packet, routers can perform several functions, including determining the next hop for the packet and checking for bit-level errors that may have happened during transmission. High-speed routers often have processing delays on the microsecond scale or less [25]. The router then sends the packet to a queue, where it may experience additional waiting time after undergoing nodal processing. Delay in the transport of a packet is a concept fundamental to packet switching. The network-wide transfer or buffering delay of a packet is equal to the sum of the store-and-forward delays it encounters in each router. Network congestion and the quantity of intermediate routers both contribute to overall packet transfer latency.

2   Literature Survey

To extend the life of wireless multihop networks in which individual nodes experience an energy deficit, Choi et al. [1] applied wireless power transfer (WPT) technology to multihop transmission. The author has developed a system model for CoWPT-based multihop transmissions and posed an optimization problem to determine the optimal WPT time of each node in order to extend the network's lifetime. The net-work's longevity is guaranteed by first making sure all nodes have the same expected lifetime. To make things easier, the author tackled the resulting linear programming (LP) problem instead of the optimization problem. Extensive simulations show that, compared to the normal WPT method, using the optimal CoWPT solution significant-ly extends the lifetime of networks in both WSN and MANET environments.

During mobile edge computing, connection failure is caused by node movement and decreased node energy, which in turn reduces the lifespan of the underlying mobile ad hoc network. Network latency grows dramatically when a route fails because route discovery must start afresh for single-path protocols. The proposed multi-path routing system is advantageous since it eliminates the need for route discovery. This study proposes LLECP-AOMDV, an ad hoc on-demand multi-path distance vector (AOMDV) routing protocol for mobile edge computing that is based on the prediction of connection lifetime and energy consumption. During the course of the path finding procedure, the energy grading method is utilized. When a node's energy gets too low, it stops participating in the route finding process. The routing selection process prioritizes paths with the lowest energy consumption and the longest predicted link lifetimes. The results of the comparisons were assessed by the author in terms of energy consumption, packet delivery rate, and end-to-end delay.

A mobile ad hoc network (MANET) is a loosely organized network of mobile, net-worked computers. Moving nodes increase route overhead and battery consumption, making mobile communication routing challenging. The field of MANET has made multiple attempts to reduce the energy consumption of nodes and the complexity of routing decisions. These concepts can improve load distribution and traffic flow while decreasing resource waste. The unique idea provided by Chandravanshi et al. [3] has the potential to yield better outcomes. Author proposes an adaptive Multipath Multichannel Energy Efficient (MMEE) routing approach in which route selection strategies are based on the expected energy usage per packet, bandwidth available, queue length, and channel utilization. Multipath reduces the probability of collisions by assigning data packets to different paths throughout a network, whereas multichannel employs a channel ideal assignment technique to reduce collisions between nodes. The multichannel approach divides the bandwidth of a link into multiple discrete channels. To reduce network collision, multiple source nodes can consume the channel bandwidth simultaneously. Through the use of a collaborative multipath multi-channel method, data can go from several sources to the same destination without experiencing collisions or congestion. The MMEE routing strategy is used to deter-mine which routes to take. The proposed MMEE improves network dependability by selecting the path using a load and bandwidth aware routing algorithm that takes into account the energy and predicted lifetime of individual nodes.

Of-floading computing from the edge is a challenging problem in 5G and 6G re-search since edge cloud services are not always available in outlying areas. Instead of investing in costly infrastructure to provide edge computing services in these areas, drones might be used instead. Due to limitations in drone range, it is challenging to provide effective edge computing services. Feng et al. [4] developed an edge computing architecture where drones with and without edge servers work cooperatively to pro-vide edge computing services to end users, with computing activities conveyed in a Mobile Ad hoc Network through multi-path and multi-hop. Extending the lifetime of a MANET while maintaining service quality is defined as a joint optimization issue involving the selection of computational drones, transmission paths, and task division schemes. Because of its ability to achieve an incomplete remedy to the problem via a greedy strategy, a Software-Defined-Network (SDN) controller is employed to investigate the issue. Taking into account the dynamic nature of the MANET, the original issue is reframed as a multi-path multi-hop task transmission problem with an arrival order constraint on the tasks to be executed.

Vehicle ad hoc networks (VANETs), UAV ad hoc networks, and wireless sensor networks are just a few examples of MANETs that benefit from the decentralized and mobile nature of the transparent architecture. There is a wealth of information available on the topic of building decentralized scheduling algorithms for topology-agnostic MANETs. Most of them do their analyses in completely delay-free settings. The requirement for MANETs to support time-sensitive traffic is only going to in-crease as more and more real-time applications switch to wireless communications. If a packet hasn't been sent within a certain amount of time, it is erased from the sys-tem regardless of its status. Compared to the regular delay-free version, this is light years ahead. To address the issue of accommodating time-sensitive traffic, Deng et al. [5] investigated distributed scheduling strategies for a topology-agnostic MANET. We compare the probabilistic ALOHA scheme to the more traditional TDMA, Chlamtac and Farag's GF sequence scheme, and a combination sequence scheme designed for a certain type of sparse network topology.

Despite named data networking's (NDN) promising benefits for MA-NETs, its deployment is hindered by MANETs' distinctive features and architectural incompatibilities. Due to the fixed nature of data routers and the prevalence of servers in NDN's role as providers, FIBs there tend to be stable. FIBs are often unreliable in MANETs because of the unpredictable behavior of mobile nodes acting as both data routers and providers. Too frequent updates to the FIB might cause data acquisition failures by causing a broadcast storm and out-of-date FIBs. In addition, NDN's reverse path-ways are extremely reliable because static data routers are used to build them. Disruptions on reverse paths are common in MANETs since they are made up of mobile nodes. The costs and time involved in data recovery are high, and data gathering failures are common as a result. With the intention of improving data collection success rates and reducing data acquisition expenses, Wang et al. [6] proposed an efficient data acquisition method for NDN-based MANET. The concept has mobile nodes retrieve information without using FIBs and then distribute it to multiple users. Data continuity and successful reception on the outbound path are also guaranteed, along with mobility support.

When a natural disaster occurs, the normal communication infrastructures are immediately destroyed. Intermittently connected MANETs are essential for providing network access in the aftermath of a disaster. Despite the fact that the majority of previous study on such networks has focused on one-to-one chat, the significance of monitoring apps has expanded in recent years. In monitoring applications, the timeliness of the data is more significant than its delay characteristics. A mathematical assessment of the age of information (AoI) was proposed by Inoue et al. [7] for MA-NETs with intermittent connectivity; this study would capture the timely nature of data collected by monitoring programs. The author used the results of the analysis to go into the basics of network architecture.

Transmission reuse, or the cooperation of the various destinations, is the primary advantage of multicast over multi unicast. Multicast's ability to increase transmission efficiency stems from this feature. This has led to extensive study of multicast in a variety of wireless contexts. When using multicast rather than multiple unicast, the implications of node mobility on transmission reuse are better known. Correlated mobility, a phenomena that faithfully replicates real-world mobility processes, is the focus of Jia et al. [8], who want to clarify the effect of mobility on transmitter reuse in mobile ad hoc networks. The multicast gain was employed by the author as a metric of transmission reuse since it represents the capacity ratio of multimedia to multi unicast under a particular delay limitation. By developing a multi-layer routing proto-col and presenting many causal scheduling techniques, we can examine the impact of different correlation degrees of node mobility on the total multicast capacity delay tradeoff. The capacity and delay advantages of multicasting are calculated and compared to the unicast scenario.

3 Proposed Model

Nodes in MANETs can function as both a source and a sink, making this type of network highly adaptable. Memory, power, and network buffers are just a few examples of scarce resources for mobile nodes. Information interchange, path selection, and routing are only few of the procedures that use up these resources. When the nodes take on the role of routers, they must communicate with one another to share and improve their routing knowledge. The network's performance, durability, and convergence can all be kept at satisfactory levels with the help of control data. Nodes rely heavily on control data to maintain an accurate routing table. Depending on the routing protocol in use, the control data may exchange multiple short packets to ascertain the accessibility of surrounding nodes, for instance.

Nodes in a MANET are characterized by their minimum requirements for transmission power, distribution, mobility, and memory. Due to the limited transmission range, path of minimum hop, minimum normalized residual energy used, minimal level absolute remaining energy used, minimum transmission energy path, throughput, bandwidth, hop count, and power to be sufficient, the wireless mobile nodes enter and leave the network dynamically in MANETs. In a typical network, information travels via several intermediate nodes before reaching its intended destination. Since each mobile ad hoc network node must contribute to network traffic that is unrelated to its own needs, a router is required to keep packets flowing smoothly. Congestion control is essential in MANETs to prevent packet loss, boost network energy, and lower overhead costs, all of which can be attributed to the high volume of traffic. Congestion happens when there are more packets trying to travel across a network than the network can handle at once. Without a fixed network, mobile nodes can talk to one another using radio links. The proposed model framework is shown in Fig. 2.

Control routing refers to the transmission of routing and control data into a network, and it is essential for network resilience. To achieve rapid convergence, it is necessary to inform all nodes in the network if there is a change in the network path due to link failure or busy nodes. Overhead refers to the portion of a transmission that consists of control routing rather than data. To maintain an appropriate degree of overhead without

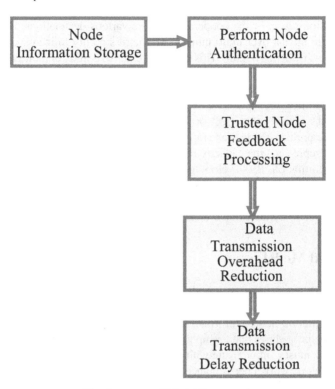

Fig. 2. Proposed Model Framework

sacrificing throughput is a key network objective. If the network's overhead is low, it will employ its maximum effective throughput at the expense of reliability. While adding more overhead improves reliability, it slows effective throughput. Because of the nodes' mobility, more control routing operations need to be performed in ad hoc networks, leading to a high level of overhead. When nodes are moved, the network's topology is constantly being reconstructed since the end-to-end path is altered. This proposal evaluates the effects of various ad hoc routing protocols by measuring certain indicators across the network. There are two main categories for metrics: reliability and overhead. Measures of reliability include throughput, packet loss, delay and round-trip time. Overhead behavior and cumulative overhead are features of the overhead set method. This research proposes a Trusted Node Feedback based Node Authentication Model with Node Transmission Analysis (NAM-NTA) model for decreasing the network overhead and delay levels in the MANET.

Initially the nodes that need to participate in ad hoc communication information is maintained in the selected node in the network that has high performance in the past data transmissions. The node information is maintained as

$$Nodeinfo[NodeList] = \sum_{n=1}^{NodeList=K} \frac{getnodeaddress(n)}{maxlimit(NodeList)} + allocenergy(n) + \tau(n)$$

$$(1)$$

$$NodeKey[K] = \prod\nolimits_{n=1}^{K} \frac{maxener(n)}{\lambda} + getTime(n) + Nodeinfo(n) \qquad (2)$$

τ is the node computational level to allocate tasks to the node. The node address is gathered for future communication. λ is the total energy allocated.

The node authentication is performed by using light weight cryptography models that is used to validate the nodes during data transmission and receiving. The process of node authentication is performed as

$$NodeAuthen[K] = \sum\nolimits_{n=1}^{K} Nodeinfo(n) + getnodeaddress(n)$$
$$+ \begin{cases} if\ NodeKey(n) == getNodeKey(n)\ return\ access \\ Otherwise \qquad\qquad\qquad\qquad\qquad remove(n) \end{cases} \qquad (3)$$

The trust factor of nodes helps in detection of malicious actions in the network. For each node, its adjacent node feedback about the current node is considered as an important parameter in node usage in data transmission. The trusted node feedback is calculated in quick packet delivery rate to avoid delay. The process is performed as

$$TfNeigh[K] = \prod\nolimits_{n=1}^{K} getmaxPDR(n+1) + enercons(n+1)$$
$$+ \max(\tau(n+1)) \begin{cases} Tf = 1\ if\ (TFNeigh(n) > Th) \\ 0 \qquad\qquad\qquad Otherwise \end{cases} \qquad (4)$$

Overhead in MANET is any unneeded consumption of time, space and energy in data transmission. Overhead is the additional storage needed for supporting data that facilitates the conveyance of a particular message from a source to a receiver. The overhead during data transmission is calculated as

$$Ovrhd[K] = \sum\nolimits_{n=1}^{K} getTime(\gamma) - getTime(\gamma+1) + \omega(n, n+1)$$
$$+ Th \begin{cases} if\ Ovrhd(n) > Th\ return\ 1 \\ otherwise \qquad\qquad return\ 0 \end{cases} \qquad (5)$$

γ is the data transmission levels of the current node to the next node. ω is the delay levels of the transmission among current and next trusted neighbor nodes.

There are several potential sources of network delays in any given communication system. The distance here between origin and the target is the most fundamental factor. Nevertheless, the information does not move straight from one location to another. Several intermediate nodes along its route will add to the transmission time. The delay level caused by malicious nodes in the network is calculated as

$$DelayL[K] = \prod\nolimits_{n=1}^{K} \frac{\tau(n, n+1) + minLoss(n)}{\max(PDR)} + \max(Ovrhd(n, n+1)) - Th \quad (6)$$

4 Results

Nodes in a mobile ad hoc network communicate wirelessly and independently. Using MANET, a temporary network can be formed without the need for a single point of control. This is because nodes in a MANET tend to move around a lot. There are many instances of broken connections and blocked pathways. If a mobile node can't determine the path to its target, it will blindly retransmit the route request packets to its adjacent nodes, resulting in a broadcast storm. A rebroadcast delay model to decide rebroadcast order, define a connectivity factor to maintain network connectivity, and establish a rebroadcast probability to maximize the use of neighbor coverage knowledge is proposed. Because of this, the overhead is decreased and the network performance is improved because to the synergy between neighbor coverage knowledge and the probabilistic approach.

This research ultimate goal is to design a energy-efficient data transmission protocol for MANET, which significantly increases the network's lifetime. Using the available protocol methods with increased transmission coverage range of relay nodes, MANET devices are able to keep the network running on a restricted battery supply. However, when compared to direct transmission protocol idea methods, cooperative communication protocol methods have not always been successful energy efficient approaches to increasing the transmission range of intermediate relay nodes with less power consumption. This research proposes a Trusted Node Feedback based Node Authentication Model with Node Transmission Analysis (NAM-NTA) model for decreasing the network overhead and delay levels in the MANET. The proposed model is compared with the traditional Lifetime Maximization in Wireless Multihop Networks (LMWMN) model and the results represent that the proposed model performance is high in delay and overhead reduction.

The nodes in the network that need to participate in data transmission will provide the node information to the monitoring authority in the network. This information helps in recognition of nodes in transmissions. The Network Node Information Storage Time Levels in milliseconds of the existing and proposed models are shown in Fig. 3.

Each node in the network will be authenticated with the unique key provided after information storage in the network. The unique key helps in recognition of malicious nodes in the network. The node authentication helps in maintaining access control only for the authorized users. The Node Authentication Accuracy Levels in percentage of the proposed and existing models are shown in Fig. 4.

The trust factors calculated in the nodes helps in consideration of only trusted nodes and neighbor feedback is considered only from the trusted nodes. The feedback helps in detection of malicious actions in the network and such activities can be avoided to increase the network lifetime. The Trusted Node Feedback Gathering Time Levels of the proposed and existing models are depicted in Fig. 5.

The sum total of all throughput, energy, storage, time, etc. consumed by all sensor nodes in the network is called overhead. The network overhead plays a major role in performance of the network. The minimum the overhead, the maximum the performance is. The proposed model as it considers the neighbor feedback and detects the malicious actions, the data transmission is also performed in active state and then nodes enters into inactive state in which the usage of the resources is reduced and balanced. The

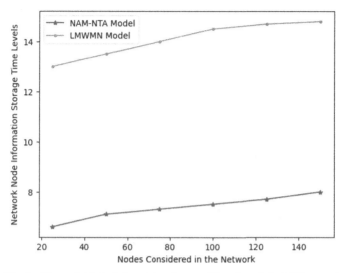

Fig. 3. Network Node Information Storage Time Levels in Milliseconds

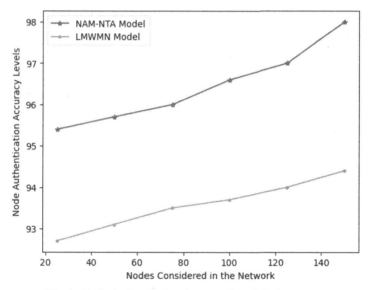

Fig. 4. Node Authentication Accuracy Levels in Percentage

Fig. 6 shows the Network Overhead Reduction Time Levels of the existing and proposed models.

A MANET network delay is a function of its design and operation. It defines the time it takes for data to go from one trusted node to another in a network. Multiplied or fractional seconds are common units of measurement. Network latency is the overall time required for a message to travel from sender to receiver, whereas propagation delay

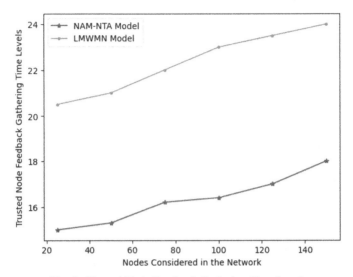

Fig. 5. Trusted Node Feedback Gathering Time Levels

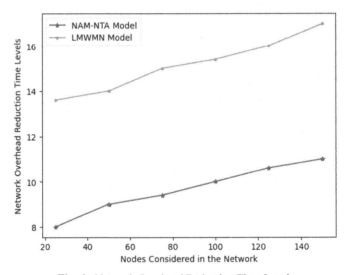

Fig. 6. Network Overhead Reduction Time Levels

is the process required the initial bit to transit over a link. The Network Delay Level Reduction Accuracy Levels of the existing and proposed models are shown in Table 1 and Fig. 7.

Table 1. Accuracy Levels

Nodes considered in the Network	Models Considered	
	NAM-NTA Model	LMWMN Model
20	96.2	90.2
40	96.3	91
60	97.3	91.3
80	97.6	91.7
100	98	92
120	98.1	92.1
140	98.3	92.4

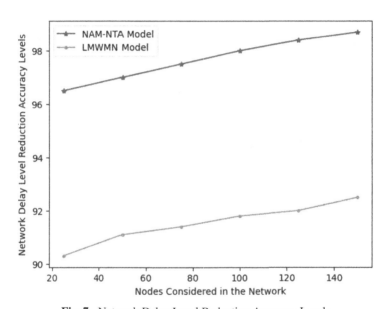

Fig. 7. Network Delay Level Reduction Accuracy Levels

5   Conclusion

The nodes in MANETs can move around freely. The nodes have complete mobility. Without a predetermined framework, these nodes are capable of dynamically self-organizing into networks of any desired topology. MANETs suffer from frequent link breakages since their nodes are constantly moving around. This resulted in several route discoveries and path failures. The time and effort required to find a new path must be taken into account. Broadcasting is an essential and efficient tool for disseminating information during the route finding process. Important network environments that greatly affect the performance of routing protocols in MANETs include power, mobility, topology, and node density. Due to their wireless nature and the fact that the nodes interact with one another via multi-hop routing, mobile ad-hoc networks are highly adaptable and independent of any central infrastructure. The majority of currently-used MANET routing methods are designed for networks in which every node has the same data transfer and processing capabilities. In terms of MANET scalability, homogeneous networks perform poorly compared to heterogeneous networks despite being simpler to model and investigate. This research proposes a Trusted Node Feedback based Node Authentication Model with Node Transmission Analysis model for decreasing the network overhead and delay levels in the MANET. The proposed model achieves 97% accuracy in overhead and delay reduction. The proposed in future can be enhanced by designing a malicious node detection models for removing such nodes for performance enhancement. The proposed model also can be extended for clusters generation and data transmission with secured node authorization models.

References

1. Choi, H.-H., Lee, K.: Cooperative wireless power transfer for lifetime maximization in wireless multihop networks. IEEE Trans. Vehic. Technol. **70**(4), 3984–3989 (2021)
2. Zhang, D.-G., et al.: A multi-path routing protocol based on link lifetime and energy consumption prediction for mobile edge computing. IEEE Access **8**, 69058–69071 (2020)
3. Chandravanshi, K., Soni, G., Mishra, D.K.: Design and analysis of an energy-efficient load balancing and bandwidth aware adaptive multipath N-channel routing approach in MANET. IEEE Access **10**, 110003–110025 (2022)
4. Feng, G., Li, X., Gao, Z., Wang, C., Lv, H., Zhao, Q.: Multi-path and multi-hop task offloading in mobile Ad Hoc networks. IEEE Trans. Vehic. Technol. **70**(6), 5347–5361 (2021)
5. Deng, L., Liu, F., Zhang, Y., Wong, W.S.: Delay-constrained topology-transparent distributed scheduling for MANETs. IEEE Trans. Vehic. Technol. **70**(1), 1083–1088 (2021)
6. Wang, X., Lu, Y.: Efficient forwarding and data acquisition in NDN-Based MANET. IEEE Trans. Mob. Comput. **21**(2), 530–539 (2022)
7. Inoue, Y., Kimura, T.: Age-effective information updating over intermittently connected MANETs. IEEE J. Select. Areas Commun. **39**(5), 1293–1308 (2021)
8. Jia, R., Lin, F., Zheng, Z.: Exploring the impact of node correlation on transmission reuse in MANETs. IEEE Access **8**, 12607–12621 (2020)
9. Veeraiah, N., et al.: Trust aware secure energy efficient hybrid protocol for MANET. IEEE Access **9**, 120996–121005 (2021)
10. Durr-e-Nayab, Zafar, M.H. and Altalbe, A.: Prediction of scenarios for routing in MANETs based on expanding ring search and random early detection parameters using machine learning techniques. IEEE Access **9**, 47033–47047 (2021)
11. Fan, R., Atapattu, S., Chen, W., Zhang, Y., Evans, J.: Throughput maximization for multi-hop decode-and-forward relay network with wireless energy harvesting. IEEE Access **6**, 582–24 (2018)
12. Noor, M.B.M., Hassan, W.H.: Current research on Internet of Things (IoT) security: a survey. Comput. Netw. **2019**(148), 283–294 (2019)
13. Arvind, S., Narayanan, V.A.: An overview of security in CoAP: attack and analysis. In: Proceedings of the 2019 5th International Conference on Advanced Computing & Communication Systems (ICACCS), pp. 655–660, Coimbatore, India (2019)
14. Surendran, S., Nassef, A., Beheshti B D.: A survey of cryptographic algorithms for IoT devices. In: Proceedings of the 2018 IEEE Long Island Systems, Applications and Technology Conference (LISAT), pp. 1–8, Farmingdale, NY, USA (2018)

15. Kumar, P., Yun, L., Guangdong, B., Andrew, P., Jin, S.D., Andrew M.: Smart grid metering networks: a survey on security, privacy and open research issues. IEEE Commun. Surv. Tutor. **21**, 2886–2927 (2019)
16. Abosata, N., Al-Rubaye, S., Inalhan, G., Emmanouilidis, C.: Internet of Things for system integrity: a comprehensive survey on security, attacks and countermeasures for industrial applications. Sensors **21**, 3654 (2021)
17. Sarenche, R., Salmasizadeh, M., Ameri, M.H., Aref, M.R.: A secure and privacy-preserving protocol for holding double auctions in smart grid. Inf. Sci. **2021**(557), 108–129 (2021)
18. Abdallah, A., Xuemin, Sherman S.: A lightweight lattice-based homomorphic privacy-preserving data aggregation scheme for smart grid. IEEE Trans. Smart Grid **9**, 396–405 (2018)
19. Khan, A., Vinod, K., Musheer, A., Saurabh, R.: LAKAF: lightweight authentication and key agreement framework for smart grid network. J. Syst. Archit. **116**, 102053 (2021)
20. Abbasinezhad-Mood, D., Nikooghadam, M.: An anonymous ECC-based self-certified key distribution scheme for the smart grid. IEEE Trans. Ind. Electron. **2018**(65), 7996–8004 (2018)
21. Grover, H.S., Kumar, D.: Cryptanalysis and improvement of a three-factor user authentication scheme for smart grid environment. J. Reliab. Intell. Environ. **2020**(6), 249–260 (2020)
22. Braeken, A., Kumar, P., Martin, A.: Efficient and provably secure key agreement for modern smart sensing communications. Energies **11**, 2662 (2018)
23. Khan, A A., Kumar, V., Ahmad, M., Rana, S., Mishra, D.: PALK: password-based anonymous lightweight key agreement framework for smart grid. Int. J. Electr. Power Energy Syst. **121**, 106121 (2020)
24. Chaudhry, S.A.: Correcting PALK: password-based anonymous lightweight key agreement framework for smart grid. Int. J. Electr. Power Energy Syst. **125**, 106529 (2021)
25. Deng, L., Gao, R.: Certificateless two-party authenticated key agreement scheme for smart grid. Inf. Sci. **2021**(543), 143–156 (2021)
26. Chaudhry, S.A., Alhakami, H., Baz, A., Al-Turjman, F.: Securing demand response management: a certificate-based access control in smart grid edge computing infrastructure. IEEE Access **8**, 101235–101243 (2020)
27. Jan, M.A., Khan, F., Alam, M., Usman, M.: A payload-based mutual authentication scheme for Internet of Things. Future Gener. Comput. Syst. **92**, 1028–1039 (2019)
28. Park, C.-S., Park, W.-S.: A group-oriented DTLS handshake for secure IoT applications. IEEE Trans. Autom. Sci. Eng. **15**, 1920–1929 (2018)
29. Abdullah, D., et al.: Super-encryption cryptography with IDEA and WAKE algorithm. J. Phys. Conf. Ser. **1019**, 012039 (2018)

Modeling the Data Object Routing in Data Aware Networking

G. K. Mohan Devarakonda[1]([✉]) [ID], Sita Rama Murthy Pilla[2] [ID], Preethi Bitra[3] [ID],
P. L. N. Prakash Kumar[1] [ID], Poodi Venkata Vijaya Durga[3] [ID], and R. Pitchai[4] [ID]

[1] Department of CSE(AI & ML), Vishnu Institute of Technology, Bhimavaram, India
krishnamohan.dg@vishnu.edu.in
[2] Department of CSE(AI & DS), Vishnu Institute of Technology, Bhimavaram, India
[3] Department of CSE, Vishnu Institute of Technology, Bhimavaram, India
[4] Department of CSE, B V Raju Institute of Technology, Narsapur, India

Abstract. Data Aware Networking (DAN) is one of the key technologies coined by the International Telecommunication for Technology. It aims to provide a system and method for managing data over the Internet. It also considers the needs of diverse users using the Internet and keeps track of their usage history to make the data proactive for them. Four key components of Data Aware networking include the documents called Data Objects which are the focal point of attention, authors who create the documents, publishers who will maintain the documents along with a catalog of documents and access rights, the End users who will be utilizing the documents in a sophisticate way provided by the concept of DAN. Considering the architecture and the way the data objects are published in data-aware networking, this paper aims to suggest a dynamic and sophisticated routing system that will study the needs of diverse users and attempt to route the data either in a proactive or reactive manner to suit the needs of the users. This paper also proposes routing algorithms to be utilized in data-aware networking for both reactive and proactive routing.

Keywords: Data Aware Networking · Data Objects · Named Data Objects · Data Driven Networking

1 Introduction to Data-Aware Networking

Depending on the use case scenarios that are being active in the field of communications and networking, the International Telecommunication Union for Technology coined the term Data Aware Networking clearly mentioning the requirements from the viewpoint of Data Forwarding, Data Routing, Mobility, Security, and Performance Management of the telecommunication networks [13, 14]. The major objective of Data Aware Networking is to come up with efficient and effective data maintenance and utilization system for consumers with the most recent data that is on demand [1]. Along with this, Data Aware Networking also attempts to create a layer over the traditional TCP/IP model to take care of the data routing and object modeling aspects.

© ICST Institute for Computer Sciences, Social Informatics and Telecommunications Engineering 2024
Published by Springer Nature Switzerland AG 2024. All Rights Reserved
P. Pareek et al. (Eds.): IC4S 2023, LNICST 536, pp. 376–382, 2024.
https://doi.org/10.1007/978-3-031-48888-7_32

Data in Data Aware Networking (DAN) may be of any kind like Text, Numbers, Images, Audio, Video, Links, and any other multimedia data. The data is termed a Data Object in DAN which is created by its author for distribution over a community of users or for the public. Such data created by the author will be given a unique name and will be made available on the internet via the publisher who takes the opportunity to maintain the data objects created by several authors [5–8]. On the other hand, the publisher also maintains the user bases who are utilizing the data objects that are available. The publisher-centered method to access data objects enables keeping track of the errant users and is also helpful to provide data on demand to the required users.

A productive Systematization of data objects makes the data objects simple to find, recover and circulate from their related DAN segments and furthermore patches up their execution and usage. This can be acknowledged by ordering and formulating the information process in like manner into different classifications of their significance [11, 12]. The routing of the data object may be reactive or proactive. In reactive routing, the data objects will be supplied to the users on demand, and in proactive routing, the history of the data object and respective users will be taken into consideration based on which the data objects will be automatically made available to the users.

2 Concept of Data-Aware Networking

The Major Contributors to the concept of Data Aware Networking are the Authors of the Data, Publishers of the Data, Data Managers, and finally the consumers of the Data who follow a sophisticated procedure to store the data files which are termed as data objects. It all starts with the creation of a data file by the author. The author will submit the same to the publisher to make the data available to the public. On receipt of the file from the author, the publisher allows a unique Id to the data file which is then called a Named Data Object. The publisher holds the responsibility to maintain a categorical catalog of data objects that he maintains and makes the catalog available to the public through the web media. The consumers on the other hand are the users in read of data. The consumers can make a request to the data manager who is responsible for maintaining the consumer base and serving the request of the consumers. On receipt of the request from the consumer regarding a particular data object, the data manager transfers the same to the publisher who in turn will reply to the consumer through the data manager with the required data object. Hence, the whole process involves the secure data transfer between Author, Publisher, Data Manager, and Consumer.

Apart from this, the concept of data-aware networking introduces the concept of networking based on data. It means that the data should be the one with which the entire networking is going to operate and it requires certain facilities to be made available in data-aware networking which are discussed in Sect. 3.

3 Requirements of Data-Aware Networking

Data Consumers in Data Aware Networking will acquire data by searching for the data using the Id or Unique Name that is given to the data object by the publisher. As there will be a huge number of data objects that relates to a particular domain, the data objects need to be categorized and kept in a particular to facilitate ease of use [9, 10]. This task can be facilitated by preparing a proper data structure for the content to be uploaded by the publisher using a suitable relational database management system. In our earlier work, we proposed a data structure to maintain a database of objects and named it a Data Object Bank so as to facilitate the categorical storage of data objects. It is also proposed to maintain the database with a suitable relational model so as to maintain the Consumer data as well as the maintenance of Consumer history. Reliability is one aspect also needs to be paid attention to build a secure and trustable system [3, 9].

Whenever the consumer is in need of data and puts the data request, we need to verify the authenticity and authorization of the user in order to prevent errant users from accessing the most secure data. At the same time, the concept of data networking ensures that the data that is being accessed by the user is the latest version and most appropriate to him. This keeps the user satisfied and helps the consumer to prevent himself from gaining wrong knowledge.

There will be some consumers who are going to access the data on a daily basis. As the recommended system is data-aware, it should keep track of the user requirements and help the user to find the data required and make it available to the consumer in time without any request. These facilities require some automated mechanism needs to be implemented while providing access to the data to the consumers based on their usage history. We may also prepare a machine learning model that used the dataset of usage history and predicts the data requirements.

In this paper, we proposed two different routing models to facilitate secure data transfer between the said entities in the data-aware networking system. One routing technique holds the responsibility to transfer data on demand and is named as On-Demand Routing. The other routing technique keeps track of consumer history and makes the proactive data transfer by making the data object available to the consumer as and when it is required. Methods for applying both routing algorithms are described in detail in the next sections.

4   Reactive Routing Algorithm for Routing in DAN

One can find different ways and algorithms to route the data objects in a variety of ways to serve the consumer base with ease and efficiency. As the focal point of attention here is the Named Data Object (NDO) and the entire functionality to be achieved is based on the, the routing here is to happen a little different than the traditional routing. We can assume this is an additional layer on the traditional layered TCP/IP network which acts based on the NDO. Hence, the routing takes two different versions based on the NDO namely Reactive Routing and Proactive Routing. The algorithm that we proposed for the reactive algorithm that is written below will provide a way to route the NDOs on demand in a very fast, effective, and efficient manner by incorporating the features of software-defined networking which is a trending domain in the era of computer science.

The goals for defining the reactive routing system include centralized control over the data and its related resources for efficient routing and distribution of resources. The conceptual diagram for the reactive routing in Data-aware networking is shown below (Fig. 1):

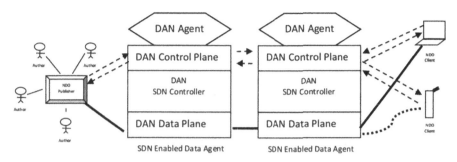

Fig. 1. SDN-Based Reactive Routing System for Data-Aware Networking

The SDN-based reactive routing system is built by adding a new layer on the conventional TCP/IP-based routing system with three agents to take control over the data-based networking namely the DAN Data Plane, DAN Control Plane, and the DAN Agent [2, 4, 15, 16]. The Control plane has the ability to be an interface between the traditional network and the DAN network and is a set of functions to receive the data from the underlying network, store and forward Named Data objects as directed by the traditional networking system. The DAN Data Plane implements physical control of underlying networks and holds the ability to change routing statistics so as to implement data-based networking. The DAN-SDN Controller will have complete control over the routing system and holds the ability to control all entities of the Data Aware Networking system.

The DAN agent holds the functionality to define the new route based on the data that is being transferred based on the earlier history of data transfer. The set of entities in the routing architecture will interact through the SDN agent which can be an integration of all entities covered in the routing system. In the mechanism of reactive routing, authors will initially create the data file and submit it to the publisher. The author may also have the possibility to publish the same data object with multiple publishers.. The publisher will maintain the catalog of data objects. The consumer data maintenance is taken care of by the DAN Agents on the other hand where the consumer will register themselves with the DAN agent. Consumers may request for a data service through the DAN agent. The DAN agent then in turn makes a request to the publisher and the publisher will be in service of data to the DAN agent with the required data object.

Proper authentication and authorization rules will be implemented for the consumer before the service is made available. The data object is then served from the DAN agent to the consumer through intermediary agents. The intermediary agents based on the statistical analysis, may keep a copy of the data object.

Algorithm for Reactive Routing of NDOs in DAN

  i. Data Created by the author will be submitted to Publisher along with relevant information.
 ii. The publisher prepares the Data Object Header and allots a unique Identity to the data object (NDO).
iii. The publisher prepares the NDO by attaching a header to the Data object.
 iv. The publisher pushes the NDO to the DAN publishing System/Database.

v. The Published data will be updated for all agents of DAN agents to be of service to consumers.

vi. Consumers may request the service of a nearby DAN Agent for the NDO.

vii. The DAN Agent searches the publisher's database for the availability of the NDO. If available, the request will be served with the required NDO, or else the request will be forwarded to the publisher.

viii. The Publisher will serve the NDO to the Consumer through the DAN agent to be on service to the consumer.

ix. DAN Agents will also copy the NDO based on the demand as per the statistical information.

5 Proactive Routing Algorithm for Data Routing in DAN

Apart from reactive routing, proactive routing holds some special responsibilities as listed below:

• It should keep track of the earlier history of the consumer data access and find the best suitable object to access next.

• It should be able to observe patterns in the data access pertaining to that particular user to serve him with the best available choice

• The data access history will also aid in timely data transfer with which the consumer will be more comfortable

As there is no initial data available for the first time proactive routing is applied, it takes the data from the reactive routing the history of data objects transferred, and the user access of particular data, dates, and times of data transfer along with all the required statistics to route the data automatically. Hence, we can presume that the proactive routing will be using the data transfer statistics from the reactive system only. The caching system in the reactive routing system will help to transfer the data objects with ease and faster as the data may not be only accessed from the server but also can be from the intermediary nodes as maintained by the reactive routing mechanism. From this point of view, the reactive routing and proactive routing are working in integration with each other and hence there is no any sort of ambiguity in the data transfer mechanisms. The step wise algorithm that we proposed for the proactive routing of NDOs in data-aware networking is given below:

Algorithm for Proactive Routing of NDOs in DAN

i. Data Created by the author will be submitted to Publisher along with relevant information.

ii. NDO Publisher prepares the header of the data object and will allot a unique ID and a unique Name to the data object.

iii. The publisher makes NDO by adding header data to the Data object

iv. The publisher will publish the document in DAN databases and updates the catalog of available data objects.

v. The Publication detail will be updated with DAN Agents as per the demand and supply statistics available with the publisher.

vi. The NDO Agents will also copy the NDO at various intermediary nodes based on the required statistical information available with the DAN database.

vii. Final Publication data will be updated with all DAN agents to be on service to the future request of Consumers.

6 Conclusion and Future Work

Both the reactive routing and proactive routing algorithms will cache the data objects at the time of data transfer anda common database of all data usage statistics will be made available with all required statistics to both algorithms. The outcome of the reactive routing system will depend entirely on the NDO and the network route will be finalized based on the availability of NDO in various intermediary nodes. This routing may also differ in different situations for the same NDO. The outcome of the proactive routing depends not only on the NDO but also on the earlier data usage statistics, and consumer base statistics, and the route will be finalized based on all the aspects of data-aware networking. Hence we may assume that the proactive routing may perform well over hundreds of data transfers and the reactive routing will start performing the best from the initial data transfer onwards.

As proactive routing takes into consideration the usage statistics of data with respect to every user, one may consider applying different machine learning techniques to classify the data statistics based on consumer's usage and predict the upcoming specifications of the same consumer. The inclusion of machine learning techniques may help the realm of data-aware networking to grow much better and more flexible and may also provide the consumers with much better data accessing platform.

References

1. Devarakonda, G.K., Pradeep,I.K., Rao, M.V.D., Krishna, Y.K.: Publishing data objects in data aware networking. In: Gupta, N., Pareek, P., Reis, M. (eds.) Cognitive Computing and Cyber-Physical Systems. EAI IC4S2022. Lecture Notes of the Institute for Computer Sciences, Social Informatics and Telecommunications Engineering, vol. 472. Springer, Cham (2023). https://doi.org/10.1007/978-3-031-28975-0_6
2. Goud, K.S., Gidituri, S.: Security challenges and related solutions in software defined networks: a survey. Int. J. Comput. Netw. Appl. (IJCNA) 9(1), 22–37 (2022)
3. Padmavathy, N.: Reliability evaluation of environmentally affected mobile Ad hoc wireless networks. In: Hura, G.S., Singh, A.K., Siong Hoe, L. (eds.) Advances in Communication and Computational Technology. ICACCT 2019. Lecture Notes in Electrical Engineering, vol. 668. Springer, Singapore. https://doi.org/10.1007/978-981-15-5341-7_98
4. Singh, J., Sunny, B.: Detection and mitigation of DDoS attacks in SDN: a comprehensive review, research challenges, and future directions. Comput. Sci. Rev. 37 (2020)
5. Yu, K., Eum, S., Kurita, T.: Information-centric networking: research and standardization status. IEEE Access 7 (2019)
6. Ravi Kumar, M., Sujatha Lakshmi, V., Sundara Krishna, Y.K.: A novel structure of data objects in data aware networking. Int. J. Eng. Res. Comput. Sci. Eng. 5(4) (2018)
7. Sujatha Lakshmi, V., Sundara Krishna, Y.K.: Contextual framework of data object in data aware networking. Int. J. Appl. Eng. Res. 13(23) (2018)

8. Li, Z., Xu, Y., Zhang, B., Yan, L.: Packet forwarding in named data networking requirements and survey of solutions. IEEE Commun. Surv. Tutor. (2018)
9. Arafath, M.S., Khan, K.U.R., Sunitha, K.V.N.: Pithy review on routing protocols in wireless sensor networks and least routing time opportunistic technique in WSN. In: 10th International Conference on Computer and Electrical Engineering, IOP Conference Series: Journal of Physics 933(2018). https://doi.org/10.1088/1742-6596/933/1/012016
10. ITU-T y.3071 Telecommunication standardization sector of Itu (03/2017) Series y: Global Information Infrastructure, Internet Protocol Aspects, Next-Generation Networks, Internet of Things and Smart Cities Future Networks Data Aware Networking (Information Centric Networking) – Requirements and Capabilities
11. Amadeo, M., Campolo, C., Quevedo, J., Corujo, D.: Information-centric networking for the internet of things: challenges and opportunities. IEEE Network, March/April (2016)
12. Shang, W., Yu, Y., Droms, R.: Challenges in IoT networking via TCP/IP architecture. NDN Technical Report Ndn-0038, February 2016
13. Itu-t y.3033 Telecommunication Standardization Sector of Itu (01/2014) Series y: Global Information Infrastructure, Internet Protocol Aspects And Next-Generation Networks Future Networks Framework of Data Aware Networking for Future Networks
14. Lopez, J.E., Arifuzzaman, M., Zhu, L.: Seamless mobility in data aware networking. In: ITU Kaleidoscope Academic Conference (2015)
15. Kandoi, R., Antikainen, M.: Denial-of-service attacks in OpenFlow SDN networks. In: 2015 IFIP/IEEE International Symposium on Integrated Network Management (IM), pp. 1322–1326. IEEE (2015)
16. Kevin, P., Bouet, M., Leguay, J.: Disco: distributed multi-domain SDN controllers. In: 2014 IEEE Network Operations and Management Symposium (NOMS), pp. 1–4. IEEE (2014)

Image Processing

Glaucoma Stage Classification Using Image Empirical Mode Decomposition (IEMD) and Deep Learning from Fundus Images

D. Shankar[1]([⊠]) [iD], I. Sri Harsha[1], P. Shyamala Madhuri[1] [iD],
J. N. S. S. Janardhana Naidu[1] [iD], P. Krishna Madhuri[1] [iD],
and Srikanth Cherukuvada[2] [iD]

[1] Department of Computer Science and Engineering, Vishnu Institute of Technology, Kovvada,
India
shankar.d@vishnu.edu.in
[2] Department of Networking and Communications, School of Computing, SRM Institute of
Science and Technology, Chennai, India

Abstract. Glaucoma is an ocular pathology characterized by the gradual deterioration of neural cells in the eye, which is attributed to elevated intra ocular pressure within the retina. Glaucoma takes the second spot in terms of its prevalence as a neurodegenerative eye disease, failure to diagnose glaucoma at an early stage can lead to complete blindness. This underlying issue requires a streamlined system that uses experienced medical experts, little equipment, and less time. To categories the stages of glaucoma, a Computer-Aided Diagnosis (CAD) method is used called Image Empirical Mode Decomposition (IEMD). Segmentation-based algorithms utilizing features like cup-to-disc ratio (CDR) and textural characteristics close to the optic disc region can distinguish glaucoma. To capture pixel variations for this investigation, the pre-processed fundus images are separated and transformed into a diverse range of intrinsic mode function (IMFs). The study employs deep learning-based framework for finding the optic nerve head from coloured fundus photographs, to transform the fundus images into a delimited region of interest (ROI) and utilizes multiple deep networks like RFCN and RCNN classifiers identified glaucoma stages separately. The CAD system serves as an automated tool for retinal image processing and demonstrates superior performance over the RFCN classifier, achieving an impressive accuracy of 96% with the RCNN classifier in classifying glaucoma stages.

Keywords: glaucoma · intraocular pressure · retina · computer-aided diagnosis · image decomposition · fundus images · optic nerve head · deep learning · glaucoma classification · optic disc · ensemble classifiers · RFCN · RCNN

P. Pareek et al. (Eds.): IC4S 2023, LNICST 536, pp. 385–402, 2024.
https://doi.org/10.1007/978-3-031-48888-7_33

1 Introduction

This study is centered on the application of convolutional neural networks (CNN) for the classification of glaucoma stages. After careful consideration, the researchers decided to adopt the RCNN (Region-based CNN) and RFCN (Region-based Fully Convolutional Network) approaches. We found that RCNN and RFCN provide the necessary flexibility, reliability, cost effectiveness, and efficiency required for the research.

Glaucoma is a sight-threatening disorder affecting the optic nerve and RNFL. It can be asymptomatic in early stages, leading to unnoticed vision changes. Fundus imaging helps diagnose and monitor glaucoma, capturing images of critical eye structures to enable early detection and intervention [1] (Fig. 1).

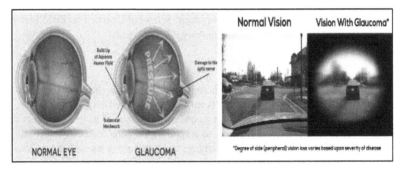

Fig. 1. Fundus Images with Normal and Glaucoma Vision

An extensive array of diverse techniques and algorithms has been developed to perform automated analysis of retinal images and extract blood vessel segments. These methods can be divided into two categories: supervised and unsupervised approaches. Blood vessels are generally visible as borders or outlines inside retinal pictures. Unsupervised methods require the use of contour models to detect and identify these vessels. This category includes techniques such as matching filter responses, morphological approaches, and deformable models. It is difficult to develop an economical and high performing segmentation approach for combined segmentation of the optic disc (OD) and optic cup (OC) [2, 3].

To solve the issues, we devised a customized Computer-Aided Diagnosis (CAD) approach utilizing retinal fundus images designed for clinical evaluation. The development of the system mainly concentrates on the use of image based empirical mode decomposition technique for image processing, RCNN and RFCN deep learning model which have low computational overhead and it is also accurate and efficient. This study aims to design a system that can accurately classify glaucoma stages based on fundus images while minimizing resource requirements. By achieving maximum accuracy, the system outcomes will assist in taking timely precautionary actions against the disease [4, 5].

2 Related Work

Segmenting the cup and disc regions in fundus images is a crucial step in detecting glaucoma. Researchers have explored various techniques to achieve accurate segmentation.

Deepak Parashar [6] a novel methodology was developed to classify glaucoma, with a focus on utilizing the 2D-TEWT for analysis. The provided images underwent preprocessing and were subsequently separated into sub-band images (SBIs) through the utilization of 2D-TEWT. These images were classified using a trained multi-class least squares support vector machine (MC-LS-SVM) classifier. Using tenfold cross-validation, the approach achieved a phenomenal classification accuracy of 93.65%.

Shubham Joshi [7] to aid in the early detection, screening, and treatment of glaucoma, the proposed CAD system makes use of three CNN models. By classifying glaucoma according to precise criteria, these models make it possible for prompt diagnosis, efficient screening, and customized therapy recommendations are used. This system identifies normal and abnormal digital fundus glaucoma images. By using the U-Net models with CNN layers it is getting accuracy of 91.11%.

Huazhu Fu [8] evaluated methods for glaucoma screening using automated processes; two deep learning-based approaches, M-Net and DENet, were examined. M-Net employs a cohesive structure to simultaneously categorize the optic disc (OD) and optic cup (OC) in a solitary stage. In contrast, DENet integrates multiple deep streams with varying levels and modules to directly predict glaucoma from fundus images.

Ratuja Shinde [9] The proposal entails the integration of a desktop application into an automated computer aided design (CAD) system that aims to facilitate detection of glaucoma through the examine of retinal fundus images. The programme incorporates the deep learning strategy known as LeNet in order to verify the input photos. An approach that relies on the brightest point is presented, and it is used to extract the area of interest (ROI). To enhance the precision of region of interest (ROI) identification, this approach uses a preprocessing stage in which an image is first gray scaled and then a Gaussian blur is applied.

Ramgopal Kashyap [10] To revolutionize glaucoma diagnosis and prediction, a cutting edge approach based on deep learning is put forward. The strategy involves the utilization of deep learning concepts to address optic cup segmentation, which entails the integration of pre trained transfer learning algorithms with the U-Net architectural style. In order to extract features, the DenseNet-201 deep convolutional neural network (DCNN) is utilized. By leveraging the power of DCNN, the presence of glaucoma can be accurately determined. The implications of this study extend to various imaging modalities, making it a promising avenue for future research.

3 Glaucoma Classification Methodology

3.1 Dataset Description

3.1.1 Training Dataset

Here, we have the complete training dataset. We can extract features and train to fit a model and so on. The findings from the learning algorithms training set gains knowledge from the experience encountered during tasks with supervised learning, where each instance includes an dependent variable in observation along with One or multiple recorded input variables. The size of the training dataset is 520.

3.1.2 Testing Dataset

It is used to test if our model is working accurately. Here, once you have obtained the model, you can use it to make predictions on the training set. The test set is an assembled group of data points used to evaluate the efficacy of the model against predefined criteria. It is vital to check that no samples from the training set made it into the exam data. It's more difficult to tell whether an algorithm has learnt to generalize from training data when the test set incorporates instances from the training set. Within our dataset, which consists of numerous records, we have allocated 80% using the data to train the mode, while the remaining 20% will be utilized for evaluating the models performance. The size of the test dataset is 130.

3.2   Pre-Processing

3.2.1   IEMD

Image Empirical Mode Decomposition is an approach to image analysis that was developed from the concept of empirical mode decomposition in Hilbert-Huang transform (HHT). This strategy was modified so that it could be applied to images. It is possible to apply IEMD to image processing jobs after increasing its scope to include two dimensions. The capacity of IEMD to efficiently separate locally overlaid frequency ranges within an image is what sets it apart as a tool with a distinctive set of capabilities (Fig. 2) and (Fig. 3).

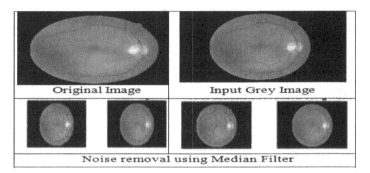

Fig. 2. Original and Grey Images

3.2.2   CLAHE

Throughout the previous few decades, there has been considerable progress in the evolution of contrast enhancement algorithms. These algorithms primarily aim to achieve two main objectives: enhancing the visual appearance of an image for improved interpretation and enhancing the efficacy of subsequent endeavors such as image segmentation, perception of object, and image analysis.

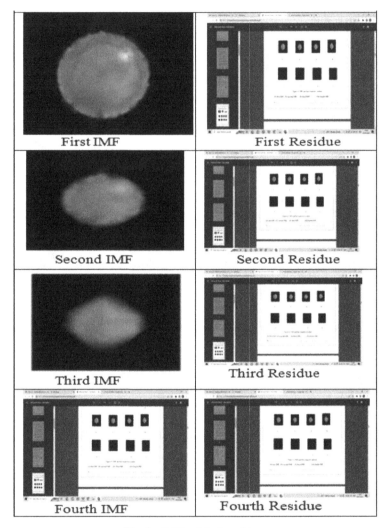

Fig. 3. IMF and their residues

Histogram adjustments, either done worldwide or regionally, are the foundation of contrast improvement approaches. To get over these worldwide restrictions, we present the Contrast Limited Adaptive Histogram Equalization (CLAHE) technique [11]. The Contrast Limited Adaptive Histogram Equalization (CLAHE) algorithm performs localized processing on image regions known as tiles, thereby mitigating the risk of excessive amplification. Additionally, the algorithm employs interpolation techniques to seamlessly merge adjacent tiles. The approach presented herein yields a notable enhancement in image contrast and is amenable to colour images, with a particular emphasis on the brightness channels in HSV images. CLAHE outperforms equalization of all channels in BGR images (Fig. 4).

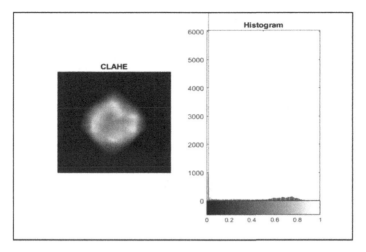

Fig. 4. CLAHE

3.2.3 Histogram

The histogram of an image is a visual representation that shows the distribution of pixel intensities. Essentially, it depicts the frequency of occurrence for each intensity value present in the image (Fig. 5) and (Fig. 6).

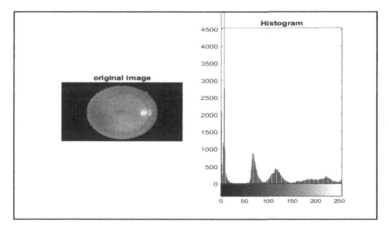

Fig. 5. Histogram of Original Eye Image

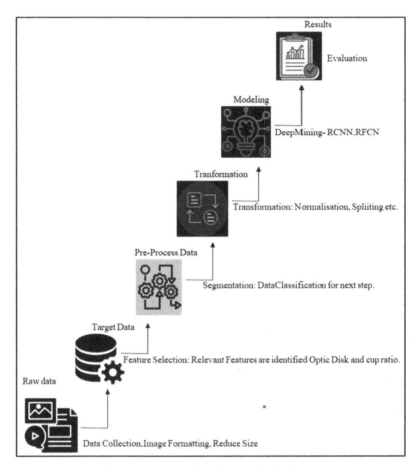

Fig. 6. Glaucoma Classification Methodology

3.2.4 RCNN

R-CNN is a cutting-edge visual object detection system that employing selective search to produce approximately 2000 region proposals per image. It then employs a pre- trained CNN to extracting 4096-dimensional feature vectors for each region of proposal. The system evaluates accuracy using the Intersection-over-Union (IoU) score, fine-tunes the CNN using selected proposals, and achieves outstanding performance without contextual rescoring or ensemble techniques [12].

In glaucoma fundus image classification, manually labeled images are used as positive examples for each class, while region proposals with IoU less than 30% act as negative examples. Bounding box regression refines proposals, and the CNN's last pooling layer features are used to train three regression models for each class. The trained R-CNN categorizes region suggestions into glaucoma, non- glaucoma, or background classes based on a 128×200 resized input, extracting 4096 dimensional feature vectors for each proposal. These feature vectors are then fed into a classification model to categorize each region proposal into one of the three classes (glaucoma, non-glaucoma, background). The region proposal is assigned the class name with the highest score, representing the most probable object type. To address overlaps and redundancies among bounding boxes, a non-maximum suppression method is used. It compares class scores and the geographic overlap of bounding boxes and keeps only the most reliable and non- overlapping bounding boxes while eliminating redundant and overlapping ones [15]. In summary, glaucoma fundus image classification involves processing the data through selective search, feature extraction, bounding box regression, and non-maximum suppression to achieve accurate and efficient glaucoma classification [16, 17] (Fig. 7).

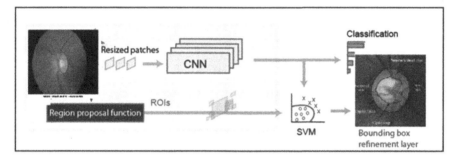

Fig. 7. RCNN Architecture

3.2.5   RFCN

A region-based detection of objects method called R-FCN makes use of fully convolutional deep networks. It offers an extremely precise and effective method of object detection. R-FCN is built as a fully CNN in contrast to earlier region-based analyzers like Fast/Faster R-CNN, which need to run a highly computational sub network for each region. As a result, there are large efficiency gains because the majority of the computation is distributed across the entire image [13, 14] (Fig. 8).

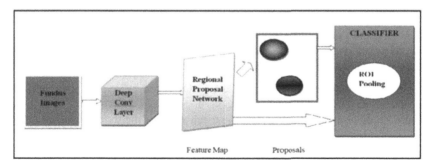

Fig. 8. RFCN Architecture

Take a look at feature map M, which has a square object and measures 5 by 5. We evenly divide the square item into three sections in order to analyze it more thoroughly. Our objective is to build a new feature map that utilizes M that only recognizes the square's top left corner. The size of the final activation map will match M's. To show that the upper left corner is present, it can only activate a subset of the grid cells. The sole activated grid cell in this instance will be the one at location [2], which is indicated in yellow. This activation clearly indicates the top left corner of a region within the map's structure (Fig. 9).

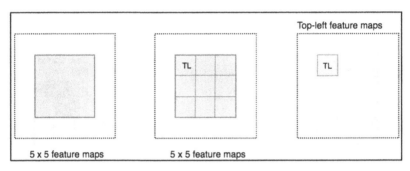

Fig. 9. Top Left Feature Maps

We can create a feature map that concentrates on the left side of the item in order to identify the top left corner of an object. We can produce nine distinct feature maps by splitting an object's area into its nine constituent pieces. Each of these maps is intended to identify a certain area of the object, such as the centre left, top left, top middle, and top right. As they are in charge of scoring and identifying sub-regions of the object according to their positions, these feature maps are known as position sensitive score maps (Fig. 10).

A proposed region of interest (ROI) is shown as a dotted red rectangle in the diagram below. We use a 3 × 3 grid to separate this ROI into smaller sections for analysis. The probability that each of these tiny regions contains the corresponding component of the thing we are particularly interested is subsequently evaluated.

The diagram on the far right shows how the analysis's findings are saved in a 3 × 3 vote array (Fig. 11).

The procedure of aligning score maps and ROIs with the vote array is commonly known as position-sensitive ROI-pooling. This technique shares similarities with the ROI pooling method employed in Fast R-CNN.

- The initial step involves the selection of the ROI region located at the top-left, which is then mapped to the corresponding top-left score map.
- The next phase involves computing the mean value of the scores obtained from the region of interest (ROI) located at the top-left quadrant. This ROI is visually depicted by the blue rectangle in the diagram.
- The findings of the research indicate that nearly 40% of the region enclosed by the blue rectangle demonstrates absence of activation, whereas the remaining 60% manifests complete activation, culminating in an average score of 0.6.

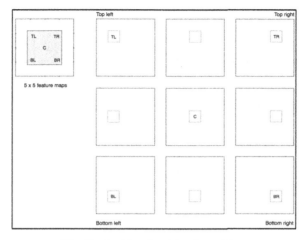

Fig. 10. Position-Sensitive Score Maps

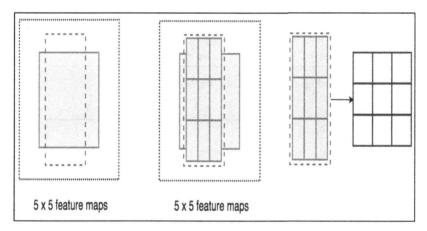

Fig. 11. Generate 9 Score Maps

- This suggests a probability of recognizing the object located in the top-left position with a level of trust of 0.6.
- The outcome of 0.6 is recorded in the array's first row and first column, denoted as [0][0].
- The aforementioned process is reiterated utilizing the top-middle Region of Interest (ROI) and its corresponding score map.
- The outcome obtained is 0.55, denoting the probability of detecting the object located at the top-middle position. This result has been saved in the array's index [0][1].

After calculating each value in the position-sensitive ROI pool, the class score is then calculated as the mean of those values (Fig. 12).

Assuming the task involves detecting C classes, we extend the number of classes to $C + 1$, which includes an additional class representing the background or non-object.

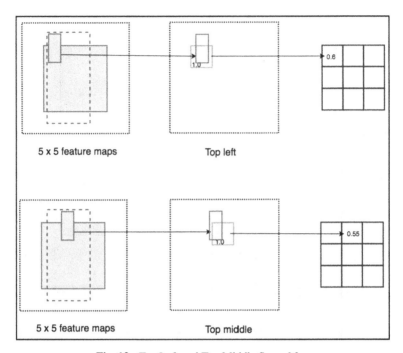

Fig. 12. Top Left and Top Middle Score Map

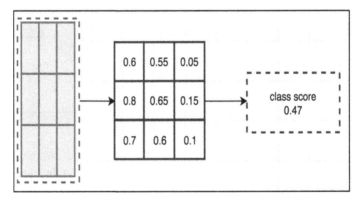

Fig. 13. Class Score

As a result, each class is associated with its own set of 3×3 score maps, resulting in a total of $(C + 1) \times 3 \times 3$ score maps. Using these individual score maps, class scores are predicted for each class. Subsequently, a softmax function is applied to these scores, yielding the probability distribution for each class (Fig. 13) and (Fig. 14).

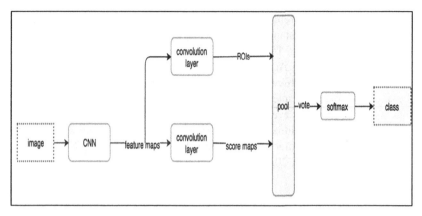

Fig. 14. Data Flow for the R-FCN

3.2.6 Training Progress

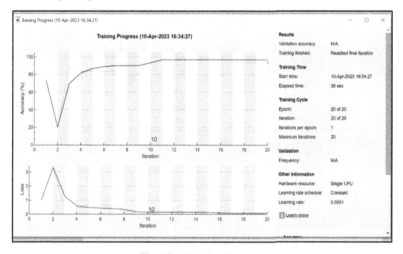

Fig. 15. Training Progress

3.2.7 Accuracy

Classification Accuracy is what we usually as commonly understood refers to the proportion of accurate predictions to the overall number of input samples However, it is important to note that this measure performs optimally only when there is an equal division of samples across each class. To illustrate, let's consider a training dataset where 98% of the samples belong to class A and only 2% belong to class B. In such a scenario, a model can easily achieve a training accuracy of 98% by simply predicting all samples as class A.

Nonetheless, if the aforementioned model were to be assessed on a test dataset comprising 60% class A instances and 40% class B instances, the accuracy of the test would diminish to 60% (Fig. 15).

Although Classification Accuracy may initially appear favorable, it can lead to a deceptive impression of achieving high accuracy. The real challenge arises when the misclassification cost of samples from the minor class is considerably high. Particularly in cases involving rare yet severe diseases, the consequences of failing to diagnose an illness in an affected individual far outweigh the cost of subjecting a healthy person to additional tests. The accuracy of test data is 93 - 96% using RCNN and 81% using RFCN (Figs. 16, 17, 18, 19, 20 and 21).

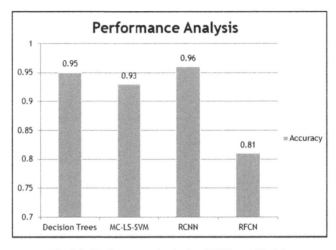

Fig. 16. Performance Analysis of Different Models

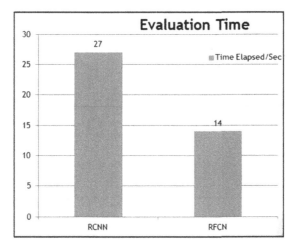

Fig. 17. Time Elapsed Comparison Analysis of Implemented Models

Epoch	Iteration	Time Elapsed (hh:mm:ss)	Mini-batch Accuracy	Mini-batch Loss	Base Learning Rate
1	1	00:00:12	74.07%	0.6322	1.0000e-04
10	10	00:00:27	96.30%	0.2039	1.0000e-04

Fig. 18. Accuracy for RFCN

Epoch	Iteration	Time Elapsed (hh:mm:ss)	Mini-batch Accuracy	Mini-batch Loss	Base Learning Rate
1	1	00:00:01	81.48%	0.9156	1.0000e-06
10	10	00:00:14	81.48%	0.8050	1.0000e-06

Fig. 19. Accuracy for RCCN

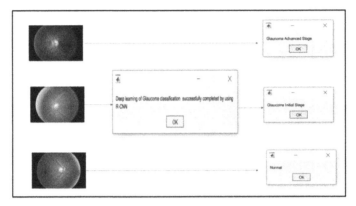

Fig. 20. Classification of glaucoma stages using RCNN

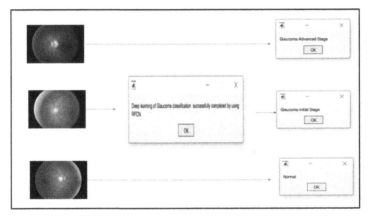

Fig. 21. Classification of glaucoma stages using RFNN

4 Conclusion

Glaucoma is an ocular condition that can lead to permanent vision loss in affected individuals. Early detection of glaucoma-related optic nerve damage is essential to prevent further deterioration. In our research, we assessed the effectiveness of R-CNN and R-FCN approaches for accurately detecting glaucoma in color fundus images. We assessed other aspects of the images, including the colour, brightness, and RGB intensity value histograms. In effect, we noticed that datasets with comparable histograms produced better results. In order to ensure its robustness over various vessel widths and luminosity circumstances, we used a supervised learning method and trained it on datasets. The accuracy curve shows that our suggested strategy outperforms existing methods. In particular, ResNet-50, a well-known deep learning architecture, performed better in our tests. By evaluating our method on the Orgia dataset, comprising 650 images, we implied that it could be a useful replacement for widespread computer-aided glaucoma screening programmes. Additionally, we assessed the performance of R-FNN and the

R-CNN. The outcomes clearly showed that R-CNN outperformed R-FCN in terms of accuracy, scoring 96.30% versus 81.48%. However, R-FCN boasted a faster processing time, taking only 14 s as opposed to R-CNN's 27 s.

5 Future Work

There is the opportunity to investigate a variety of neural network topologies in order to improve the results of vascular segmentation by lowering the number of instances of false-positive detections. In the course of our ongoing research, we are attempting to design innovative structures that are tailored to the detection of glaucoma on vast databases.

References

1. Hagiwara, Y., et al.: Computer-aided diagnosis of glaucoma using fundus images: a review. Comput. Methods Programs Biomed. **165**, 1–12 (2018)
2. Diaz-Pinto, A., Colomer, A., Naranjo, V., Morales, S., Xu, Y., Frangi, A.F.: Retinal image synthesis and semi-supervised learning for glaucoma assessment. IEEE Trans. Med. Imaging **38**, 2211–2218 (2019). https://doi.org/10.1109/tmi.2019.2903434
3. Aloudat, M., Faezipour, M., El-Sayed, A.: High intraocular pressure detection from frontal eye images: a machine learning based approach. PubMed (2018). https://doi.org/10.1109/embc.2018.8513645
4. Saha, S.K., Fernando, B., Cuadros, J., Xiao, D., Kanagasingam, Y.: Automated quality assessment of colour fundus images for diabetic retinopathy screening in telemedicine. J. Digit. Imaging **31**, 869–878 (2018). https://doi.org/10.1007/s10278-018-0084-9
5. Maheshwari, S., Pachori, R.B., Kanhangad, V., Bhandary, S.V., Acharya, U.R.: Iterative variational mode decomposition based automated detection of glaucoma using fundus images. Comput. Biol. Med. **88**, 142–149 (2017). https://doi.org/10.1016/j.compbiomed.2017.06.017
6. Parashar, D., Agrawal, D.K.: Classification of glaucoma stages using image empirical mode decomposition from fundus images. J. Digit. Imaging **35**(5), 1283–1292 (2022). https://doi.org/10.1007/s10278-022-00648-1
7. Joshi, S., Partibane, B., Hatamleh, W.A., Tarazi, H., Yadav, C.S., Krah, D.: Glaucoma detection using image processing and supervised learning for classification. J. Healthcare Eng. **2022**, 2988262 (2022). https://doi.org/10.1155/2022/2988262
8. Fu, H., Cheng, J., Xu, Y., Liu, J.: Glaucoma detection based on deep learning network in fundus image. Deep Learning and Convolutional Neural Networks for Medical Imaging and Clinical Informatics. 119–137 (2019). https://doi.org/10.1007/978-3-030-13969-8_6
9. Shinde, R.: Glaucoma detection in retinal fundus images using U-Net and supervised machine learning algorithms. Intell.-Based Med. **5**, 100038 (2021). https://doi.org/10.1016/j.ibmed.2021.100038
10. Kashyap, R., Nair, R., Gangadharan, S.M.P., Botto-Tobar, M., Farooq, S., Rizwan, A.: Glaucoma detection and classification using improved U-Net deep learning model. Healthcare **10**, 2497 (2022). https://doi.org/10.3390/healthcare10122497
11. Agarwal, T.K., Tiwari, M., Lamba, S.S.: Modified Histogram based contrast enhancement using Homomorphic Filtering for medical images. In: 2014 IEEE International Advance Computing Conference (IACC), pp. 964–968 (2014). https://doi.org/10.1109/iadcc.2014.6779453

12. Li, L., et al.: A large-scale database and a CNN model for attention-based glaucoma detection. IEEE Trans. Med. Imaging **39**(2), 413–424 (2019). https://doi.org/10.1109/tmi.2019.2927226
13. Shelhamer, E., Long, J., Darrell, T.: Fully convolutional networks for semantic segmentation. IEEE Trans. Pattern Anal. Mach. Intell. **39**, 640–651 (2017). https://doi.org/10.1109/TPAMI.2016.2572683
14. Shankar, D., George, G.V.S., JNSS, J.N., Madhuri, P.S.: Deep analysis of risks and recent trends towards network intrusion detection system. Int. J. Adv. Comput. Sci. Appl. 14, (2023). https://doi.org/10.14569/ijacsa.2023.0140129
15. Shankar, D., George, G.V.S., Kanya, N.: OptiBiNet_GRU: robust network intrusion detection system using optimum bi-directional gated recurrent unit. Int. J. Intell. Eng. Syst. **16**, 75–91 (2023). https://doi.org/10.22266/ijies2023.0630.06
16. Shanmugam, P., Raja, J., Pitchai, R.: An automatic recognition of glaucoma in fundus images using deep learning and random forest classifier. Appl. Soft Comput. **109**, 107512 (2021). https://doi.org/10.1016/j.asoc.2021.107512
17. Sandhya, M., Morampudi, M.K., Grandhe, R., Kumari, R., Banda, C., Gonthina, N.: Detection of Diabetic Retinopathy (DR) severity from fundus photographs: an ensemble approach using weighted average. Arab. J. Sci. Eng. (2022). https://doi.org/10.1007/s13369-021-06381-1

Implementation and Performance Evaluation of Asymmetrical Encryption Scheme for Lossless Compressed Grayscale Images

Neetu Gupta[1]([✉]), Hemant Kumar Gupta[2], K. Swapna[3], Kommisetti Murthy Raju[4], and Rahul Srivastava[2]

[1] Department of Computer Science and Engineering, Manipal University Jaipur, Jaipur, Rajasthan, India
neetu.gupta@jaipur.manipal.edu

[2] Department of Electronics and Communication Engineering, Arya College of Engineering & I.T., Jaipur, Rajasthan, India
rahu_79@rediffmail.com

[3] Department of Electronics and Communication Engineering, Vaagdevi College of Engineering, Warangal, Telangana, India
swapna_k@vaagdevi.edu.in

[4] Department of Electronics and Communication Engineering, Shri Vishnu Engineering College for Women, Bhimawaram, Andhra Pradesh, India
Venkateswara.103@svecw.edu.in

Abstract. The employment of compression and encryption methods enables the transmission of images across a communication link with less bandwidth consumption and resistance against differential assaults. In this study, the gray scale images are compressed using the Huffman lossless compression approach. Each pixel is given a unique prefix code with a configurable length in this. The frequency of occurrence of characters has an inverse relationship with prefix code length. Asymmetrical RSA encryption is used to encrypt compressed images. In the RSA encryption technique, the encryption key is kept in the open as opposed to the decryption key, which is kept private. Analysis of the compression parameters, together with correlation coefficient and entropy analysis, is used to convey the effectiveness of suggested techniques. Five 512×512 grayscale test images are used as text images and to verify the results.

Keywords: Huffman Lossless Compression · RSA asymmetrical encryption · Correlation Coefficient analysis · Grayscale Images

1 Introduction

Compression techniques are used during transmission to efficiently convey the information via a medium of communication with a smaller bandwidth [1–4]. The redundant substance from the original information source is removed during the compression process so as to reduce the amount of the information. Lossy compression methods have

P. Pareek et al. (Eds.): IC4S 2023, LNICST 536, pp. 403–413, 2024.
https://doi.org/10.1007/978-3-031-48888-7_34

a high rate of compression, they may also exclude some crucial information, whereas lossless compression approaches keep the information's integrity and certainty while compressing it enough [5, 6]. Another crucial factor is protecting the data during transmission from different unethical assaults. During transmission, encryption methods are combined with the data to ensure data secrecy [7]. Both symmetrical and asymmetric encryption methods are possible. Key creation is typically used to accomplish encryption. The transmitter key and receiver key in symmetrical encryption processes are identical, which makes it possible to intercept information in transit [8]. Using an asymmetrical encryption technique, where the receiver key is private but the sender key is public, can enhance security characteristics.

Both the compression and encryption processes are crucial for secure data transfer on lowered bandwidth. It is possible to use the compression technique either after or before the encryption. When encryption is applied to an un-compressed image, execution time is prolonged and the encryption process is misused [9, 10]. By adopting a compression strategy before encrypting an image, the likelihood that hackers will be able to decode it is decreased [9, 11, 12].

In this study, the RSA asymmetric encryption method is used after the Huffman lossless compression algorithm. The compression efficiency metrics Peak signal to noise ratio (PSNR), Mean square error (MSE) and Compression ratio (CR) are investigated to assess the performance of image reconstruction by comparing them to existing methods. The encryption efficiency of the proposed technique is assessed using entropy and coefficient of correlation studies. Five typical grayscale test images with a 512×512 pixel size are used for the research experiment.

2   Related Work

M. Yassein et al.'s study on several encryption methods in the symmetrical and asymmetrical categories was published in 2017 [13]. The asymmetrical encryption algorithm RSA is contrasted with the symmetric encryption techniques 3DES, AES, Blowfish, and DES. The RSA encryption approach offers higher protection against various assaults, as shown by the authors' analysis of the symmetrical and asymmetrical encryption algorithms using various metrics for evaluating encryption performance. Galla et al. [14] executed RSA technique for image data encryption in 2016. The authors have shown how the RSA method depends on number factorization. Jumgekar et al. [15] also developed and illustrated basic cryptanalysis in 2013 and explained the implementation of RSA method through the production of public and private keys. A RSA encryption technique was introduced by Jonsson et al. [16] in 2002, and it was compared to the incomplete RSA algorithm. Authors have demonstrated that security characteristics based on pseudo-random functions are diminished by TLS-based algorithms. By modernizing the computing procedure, Katz et al. [17] demonstrated the contemporary cryptanalysis method in 2014, which lowers the drawbacks of complete secrecy. Authors have employed a private key encryption technique and message signal authentication.

S. Han et al. [18] solved the neural network restrictions in 2016 by demonstrating the deep compression technique. Pruning, quantization, and Huffman coding are all components of this deep compression technique. The deep compression strategy, according to the authors, lowers the need for storage without compromising the accuracy of the rebuilt image. To improve the compression effectiveness and privacy of text-based information, E. Satir et al. [19] published a lossless Huffman compression approach in 2014. A lossless image compression technique comparable to bzip2 was devised in 2012 by Y. Zhang et al. [20] with the goal of parallelizing the development of Huffman coding, BWT and MTF. To demonstrate the advantages and disadvantages of the recommended algorithms, a performance study is also conducted.

Using LZW and Run length encoding, M. Sharma [21] examined Huffman compression approach in 2010. The author shows that Huffman coding has a higher compression ratio and efficiency than other compression techniques.

According to a survey of the literature, Huffman coding is the most popular lossless compression method, which encourages future study to increase the reconstructed image compression ratio and effectiveness. This research paper goal is to apply asymmetrical encryption on lossless compressed grey scale photos and examine the results.

3 Fundamental Information and Proposed Model

A form of lossless compression method based on how frequently pixels occur in visual data is called Huffman coding (Fig. 1).

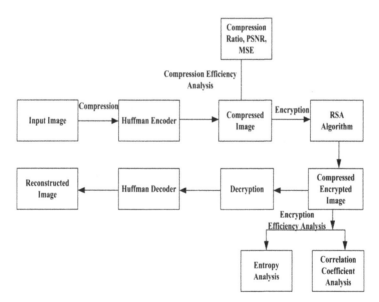

Fig.1. Fundamental structural representation of suggested system paradigm

In the Huffman coding, Image pixels are placed according to their frequency in decreasing order. Once more, each pixel is ordered in decreasing order of frequency of occurrence after merging the frequencies of the two pixels with the lowest frequencies. This procedure will continue until there are only two frequencies of pixels left. The last two pixel frequencies are stored by giving higher frequency pixels a value of 0 and lower frequency pixels a value of 1.

Algorithm for Compression

 # Every pixel value is shown through 8 bits, each of which has a symbol and values ranging from 0 to 255.

➤ *Use I = imread(file) to read relevant input image.*

➤ *Use I = rbg2gray(file) to convert a colour image to a grayscale image if the original image was colour.*

➤ *Use [m,n] = size(I) to calculate the image's size.*

➤ *Utilise the programme to calculate the cumulative probability.*

➤ *Sum (k(:)) = count (cnt)*

➤ *Count (cnt) = Pro(cnt)/Total number of count*

➤ *Sigma plus Pro(cnt) equals Cumpro(cnt)*

➤ *Use dict = huffmandict(symbols, pro) to invoke the Huffman code dictionary.*

➤ *A vector is created from an array of symbols using the formula newvec (vecsize) = I (m,n)*

➤ *The statement hcode = huffmanenco(newvec, dict) is used to conduct Huffman encoding.*

➤ *Dhsig1 = huffmandeco(hcode,dict) performs the Huffman decoding.*

➤ *Calculate MSE, PSNR and compression ratio.*

RSA generates two unique keys. A public key has been allotted to the transmitting side of the communication. The recipient is given a private key in addition to another key. The choice of two enormous prime integers is key to the security of this procedure. The three essential components of the RSA encryption technique are key creation, encryption, and decryption.

Algorithm for Encryption
 # Encryption and decryption keys are both produced during key production.

➤   *Using the primer testing method, find two distinct integer prime values with same bit length, g and h.*

➤   *Use n = g h to calculate the key length in bits.*

➤   *Make sure that the encryption key e is chosen so that it cannot be a combination of 1e(g,1,h) and (g,1,h). A public key can be formed by (n,e).*

➤   *Choose the private decryption key d such that 1d(g-1)(h-1) and (d e)mod (g-1)(h-1) = 1.*

➤   *Divide the picture I into a series of blocks where each block meets the condition 0 Ii n. Put these blocks into an encryption using the equation E = Ie mod (n).*

➤   *Using the formula D = Ed mod (n), decryption can be performed of mage E.*

4   Evaluation of the Effectiveness of Compression and Encryption

4.1 Parameters for Evaluating Compression Efficiency

To evaluate the efficacy of the Huffman compression approach, PSNR and MSE are used. For an image to be successfully reconstructed, the PSNR must be high, or over 30 dB, and the MSE must be as low as possible.

$$\text{PSNR} = log_{10}\left[\frac{\{J \times K\}^2\}}{MSE} \right] \tag{1}$$

$$\frac{1}{J \times K}\sum_{X=1}^{J}\sum_{Y=1}^{K}[u(x, y) - v(x, y)]^2 \tag{2}$$

where $u(x, y)$ and $v(x, y)$ denotes the uncompressed and compressed pixel respectively.

4.2 Parameters for Evaluating Encryption Efficiency

4.2.1 Analysis Based on Entropy

The average amount of information in a lengthy string of pixels in an image data is referred to as entropy.

$I(X_i)$ has m distinct symbols, and the mean value or entropy is given by

$$H(X) = \sum_{i=1}^{j} P(X_i)I(X_i) \tag{3}$$

$$H(X) = \sum_{i=1}^{j} P(X_i)\log_2 P(X_i) \tag{4}$$

4.2.2 Analysis Based on Correlation Coefficient

The correlation coefficient between two neighboring pixels on the horizontal, vertical, and diagonal axes is used to describe the correlation between the original and encrypted image. This is how the correlation coefficient is shown:

$$C = \frac{\sum_{i=1}^{j}(x_i - \overline{x})(y_i - \overline{y})}{\sqrt{\sum_{i=1}^{j}(x_i - \overline{x})^2} \times \sqrt{\sum_{i=1}^{j}(y_i - \overline{y})^2}} \tag{5}$$

where $\overline{x}$ and $\overline{y}$ can be expressed as

$$\overline{x} = \frac{1}{K}\sum_{i=1}^{k} x_i \text{ and } \overline{y} = \frac{1}{K}\sum_{i=1}^{k} y_i \tag{6}$$

The original image's neighboring pixels must be significantly connected with one another in the horizontal, vertical, and diagonal dimensions. As a result, the correlation coefficient must be high. However, an encrypted image should have little to no pixel correlation and a low correlation coefficient.

5 System Environments

MATLAB 2018 is used in this experimental investigation to simulate the outcomes. Operating system utilized is Windows 10. 512×512 standardized grayscale test pictures. To authenticate the outcomes, we use the bmp format from the SIPI data store.

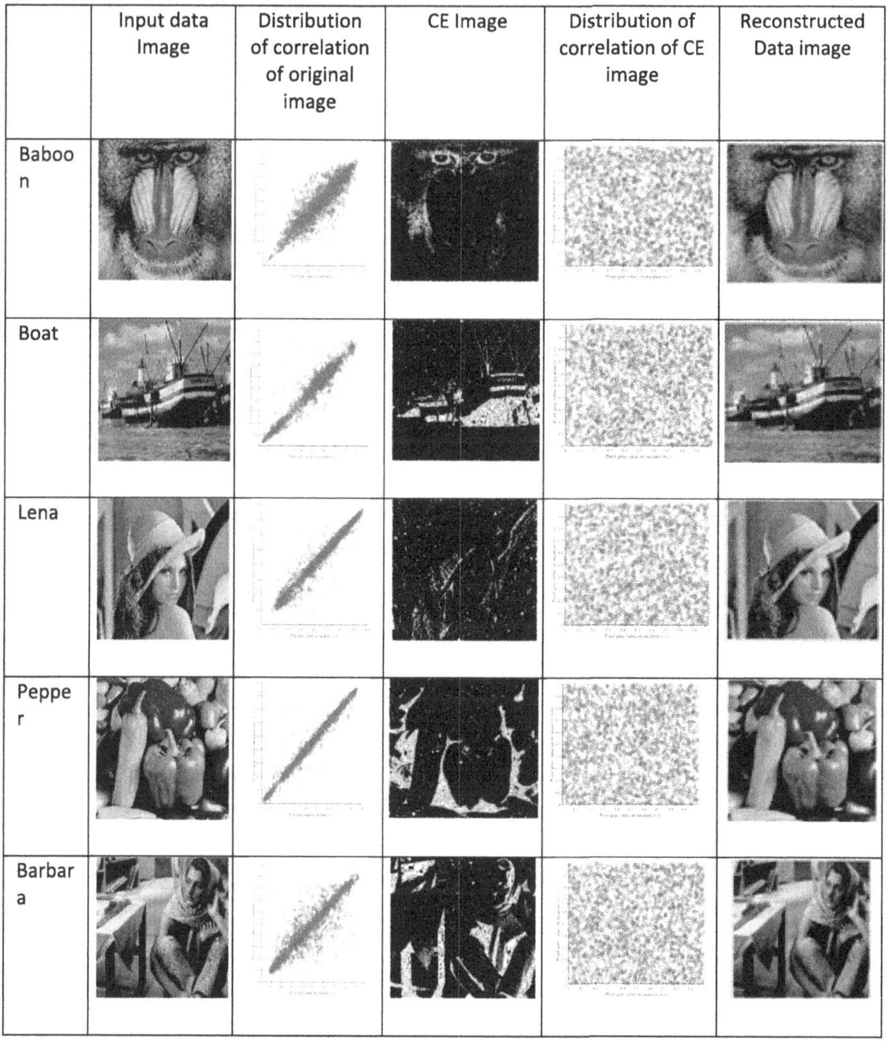

	Input data Image	Distribution of correlation of original image	CE Image	Distribution of correlation of CE image	Reconstructed Data image
Baboon					
Boat					
Lena					
Pepper					
Barbara					

Fig. 2. Representation of pictorial outputs during proposed compression and encryption process

6 Results of Proposed Model

Table 1 displays results for compression efficiency parameters. Existing techniques are also contrasted. The compression efficiency is represented by CR, PSNR, and MSE. Graphical depiction of compression efficiency characteristics is shown in Fig. 3.

Columns 2 and 3 of Fig. 2 display the input image and distribution of the associated correlation coefficients, respectively. Columns 4 and 5 respectively display the compressed encrypted (CE) picture and the distribution of the related correlation coefficients, while column 6 displays the reconstructed image following decryption and decompression (Table 2).

Table 1. Results obtained for proposed CE technique in terms of MSE, CR and PSNR

Title	CR		PSNR in dB		MSE	
	Proposed scheme	Existing Studies	Proposed Scheme	Existing Studies	Proposed Scheme	Existing Studies
Baboon	2.9045	NA	29.3446	21.23 [22]	44.531	126.83 [22]
Boat	5.3956	NA	34.5563	29.68 [23]	16.2547	NA
Lena	7.8546	5.83 [24]	36.0938	29.57 [25]	7.3748	163.56 [26]
Pepper	7.4567	3.99 [27]	38.1706	31.26 [27]	5.5758	37.760 [27]
Barbara	4.8568	NA	34.3487	21.81 [22]	15.235	289.76 [22]

Table 2. Results obtained for proposed encryption scheme in terms of entropy and Correlation coefficient

Title	V-Direction		D-Direction		H-Direction		Value of Entropy	
	Proposed Model	Existing Studies	Proposed Model	Existing Studies	Proposed Model	Existing Studies	Proposed Model	Existing Studies
Baboon	.0712	.0093 [28]	.0687	−.0251 [28]	.0886	−.0225 [28]	3.8652	7.9987 [1]
Boat	.2294	−.0064 [1]	.1920	.00007 [1]	.2176	.010 [1]	3.8863	7.9989 [1]
Lena	.2110	.0240 [28]	.1005	−.0411 [28]	.1352	−.0094 [28]	3.7257	7.9989 [1]
Pepper	.2521	.0093 [28]	.2144	−.0251 [28]	.2470	−.0225 [28]	3.9000	7.9988 [1]
Barbara	.2894	.0079 [1]	.2426	.0200 [1]	.2666	.0122 [1]	3.9308	7.9901 [1]

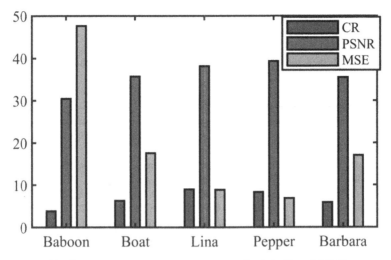

Fig. 3. Comparison chart among values of MSE, CR and PSNR

7 Conclusion and Future Scope

In this study, image data is compressed and encrypted using the Huffman lossless compression algorithm and RSA asymmetrical techniques. According to a review of the data, the dynamic range of the PSNR of the Huffman compression approach is lying between 30 to 40 dB, which is beyond the minimum requirement of PSNR for successful compression while it offers a sizable compression ratio. MSE, with a range of 6.85 to 47.64, is a less sophisticated algorithm than other cutting-edge ones. The low correlation coefficients of encrypted images show that the RSA encryption method is resilient to numerous attacks and the entropy being in the range of 3.72 to 3.93.

References

1. Tong, X.J., Chen, P., Zhang, M.: A joint image lossless compression and encryption method based on chaotic map. Multimedia Tools Appl. **76**(12), 13995–14020 (2017)
2. Singh, R.K., Kumar, B., Shaw, D.K., Khan, D.A.: Level by level image compression-encryption algorithm based on quantum chaos map. J. King Saud Univ. Comput. Inf. Sci. **33**(7), 844–851 (2018)
3. Gupta, N., Vijay, R., Gupta, H.K.: Performance analysis of DCT based lossy compression method with symmetrical encryption algorithms. EAI Endorsed Trans. Energy Web **7**(28), 1–11 (2020)
4. Daubechies, I., Barlaud, M., Mathieu, P.: Image coding using wavelet transform. IEEE Trans. Image Process. **1**(2), 205–220 (1992)
5. Maniccama, S.S., Bourbakis, N.G.: Lossless image compression and encryption using SCAN. Pattern Recogn. **34**(6), 1229–1245 (2001)
6. Zhang, X.: Lossy compression and iterative reconstruction for encrypted image. IEEE Trans. Inf. Forensics Secur. **6**(1), 53–58 (2011)

7. Al-Khasawneh, M.A., Shamsuddin, S.M., Hasan, S., Bakar, A.A.: An improved chaotic image encryption algorithm. In: International Conference on Smart Computing and Electronic Enterprise ICSCEE 2018, pp. 1–8, Shah Alam, Malaysia (2018)
8. Carpentieri, B.: Efficient compression and encryption for digital data transmission. Secur. Commun. Netw. **2018**, 1–9 (2018)
9. Setyaningsih, E., Wardoyo, R.: Review of image compression and encryption techniques. Int. J. Adv. Comput. Sci. Appl. **8**(2), 83–94 (2017)
10. Gupta, N., Vijay, R., Gupta, H.K.: Performance evaluation of symmetrical encryption algorithms with wavelet based compression technique. EAI Endorsed Trans. Scalable Inf. Syst. **7**(28), 1–14 (2020)
11. Sharma, M., Gandhi, S.: Compression and encryption : an integrated approach. Int. J. Eng. Res. Technol. **1**(5), 1–7 (2012)
12. Gupta, N., Vijay, R.: Hybrid image compression-encryption scheme based on multilayer stacked autoencoder and logistic map. China Commun. **19**(1), 238–252 (2022)
13. Yassein, M.B., Aljawarneh, S., Qawasmeh, E., Mardini, W., Khamayseh, Y.: Comprehensive study of symmetric key and asymmetric key encryption algorithms. In: Proceedings of 2017 International Conference on Engineering and Technology ICET 2017, pp. 1–7 Antalya Turkey (2018)
14. Galla, L.K., Koganti, V.S., Nuthalapati, N.: Implementation of RSA. In: 2016 International Conference on Control Instrumentation Communication and Computational Technologies ICCICCT 2016, pp. 81–87 Kumaracoil, India, (2016)
15. Jamgekar, R.S., Joshi, G.S.: File encryption and decryption using secure RSA. Int. J. Emerging Sci. Eng. **1**(4), 11–14 (2013)
16. Jonsson, J., Kaliski, B.S.: On the security of RSA encryption in TLS, Lect. Notes Comput. Sci. (including Subser. Lect. Notes Artif. Intell. Lect. Notes Bioinformatics), vol. 2442, pp. 127–142 (2002)
17. Katz, J., Lindell, Y.: Introduction to Modern Cryptography (2014)
18. Han, S., Mao, H., Dally, W.J.: Deep compression: compressing deep neural networks with pruning, trained quantization and Huffman coding. In: 4th International Conference on Learning Representations, ICLR 2016 - Conference Track Proceedings (2016)
19. Satir, E., Isik, H.: A Huffman compression based text steganography method. Multimedia Tools Appl. **70**(3), 2085–2110 (2014)
20. Patel, R.A., Zhang, Y., Mak, J., Davidson, A., Owens, J.D.: Parallel lossless data compression on the GPU. In: 2012 Innovative Parallel Computing, InPar 2012 (2012)
21. Sharma, M.: Compression using huffman coding. Int. J. Comput. Sci. Netw. Secur. **10**(5), 133–141 (2010)
22. Raja, S.P., Suruliandi, A.: Performance evaluation on EZW & WDR image compression techniques. In: IEEE International Conference on Communication Control and Computing Technologies, ICCCCT 2010, pp. 661–664 (2010)
23. Agarwal, C., Mishra, A., Sharma, A.: A novel gray-scale image watermarking using hybrid Fuzzy-BPN architecture. Egyptian Inform. J. **16**(1), 83–102 (2015)
24. Praisline Jasmi, R., Perumal, B., Pallikonda Rajasekaran, M.: Comparison of image compression techniques using Huffman coding, DWT and fractal algorithm. In: International Conference on Computer Communication and Informatics, ICCCI 2015, pp. 1–5 (2015)
25. Zhou, N., Pan, S., Cheng, S., Zhou, Z.: Image compression-encryption scheme based on hyper-chaotic system and 2D compressive sensing. Optical Laser Technol. **82**, 121–133 (2016)
26. Dang, P.P., Chau, P.M.: Image encryption for secure Internet multimedia applications. IEEE Trans. Consumer Electron. **46**(3), 395–403 (2000)

27. Hu, F., Pu, C., Gao, H., Tang, M., Li, L.: Image compression and encryption scheme based on deep learning. Nauk. Visnyk Natsionalnoho Hirnychoho Universytetu **6**, 142–148 (2016)
28. Zhang, Y.: The unified image encryption algorithm based on chaos and cubic S-Box. Inf. Sci. (Ny) **450**, 361–377 (2018)

Optimization of Single Image Dehazing Based on Stationary Wavelet Transform

M. Ravi Sankar[1] , P. Rama Krishna[2] , A. Yamini[1], Ch. Manikanta[1],
R. Rupa Swathika[1], Y. Tanuja Tulasi[1], and B. Elisha Raju[3]($\boxtimes$)

[1] Department of ECE, Sasi Institute of Technology and Engineering, Tadepalligudem, India
[2] Department of IoT, Seshadri Rao Gudlavalleru Engineering College, Gudlavalleru, India
[3] Department of ECE, Vishnu Institute of Technology, Bhimavaram, India
elisharaju.b@vishnu.edu.in

Abstract. Dehazing of images holds a crucial significance in the domains of artificial intelligence and picture dispensation, primarily due to the detrimental impact of haze on visual quality, thereby impeding the effectiveness of subsequent tasks. Over the past years, the stationary wavelet transform (SWT) has gained prominence as a potent tool for image dehazing, owing to its capability to capture both frequency and location information effectively. The objective of this study is to enhance the visual quality of a dehazed image by leveraging the multi-level stationary wavelet transform (SWT). This approach facilitates reducing image dimensions without compromising image quality. Using the advantage of SWT, an efficient dehazing methodology based on sub band image model has been implemented in this work. The efficiency of proposed methodology has been evaluated in terms of PSNR, SSIM, and MMSE. This study includes a comparative analysis between DHWT and SWT concerning the mentioned parameters. The investigational results clearly demonstrate that the proposed method delivers outstanding visual quality after dehazing. Compared to DHWT, the SWT-based dehazing method achieves a remarkable 11.13% improvement in PSNR, 13.93% enhancement in SSIM, and a significant 40% reduction in MSE.

Keywords: Image dehazing · Stationary wavelet transform · Haar wavelet transform · Enhancement · Efficiency

1 Introduction

Image dehazing is a crucial image enhancement technique that aims to improve the visibility and quality of hazy or foggy images [1]. It addresses the problem of image degradation caused by atmospheric scattering, which occurs when light interacts with particles and molecules in the atmosphere, leading to reduced contrast, colour distortion, in the taken images [2]. The existence of haze in images can significantly impact various applications, such as surveillance, outdoor photography, computer vision, and remote sensing [3]. It not only affects the aesthetics of images but also hinders the performance of image analysis algorithms and visual perception. Dehazing includes

P. Pareek et al. (Eds.): IC4S 2023, LNICST 536, pp. 414–421, 2024.
https://doi.org/10.1007/978-3-031-48888-7_35

reducing or eliminating haze effects to produce clearer, more aesthetically pleasing photographs [4]. Various picture dehazing algorithms have been developed to accomplish this, ranging from conventional methods based on single-image processing techniques to sophisticated methods utilizing numerous images and advanced machine learning models. Estimating the amount of haze or ambient light present in the picture is the main obstacle in image dehazing [5]. Once the atmospheric light has been calculated, the haze can be successfully eliminated or diminished, revealing the real details of the scene. Taking use of developments in computer vision, deep learning [6, 7], and image processing [8–11], image dehazing systems [12, 13] are still developing.

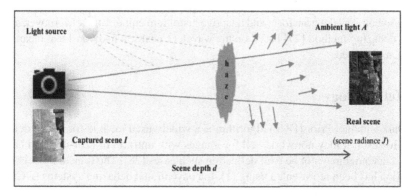

Fig. 1. Haze conception model [13].

Decomposing the hazy image into several frequency bands is the fundamental concept behind image dehazing using wavelet transform [14]. The low-frequency sub bands, which hold the information about the global lighting, are supposed to be haze-affected, but the high-frequency sub bands, which contain the fine details of the image, are believed to be haze-free [15]. The haze is removed from the image by applying a dehazing algorithm to the low-frequency sub bands. After the dehazing procedure, the wavelet transform can also be employed for enhancement. As a result, wavelet transform image dehazing is a potent method for reducing haze from photos while maintaining their fine details and edges. It is a potential strategy for a variety of applications, including remote sensing, surveillance, and outdoor photography [16].

2 Literature Review

Image dehazing stands as a fundamental and crucial task in the realm of image processing and computer vision. Over the years, substantial research efforts have been devoted to developing effective algorithms and techniques in this domain. This paper [17] presents a single image dehazing based on deep learning. In [18], the DCP algorithm is presented for single image dehazing, leveraging the observation that pixels in non-hazy regions tend to have very low values in at least one-color channel. Additionally, the authors of [19] propose an image dehazing method utilizing the discrete wavelet transform. Furthermore, in [20], a multi-scale convolutional neural network for single image dehazing is proposed, utilizing a multi-scale approach to capture both local and global information within the image.

In [21], the authors propose an image dehazing method utilizing the stationary wavelet transform. In [22], a deep learning-based approach for single image dehazing is presented, featuring a multi-scale multi-feature fusion network to capture both local and global information in the image. A novel image dehazing method is introduced in [23], which employs the discrete wavelet transform. Furthermore, in the same paper, another image dehazing method utilizes the stationary wavelet transform [24]. In this paper [25], another image dehazing method is introduced, utilizing wavelet-based local contrast enhancement. Additionally, this paper presents an image dehazing method [26] based on wavelet transform and adaptive histogram equalization. Moreover, a novel image dehazing method [27] based on the wavelet transform and non-local means filter has been proposed.

3   Methodology

The Dark Channel Prior (DCP) algorithm is a widely used method for image dehazing [28]. However, it may not work well for images with uniform or near-uniform haze, as the dark channel may not be well defined in such cases [28]. The typical model of haze formation has been represented using (1). In conventional dehazing systems (DCP), the dark channel is epitomized using (2).

$$I_{haze}(x) = J_{haze-free}(x) \times e^{-\beta d(x)} + A \times (1 - e^{-\beta d(x)}) \tag{1}$$

$$J^{dark}(x) = \frac{min}{y \in \Omega(x)} \left(\frac{min}{c \in \{r, g, b\}} J^c(y) \right) \tag{2}$$

3.1   Image Model

The clear image, which symbolizes the light reflected from things' surfaces, is the first part. The scattering transmission, often known as haze, is the second element. Equation (3) can be used to express a picture's optical model.

$$I_c = J_c \odot t + a_c \times (1 - t), where \ c = 1, 2, 3 \ldots \tag{3}$$

3.2 Sub-band Image Model

Given that the optical image model implies that the transmission rate is uniformly distributed throughout the environment, the frequency response of haze in images should mostly be distributed across the low-frequency sub-band [29]. The light distribution t can be expressed by using (4). Exploiting the SWT, which can be expressed by Eq. (5). Furthermore, the corresponding reconstruction algorithm, known as IDSWT (Inverse Discrete Stationary Wavelet Transform), is represented by Eq. (6).

$$t = e^{-\beta d} \tag{4}$$

$$
\begin{cases}
A_{j,k_1 k_2} = \sum_{n_1} \sum_{n_2} h_0^{\uparrow 2^j}(n_1 - 2k_1) h_0^{\uparrow 2^j}(n_2 - 2k_2) A_{j-1,n_1,n_2} \\
D^1_{j,k_1 k_2} = \sum_{n_1} \sum_{n_2} h_0^{\uparrow 2^j}(n_1 - 2k_1) g_0^{\uparrow 2^j}(n_2 - 2k_2) A_{j-1,n_1,n_2} \\
D^2_{j,k_1 k_2} = \sum_{n_1} \sum_{n_2} g_0^{\uparrow 2^j}(n_1 - 2k_1) h_0^{\uparrow 2^j}(n_2 - 2k_2) A_{j-1,n_1,n_2} \\
D^3_{j,k_1 k_2} = \sum_{n_1} \sum_{n_2} g_0^{\uparrow 2^j}(n_1 - 2k_1) g_0^{\uparrow 2^j}(n_2 - 2k_2) A_{j-1,n_1,n_2}
\end{cases} \tag{5}
$$

$$
\begin{aligned}
\Big(A_{j-1,n_1,n_2} \\
= \tfrac{1}{4} \sum_{i=0}^{3} \Big\{ &\sum_{k_1} \sum_{k_2} h_1(n_1 - 2k_1 - i) h_1(n_2 - 2k_2 - i) A_{j,k_1 k_2} \\
+ &\sum_{k_1} \sum_{k_2} h_1(n_1 - 2k_1 - i) g_1(n_2 - 2k_2 - i) D^1_{j,k_1 k_2} \\
+ &\sum_{k_1} \sum_{k_2} g_1(n_1 - 2k_1 - i) h_1(n_2 - 2k_2 - i) D^2_{j,k_1 k_2} \\
+ &\sum_{k_1} \sum_{k_2} g_1(n_1 - 2k_1 - i) g_1(n_2 - 2k_2 - i) D^3_{j,k_1 k_2} \Big)
\end{aligned} \tag{6}
$$

4 Simulation Outcomes

4.1 Subjective Analysis

The term subjective analysis refers to an analysis that bases its quality evaluation on the observers' perceptions of the image. For a variety of foggy images, the subjective analysis has been done (see Fig. 2).

4.2 Objective Analysis

The proposed methodology has been subjected to a thorough objective analysis, evaluating various metrics as mentioned in Table1. The comprehensive overview of the conducted analysis has been depicted in Table 1.

(a)Input Image	(b)Output Image	PSNR	SSIM
		69.7440	0.9503
		63.2154	0.8546
		65.9343	0.8657
		67.3452	0.9612

Fig. 2. (a) Input haze image (b) SWT based dehazed image.

| | 68.8765 | 0.9406 |

Fig. 2. (*continued*)

Table 1. Performance metrics for DHWT and SWT

Image Name	Metric	DHWT	SWT(Proposed)
Image-1	MSE	0.0087	0.0073
	PSNR	62.7544	69.7440
	SSIM	0.8341	0.9503
Image-2	MSE	0.0065	0.0039
	PSNR	61.4634	63.2154
	SSIM	0.8535	0.8546
Image-3	MSE	0.0079	0.0098
	PSNR	64.7544	65.9343
	SSIM	0.7986	0.8657
Image-4	MSE	0.0092	0.0083
	PSNR	63.8243	67.3452
	SSIM	0.8345	0.9612
Image-5	MSE	0.0079	0.0069
	PSNR	61.4570	68.8765
	SSIM	0.8845	0.9406

5 Conclusion

The main goal of this study is to use the multi-level stationary wavelet transform (SWT) to improve the visual quality of dehazed photographs. This method enables image size reduction without sacrificing image quality. This work makes use of SWT to build an effective dehazing mechanism based on the sub-band image model. The suggested methodology's efficiency has been assessed utilizing performance indicators such as PSNR, SSIM, and MMSE. Based on these characteristics, a comparison of DHWT (Discrete Haar Wavelet Transform) and SWT was performed. The simulation results clearly show that the suggested approach provides excellent visual quality after dehazing. When compared to DHWT, the SWT-based dehazing method delivers a stunning 11.13% increase in PSNR, a 13.93% increase in SSIM, and a significant 40% decrease in MSE. These findings demonstrate the efficacy and superiority of the SWT-based technique to image dehazing.

References

1. Lu, J., Dong, C.: DSP-based image real-time dehazing optimization for improved dark-channel prior algorithm. J. Real-Time Image Proc. **17**, 1675–1684 (2020)
2. Golts, A., Freedman, D., Elad, M.: Unsupervised single image dehazing using dark channel prior loss. IEEE Trans. Image Proc. **29**, 2692–2701 (2020)
3. Zhu, Q., Mai, J., Shao, L.: A fast single image haze removal algorithm using color attenuation prior. Image Process. IEEE Trans. **24**(11), 3522–3533 (2015)
4. Cai, B., et al.: Dehazenet: an end-to-end system for single image haze removal. IEEE Trans. Image Process. **25**(11), 5187–5198 (2016)
5. Gibson, K.B., Vo, D.T., Nguyen, T.Q.: An investigation of dehazing effects on image and video coding. IEEE Trans. Image Process. **21**(2), 662–673 (2012)
6. Vijjapu, A., Vinod, Y.S., Murty, S.V.S.N., Raju, B.E., Satyanarayana, B.V.V., Kumar, G.P.: Steganalysis using Convolutional Neural Networks-Yedroudj Net. In: International Conference on Computer Communication and Informatics (ICCCI), Coimbatore, India, pp. 1–7 (2023)
7. Sharmila, S., Thanga Revathi, S.K., Sree, P.: Convolution neural networks based lungs disease detection and severity classification. In: International Conference on Computer Communication and Informatics (ICCCI), Coimbatore, India, pp. 1–9 (2023)
8. Ramesh Chandra, K., Prudhvi Raj, B., Prasannakumar, G.: An efficient image encryption using chaos theory. In: International Conference on Intelligent Computing and Control Systems (ICCS), Madurai, India, pp. 1506–1510 (2019)
9. Raju, E.B., Sankar, R.M., Kumar, S.T., Chandra, R.K., Durga, B.V., Kumar, P.G.: Modified encryption standard for reversible data hiding using AES and LSB steganography. In: International Conference on Computer Communication and Informatics (ICCCI), Coimbatore, India, pp. 1–5 (2023)
10. Ramesh Chandra, K., Donga, M., Budumuru, P.R.: Reversible data hiding using secure image transformation technique. In: Suma, V., Chen, J.I.Z., Baig, Z., Wang, H. (eds.) Inventive Systems and Control. Lecture Notes in Networks and Systems, vol. 204. Springer, Singapore (2021)
11. Elisha Raju, B., Ramesh Chandra, K., Budumuru, P.R.: A two-level security system based on multimodal biometrics and modified fusion technique. In: Karrupusamy, P., Balas, V.E., Shi, Y. (eds.) Sustainable Communication Networks and Application. Lecture Notes on Data Engineering and Communications Technologies, vol. 93. Springer, Singapore (2022)
12. Ravi Sankar, M., et al.: Performance evaluation of multiwavelet transform for single image dehazing. In: Gupta, N., Pareek, P., Reis, M. (eds.) Cognitive Computing and Cyber Physical Systems. Lecture Notes of the Institute for Computer Sciences, Social Informatics and Telecommunications Engineering, vol. 472. Springer, Cham (2023)
13. Sravanthi, I., et al.: Performance evaluation of fast DCP algorithm for single image dehazing. In: Gupta, N., Pareek, P., Reis, M. (eds.) Cognitive Computing and Cyber Physical Systems, IC4S 2022, Lecture Notes of the Institute for Computer Sciences, Social Informatics and Telecommunications Engineering, vol. 472. Springer, Cham (2023)

14. Lee, S., Yun, S., Nam, J.H., et al.: A review on dark channel prior based image dehazing algorithms. J. Image Video Proc. **2016**, 4 (2016)
15. Anwar, M.I., Khosla, A.: Vision enhancement through single image fog removal. Eng. Sci. Technol. Int. J. **20**(3), 1075–1083 (2017)
16. Badhe, M.V., Prabhakar, L.R.: A survey on haze removal using image visibility restoration technique. Int. J. Comput. Sci. Mob. Comput. **5**(2), 96–101 (2016)
17. Hodges, C., Bennamoun, M., Rahmani, H.: Single image dehazing using deep neural networks. Pattern Recogn. Lett. **128**, 70–77 (2019)
18. He, K., Sun, J., Tang, X.: Single image haze removal using dark channel prior. IEEE Trans. Pattern Anal. Mach. Intell. **33**(12), 2341–2353 (2010)
19. Portilla, J., Strela, V., Wainwright, M.J., Simoncelli, E.P.: Image denoising using scale mixtures of Gaussians in the wavelet domain. IEEE Trans. Image Process. **12**(11), 1338–1351 (2003)
20. Sharmila, K.S., Asha, A.V.S., Archana, P., Chandra, K.R.: Single image dehazing through feed forward artificial neural network. In: Gupta, N., Pareek, P., Reis, M. (eds) Cognitive Computing and Cyber Physical Systems. IC4S 2022. Lecture Notes of the Institute for Computer Sciences, Social Informatics and Telecommunications Engineering, vol. 472. Springer, Cham (2023)
21. Khmag, A., Al-Haddad, S.A.R., Ramli, A.R., Kalantar, B.: Single image dehazing using second-generation wavelet transforms and the mean vector L2-norm. Vis. Comput. **34**, 675–688 (2018)
22. Zhang, C., Wu, C.: Multi-scale attentive feature fusion network for single image dehazing. In: International Joint Conference on Neural Networks, pp. 1–7 (2022)
23. Gibson, K.B., Nguyen, T.Q.: On the effectiveness of the dark channel prior for single image dehazing by approximating with minimum volume ellipsoids. In: IEEE International Conference on Acoustics, Speech and Signal Processing, pp. 1253–1256 (2011)
24. Xie, B., Guo, F., Cai, Z.: Improved single image dehazing using dark channel prior and multi-scale retinex. In: International Conference on Intelligent System Design and Engineering Application, pp. 848–851 (2010)
25. Park, D., Han, D.K., Ko, H.: Single image haze removal with WLS-based edge-preserving smoothing filter. In: IEEE International Conference on Acoustics, Speech, and Signal Processing, pp. 2469–2473 (2013)
26. Gibson, K.B., Nguyen, T.Q.: Fast single image fog removal using the adaptive Wiener filter. In: IEEE International Conference on Image Processings, pp. 714–718 (2013)
27. Zhang, Q., Li, X.: Fast image dehazing using guided filter. In: IEEE 16th International Conference on Communication Technology, pp. 182–185 (2015)
28. Berman, D., Treibitz, T., Avidan, S.: Non-local image dehazing. In: Computer Vision and Pattern Recognition, pp. 1674–1682 (2016)
29. Iwamoto, Y., Hashimoto, N., Chen, Y.-W.: Fast dark channel prior based haze removal from a single image. In: 14th International Conference on Natural Computation, Fuzzy Systems and Knowledge Discovery, pp. 458–461 (2018)

A Chest X-Ray Image Based Model for Classification and Detection of Diseases

Srinivas Yallapu[1,2]([⊠]) [iD] and Aravind Kumar Madam[3] [iD]

[1] Bharatiya Engineering Science and Technology Innovation University, Gownivaripalli, Gorantla, Andhra Pradesh 515231, India
yallapu.srinivas@gmail.com
[2] Department of ECE, Vishnu Institute of Technology, Bhimavaram, Andhra Pradesh 534202, India
[3] West Godavari Institute of Science and Engineering, East Godavari District, Prakashraopalem, Andhra Pradesh, India

Abstract. Radiography holds significant importance in the medical field, allowing for efficient and widespread access, primarily due to the predominant utilization of thoracic imaging equipments within healthcare infrastructure. However, radiologist's ability to understand radiography images is constrained by their inability to recognize the fine visual details present in the images. There have been numerous studies published in the literature describe machine learning (ML) recent advance models that use support vector machines do differentiate between COVID-19 and non COVID-19 cases by means of open-access chest radiograph databases. The AI can be familiar with characteristics observed in chest X-rays tasks that are typically beyond the scope of a radiologist. They did, however, produce poor categorization performance. Deep - learning techniques in artificial intelligence (AI) are high-performance classifiers engage in recreation a most important significance in the identification of the disease through analysis of chest radiograph metaphors. The major goal of this learning is to examine the enhancement of pre-trained ConvNet (CNNs) as XCOVNet for COVID-19 classification by means of chest radiograph, considering the abundance of newly developed deep learning models specifically designed for this task with more accuracy of 99.51% than previous methods.

Keywords: Radiography · Deep-Learning · Convolutional Neural Networks (ConvNet or CNN) · Artificial Intelligence (AI) · Machine Learning (ML)

1 Introduction

Since the pandemic disease (COVID-19) originating in Wuhan, China in December 2019, it has impacted billions of people. The virus caused an outbreak that spread quickly. The virus that caused COVID-19 sickness was identified after extensive investigation, and it was shown to be a member of a large family of respiratory viruses that may also cause diseases such as severe acute respiratory syndrome (SARS-CoV) and Middle East

P. Pareek et al. (Eds.): IC4S 2023, LNICST 536, pp. 422–432, 2024.
https://doi.org/10.1007/978-3-031-48888-7_36

Respiratory Syndrome (MERS-CoV) are among the included. Viral pneumonitis can arise from the novel SARS-CoV-2 virus. In some states; there is very high mortality rate among the populace. The number of fatalities worldwide is rising daily. Therefore, it is essential to create a method that can accurately, quickly, and affordably diagnose viral pneumonia. This will be the first step in implementing other preventive measures, including as isolation, contact tracing, and treatment, to end the outbreak. Real-time polymerase chain reaction (RT-PCR) testing for the detection of viral nucleic acid is one well-liked technique for virus detection. This exam has a lot of restrictions and is quite sensitive. For instance, it cannot find corona vi-ruses that were already present when DNA sequence samples were taken [1].

However, only with the assistance of specialized knowledge is perfect and quick analysis of an X-Ray image possible. The typical diagnosis of pneumonia is based on the presence of fever, chills, and a dry cough; however, because many asymptomatic patients test positive, it is imperative to enhance the screening procedure by using X-ray metaphors and testing as many people as possible as quickly as feasible. The screening procedure becomes a challenging work as a result of an increase in cases and a decrease in the number of professionals available to make diagnoses. Therefore, for a quick and precise diagnosis, clinicians must rely on machine learning algorithms. The X-ray images have already been examined using a number of machine learning techniques [2]. Support vector methods (SVMs), a type of conventional approach, have a number of drawbacks. Their performance has declined over time and is no longer seen as meeting realistic criteria. Additionally, their development takes a long time. Deep learning techniques have produced significant improvements in the classification of medical images, and they are now a useful tool for clinicians to examine the photos and identify the issue. They can now do many of the activities currently associated with medical image analysis, including pathological abnormality recognition, staging, and explanation. CNN is a common method for image analysis that has achieved great success in the medical industry [3].

In order to evaluate chest images, identify prevalent thoracic disorders, and distinguish between viral and non-viral pneumonia, Deep Convolutional Networks (DCNNs) are being developed [4, 5]. Though many common viruses, including influenza A/B, only viruses that cause viral pneumonia significantly alter X-Ray images. This implies that each incidence of viral pneumonia will have a unique visual presentation. Another issue is locating a dataset with positive sample sizes. It is imperative to create a representation that can circumvent these pathological anomalies and accurately identify the virus. Since 2012, these techniques have been applied in the medical field and have demonstrated noticeably superior performance to earlier techniques. CNN with 121 layers that was skilled using a dataset comprising 112,120 frontal-view chest X-ray images from the ChestX-ray 14 dataset outperformed four radiologists on average [6]. CNN differs from traditional machine learning techniques since it can learn automatically from domain-specific images. Healthcare systems around the world are being overloaded by the COVID-19 patient population's exponential growth. Every patient with a respiratory ailment cannot be checked using traditional methods (RT-PCR) due to a lack of testing kits. High risk patients may benefit from being quarantined while test results are awaiting if suspected COVID-19 infections are found on chest X-rays. In this work, a methodology has been developed using chest X-rays to priorities the patient selection for additional RT-PCR testing. In a hospital context where the current systems are having trouble making decisions, this could be helpful.

2 Literature Review

The research on utilizing machine learning algorithms for disease diagnosis and classification using chest X-ray metaphors have been dominated from last decade. Numerous authors have utilized and developed various algorithms for training and testing, and the reported performance metrics. In today the radiologist's data increased drastically by thin-slice chest scans, which have become necessary in thoracic radiology. Because of this it is important to automate the processing of such data; as a result it has led to a fast escalating study field in medical imaging. Various pulmonary structures can be segmented, chest images can be registered, and applications for the detection, classification, and quantification of chest anomalies addressed in [7].

Recent research indicates that several physiologic mechanisms used by pulmonary vascular disorders can specially affect arteries or veins. Physicians by hand examine the patients' chest computed tomography (CT) images in search of anomalies to recognize changes in the two vascular trees. The classification of veins and arteries from chest computed tomography (CT) images has been proposed in [8]. The authors of [9] described a methodology that identifies disease classification issue for the chest X-ray (CXR). In databases of medical, general illnesses are regularly overrepresented while unusual medical conditions are underrepresented. As a result of privacy issues it is difficult to unite massive datasets across health care facilities. To deal with these issues, the authors of [10] recommended synthesizing pathology in medical imagery.

Based on a modestly large labeled dataset, a deep convolutional generative adversarial network (DCGAN) has been proposed to generate synthetic chest X-rays. This improved performance is partly due to the dataset's balance using DCGAN synthesized images, which preferentially enhance classes that lack example photos. To provide more useful picture representations, the authors of [11] have considered the modeling of relations of image-level. Therefore, it was suggested in [12] that a distance map have to be estimate. And use it to compel a more comprehensive and classification-appropriate semantic segmentation prediction. The recent pandemic COVID 19 affected many of the humans, from which many of them get examined by the different methods like CT and CXR. These methods created the datasets for the researchers to get the abundant data in the format of images for their evaluation of the biomedical features. Due to privacy issues, small collection of data is available for the researchers.

A comprehensive collection of COVID-19 CXR and CT metaphors take out from publications in the Pub Med Central Open Access (PMC-OA) Subset, relevant to COVID-19; can be accessed through [13]. The authors of [14, 15] proposed a neural network in order to improve pneumothorax detection by incorporating both frontal and lateral X-ray data. This network has two inputs and three outputs. Accurate and automated interpretation of medical images, including tasks like segmentation, detection, and classification, is crucial for current illness analysis and prediction. Computer-aided diagnosis (CAD) systems play a significant role in providing fast and reliable diagnoses for a wide range of medical conditions. Over the past decade, deep learning (DL)-based CAD systems have shown remarkable performance in various healthcare applications.

To address this gap, the authors in [16] have introduced the Hercules model, an innovative a specialized uncertainty-aware hierarchical considerate multilevel feature fusion model tailored for medical image categorization. In [17] the authors present pre-processing techniques proposed by various researchers to address the issue of artifacts in metaphors attained from automated diagnostic equipments. Systems so far we discussed show the value of further research in this field for enhancing medical diagnosis and patient outcomes by demonstrating the potential of DL and ensemble representations for the categorization and detection of diseases using chest X-ray images [18–24].

Table 1. Performance comparison of literature

Ref. No	Method	Accuracy	Gap
[8]	3-D Convolutional neural network (CNN)	94%	Need to increase Accuracy
[9]	ConsultNet	82.2%	Disease recognition problem in case of saliency detection
[25]	Deep learning method	99%	Restricted only to analyze diseases by doctors who are deals outpatients
[26]	Ensemble Bootstrap aggregating training and Multiple NN methods	93%	Need to increase Accuracy
[27]	Capsule Networks	95.7%	New versions of covid not detected

In [25] authors proposed classification of chest X-ray view, in this method they achieved accuracy of 97%. Authors applied Deep learning to the classification of COVID-19 disease using X-ray images and got accuracy of 93% in [26]. Authors' proposed in [27] capsule networks for corona disease separation and got accuracy of 95.7%. The following Table 1 depicts the Performance comparison of some of the literature. Therefore, it can be concluded that additional feature extraction operations are not necessary for multiple classifications. The training complexity hinders the analysis of image properties, leading to poor performance metrics such as low accuracy [28–32].

3   Methodology

The proposed system suggested the solution which uses chest X-ray metaphors to classify and identify diseases using a deep learning-based model. The system classifies the input photos into several disease categories and given to CNN architecture by extracting features from input [15]. The method begins by gathering chest X-ray dataset metaphors, which contains metaphors of various illnesses like pneumonia, COVID-19, and tuberculosis. The dataset is then preprocessed to accommodate missing data and normalize the image intensities. The preprocessed dataset is then separated into three sets: training, validation and testing. The validation set is utilized to adjustment of hyper parameter and selection of model, while the CNN model trained by training set. The testing set is used to assess the trained model performance.

3.1 Block Diagram

For the categorization and detection of diseases from chest X-ray a block schematic has been developed and is depicted in Fig. 1. The system's brain is a deep learning model, which examines chest X-ray metaphors to find patterns that can be used to identify thoracic disorders like pneumonia, TB, and COVID-19. CNN used to train the model using a sizable dataset of chest X-ray metaphors labeled with various dis-eases. Before the chest X-ray metaphors are put into the DL model, the pre-processing part of the system is in charge of cleaning, normalizing, and improving them. The data management component of the system is responsible for storing, retrieving, and managing the large dataset of chest X-ray images with disease labels. Therefore, the proposed block diagram significance for integrating different components to create a robust and precise system for the classification and detection of thoracic diseases from chest X-ray images.

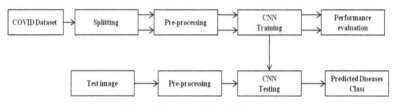

Fig. 1. Proposed Block Diagram

3.2 Deep COVID NET CNN Architecture

The below Fig. 2 shows the Deep COVID NET CNN Architecture which describes four Convolutional layers sequential model architecture for classify X-ray images. The first layer consists of 32 filters, second layer consists of 64 filters, third layer consists of 64 filters and the final layer consists of 128 filters.

4 Simulation Results

In below Fig. 3 detected 820 dataset processed images. The next step is to generate CNN model for given images and get the following window.

In below Fig. 4 represents newly developed deep learning models specifically designed with more accuracy of 99.51% than previous methods.

The Fig. 5 shows no of images have different sizes are suppressed at different layers.

In first layer 62×62 image size was used and in second layer 62×62 and goes on. Then XCOVNet model is willing to ready to accept test images and will predict disease in that image. The graph in Fig. 6 shows the relationship between epoch/iteration accuracy

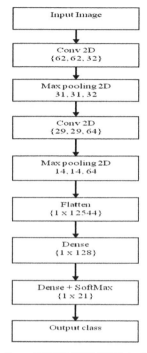

Fig. 2. Deep COVID NET CNN Architecture

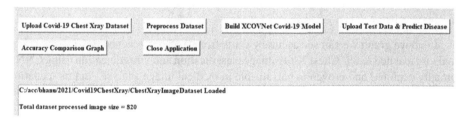

Fig. 3. XCOVNet Covid-19 Building Model

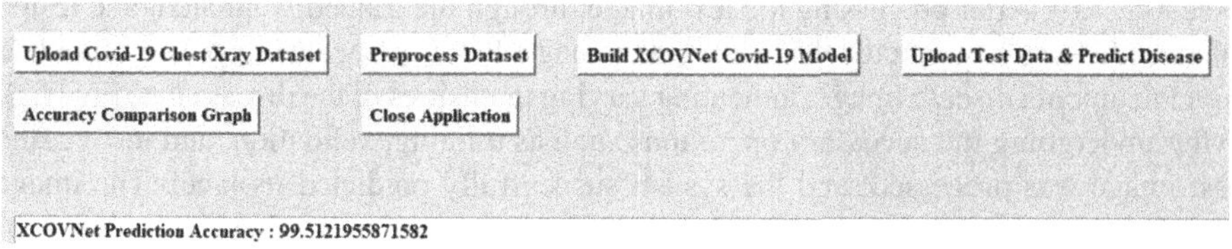

Fig. 4. XCOVNet Prediction accuracy of CNN Model

and loss values. Here consider 20 iterations to build XCOVNet. As number of iterations increase Accuracy will increase whereas Loss will decrease.

```
Model: "sequential_1"

Layer (type)                    Output Shape              Param #
=================================================================
conv2d_1 (Conv2D)               (None, 62, 62, 32)        896
max_pooling2d_1 (MaxPooling2    (None, 31, 31, 32)        0
conv2d_2 (Conv2D)               (None, 29, 29, 64)        18496
max_pooling2d_2 (MaxPooling2    (None, 14, 14, 64)        0
flatten_1 (Flatten)             (None, 12544)             0
dense_1 (Dense)                 (None, 128)               1605760
dense_2 (Dense)                 (None, 21)                2709
=================================================================
Total params: 1,627,861
Trainable params: 1,627,861
Non-trainable params: 0

None
```

Fig. 5. Details of proposed CNN Layers

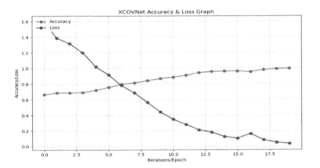

Fig. 6. Loss and Accuracy graph

In above graph we can see accuracy starts from 0.65 and reached to 1.0% accuracy and loss reached to 0%.Chest X-ray image classification and virus detection using CNNs broadly explored and proven to be valuable in medical image analysis. In this scenario, a CNN can be trained for classification of chest X-ray metaphors into different disease categories or detect the presence of specific diseases. Figure 7.a shows the pneumonia decease detection.

After processed the test image (operations like trained, validated, and tested), predicted the images and labeled as like this, this is the Pneumonia disease predicted image (see Fig. 7.b). After processing the test image through the trained, validated, and tested phases, the system predicted the label for the image. In this case, the image was classified as a Pneumonia disease image; indicating varying intensities of the disease (see Fig. 7.c). After undergoing the necessary operations such as training, validation, and testing, the test image was processed, and the system successfully predicted its label. The image was classified as Pneumonia - Viral - COVID-19 disease, indicating the presence of COVID-19 with varying intensities of the disease (see Fig. 7.d).After going through the stages of training, validation, and testing, the test image was processed, and the system made a prediction and assigned a label.

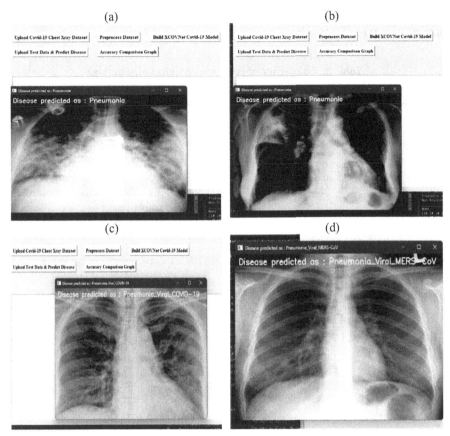

Fig. 7. Identification of various diseases: a) Low severity pneumonia disease, b)High severity pneumonia disease, c) Pneumonia-Viral-COVID-19 disease, d) Pneumonia-Viral-MERS-Cov disease

5 Conclusion

In this work, a model has been created to identify various diseases like COVID-19, pneumonia, Pneumonia-Viral-COVID-19, Pneumonia-Viral-MERS-Covinfection from chest X-ray metaphors. A dataset of 392 X-ray patient metaphors has been utilized with positive and negative COVID results. The proposed system utilizes a deep learning-based model and chest X-ray metaphors for classification and identification of diseases. The system employs CNN architecture to extract features from the input photos and categorize them into various disease categories. The process involves gathering chest X-ray images dataset, preprocessing it to handle missing data and normalize image intensities, and then dividing it into training, validation, and testing sets. The validation set is utilized to adjustment of hyper parameter and selection of model, while the CNN model is trained by training set. Finally, the performance of the trained model is evaluated using the testing set, providing a comprehensive assessment of the system's effectiveness in disease classification. The following Table 2 describes the performance of proposed method with existed methods. The proposed method gives the accuracy of 99.51% which is more than existed methods.

Table 2. Comparison of performance with existed methods.

Ref. No	Method	Accuracy	Research Gap/Future work
[8]	3-D CNN	94%	Need to increase Accuracy
[9]	ConsultNet	82.2%	Disease recognition problem in case of saliency detection
[25]	Deep learning method	99%	Restricted only to analyze diseases by doctors who are deals outpatients
[26]	Ensemble Bootstrap aggregating training and Multiple NN methods	93%	Need to increase Accuracy
[27]	Capsule Networks	95.7%	New versions of covid not detected
Proposed	XCOVNet	99.51%	By doing experiments on CT scan image data of Chest for finding of COVID-19 and mixing both models to recognize the severity level

References

1. Kim, H., Hwang, J., Kim, J.H., Lee, S., Kang, M.: Sensitive detection of multiple fluorescence probes based on Surface-enhanced Raman Scattering (SERS) for MERS-CoV. In: IEEE 14th International Conference on Nano/Micro Engineered and Molecular Systems (NEMS), Bangkok, Thailand, pp. 498–501 (2019)
2. Bhat, A., Bhardwaj, U., Singla, M., Garg, K.: Prediction of COVID 19 using chest x-ray images through CNN optimised using genetic algorithm. In: 2022 2nd International Conference on Intelligent Technologies (CONIT), Hubli, India, pp.1–8 (2022)
3. Lauer, S.A., et al.: The incubation period of coronavirus disease 2019 (COVID-19) from publicly reported confirmed cases: estimation and application. Ann. Intern. Med. **172**(9), 577–582 (2020)
4. Holshue, M.L., et al.: First case of 2019 novel coronavirus in the United States. N. Engl. J. Med. **382**, 929–936 (2020)
5. WHO Coronavirus Disease (COVID-19) Dashboard (2021)
6. Yicheng, F., Huangqi, Z., Jicheng, X., Minjie, L., Lingjun, Y., Peipei, P., Wenbin, J.: Sensitivity of chest CT for covid-19: comparison to RT-PCR. Radiology, 200–432 (2020)
7. Sluimer, I., Schilham, A., Prokop, M., Van Ginneken, B.: Computer analysis of computed tomography scans of the lung: a survey. IEEE Trans. Med. Imaging **25**(4), 385–405 (2006)
8. Nardelli, P., et al.: Pulmonary artery-vein classification in CT images using deep learning. IEEE Trans. Med. Imaging **37**(11), 2428–2440 (2018)

9. Guan, Q., Huang, Y., Luo, Y., Liu, P., Xu, M., Yang, Y.: Discriminative feature learning for thorax disease classification in chest x-ray images. IEEE Trans. Image Process. **30**, 2476–2487 (2021)

10. Salehinejad, H., Colak, E., Dowdell, T., Barfett, J., Valaee, S.: Synthesizing chest x-ray pathology for training deep convolutional neural networks. IEEE Trans. Med. Imaging **38**(5), 1197–1206 (2019)

11. Mao, C., Yao, L., Luo, Y.: ImageGCN: multi-relational image graph convolutional networks for disease identification with chest x-rays. IEEE Trans. Med. Imaging **41**(8), 1990–2003 (2022)

12. Hu, J., Zhang, C., Zhou, K., Gao, S.: Chest x-ray diagnostic quality assessment: how much is pixel-wise supervision needed? IEEE Trans. Med. Imaging **41**(7), 1711–1723 (2022)

13. Peng, Y., Tang, Y., Lee, S., Zhu, Y., Summers, R.M., Lu, Z.: COVID-19-CT-CXR: a freely accessible and weakly labeled chest x-ray and CT image collection on covid-19 from biomedical literature. IEEE Trans. Big Data **7**(1), 3–12 (2021)

14. Luo, J.X., Liu, W.F., Yu, L.: Pneumothorax recognition neural network based on feature fusion of frontal and lateral chest x-ray images. IEEE Access **10**, 53175–53187 (2022)

15. Chen, B., Li, J., Lu, G., Yu, H., Zhang, D.: Label co-occurrence learning with graph convolutional networks for multi-label chest x-ray image classification. IEEE J. Biomed. Health Inform. **24**(8), 2292–2302 (2020)

16. Abdar, M., et al.: Hercules: deep hierarchical attentive multilevel fusion model with uncertainty quantification for medical image classification. IEEE Trans. Industr. Inf. **19**(1), 274–285 (2023)

17. Budumuru, P.R., Varma, A.K.C., Satyanarayana, B.V.V., Srinivas, Y., Raju, B.E., Parasanna Kumar, G.: Preprocessing analysis of medical image: a survey. In: 2022 4th International Conference on Advances in Computing, Communication Control and Networking (ICAC3N), Greater Noida, India, pp. 2369–2375 (2022)

18. Shaik, A.R., Chandra, K.R., Raju, B.E., Budumuru, P.R.: Glaucoma identification based on segmentation and fusion techniques. In: 2021 International Conference on Advances in Computing, Communication, and Control (ICAC3), Mumbai, India, pp. 1–4 (2021)

19. Chandra, K.R., Donga, M., Budumuru, P.R.: Reversible data hiding using secure image transformation technique. In: Suma, V., Chen, J.I.Z., Baig, Z., Wang, H. (eds.) Inventive Systems and Control. Lecture Notes in Networks and Systems, Springer, Singapore, 204 (2021)

20. Sharmila, K.S., Asha, A.V.S., Archana, P., Chandra, K.R.: Single image dehazing through feed forward artificial neural network. In: Gupta, N., Pareek, P., Reis, M. (eds.) Cognitive Computing and Cyber Physical Systems. IC4S 2022. Lecture Notes of the Institute for Computer Sciences, Social Informatics and Telecommunications Engineering, Springer, Cham, 472 (2023)

21. Budumuru, P.R., Shaik, A.R., Satyanarayana, B.V.V., Manikanta, S.P., Sharmila, K.S., Durga Prasad, D.: Normalized algorithm with image processing methods for estimation of crack length. In: 2022 6th International Conference on Electronics, Communication and Aerospace Technology, Coimbatore, India, pp. 1436–1439 (2022)

22. Sharmila, S., Thanga Revathi, S.K., Sree, P.: Convolution Neural Networks based lungs disease detection and Severity classification. In: International Conference on Computer Communication and Informatics (ICCCI), Coimbatore, India, pp. 1–9 (2023)

23. Kiruban, M., Jayamani, R., Ramu, P.: Removal of salt and pepper noise from SAR images using optimized APCNN in Shearlet transform domain. Arab. J. Geosci. **14**, 458 (2021)

24. Sakthimohan, M., Reddy, P.G.K., Narendra, T., Venkatesh, B., Elizabeth, R.G.: Leaf health monitoring and disease detection using image processing. In: 2nd International Conference on Advance Computing and Innovative Technologies in Engineering (ICACITE), Greater Noida, India, pp. 773–777 (2022)

25. Xue, Z., et al.: Chest x-ray image view classification. In: IEEE 28th International Symposium on Computer-Based Medical Systems, pp. 66–71 (2015)

26. Asif, S., Wenhui, Y., Jin, H., Jinhai, S.: Classification of COVID-19from chest X-ray images using deep convolutional neural network. In: IEEE 6th international conference on computer and communications (ICCC), pp. 426–433 (2020)

27. Afshar, P., Heidarian, S., Naderkhani, F., Oikonomou, A., Plataniotis, K.N., et al.: Covid-caps: a capsule network-based framework for identification of covid-19 cases from x-ray images. Pattern Recogn. Lett. **138**, 638–643 (2020)

28. Ravi Sankar, M., et al.: Performance evaluation of multiwavelet transform for single image dehazing. In: Gupta, N., Pareek, P., Reis, M. (eds.) Cognitive Computing and Cyber Physical Systems. Lecture Notes of the Institute for Computer Sciences, Social Informatics and Telecommunications Engineering, vol. 472. Springer, Cham (2023)

29. Sravanthi, I., et al.: Performance evaluation of fast DCP algorithm for single image dehazing. In: Gupta, N., Pareek, P., Reis, M. (eds.) Cognitive Computing and Cyber Physical Systems, IC4S 2022,Lecture Notes of the Institute for Computer Sciences, Social Informatics and Telecommunications Engineering, vol. 472. Springer, Cham (2023)

30. Elisha Raju, B., Ramesh Chandra, K., Budumuru, P.R.: A two-level security system based on multimodal biometrics and modified fusion technique. In: Karrupusamy, P., Balas, V.E., Shi, Y. (eds.) Sustainable Communication Networks and Application. Lecture Notes on Data Engineering and Communications Technologies, vol. 93. Springer, Singapore (2022)

31. Vijjapu, A., Vinod, Y.S., Murty, S.V.S.N., Raju, B.E., Satyanarayana, B.V.V., Kumar, G.P.: Steganalysis using Convolutional Neural Networks-Yedroudj Net. In: International Conference on Computer Communication and Informatics (ICCCI), Coimbatore, India, pp. 1–7 (2023)

32. Satyanarayana, B.V.V., Kumar, G.P., Varma, A.K.C., Dileep, M., Srinivas, Y., Budumuru, P.R.: Alzheimer's disease detection using ensemble of classifiers. In: Gupta, N., Pareek, P., Reis, M. (eds.) Cognitive Computing and Cyber Physical Systems. IC4S 2022. Lecture Notes of the Institute for Computer Sciences, Social Informatics and Telecommunications Engineering, Springer, Cham. 472, pp. 55–65 (2022)

Enlighten GAN for Super-Resolution Images from Surveillance Car

Pallavi Adke[1], Ajay Kumar Kushwaha[2]([✉]), Pratik Kshirsagar[1], Mayur Hadawale[1], and Prajwal Gaikwad[1]

[1] Pimpri Chinchwad College of Engineering and Research, Pune, India
[2] Bharati Vidyapeeth (Deemed to be University) College of Engineering, Pune, India
akkushwaha@bvucoep.edu.in

Abstract. Law enforcement and security officers utilize surveillance cars to monitor and investigate suspicious activities, including traffic violations and neighbourhood security. This paper delves into the comprehensive utilization of surveillance cars in modern society. By designing and developing an upgraded system, we aim to address ethical and legal concerns surrounding surveillance practices. The primary objective is to obtain super-resolution images from captured footage using Generative Adversarial Networks (GANs) for image enhancement. GANs, a type of Neural Network, enable the creation of high-quality images from existing low-resolution data. This study explores the application of GANs to enhance image quality in the context of surveillance cars. The proposed system leverages GANs to generate high-resolution images from the low-resolution ones captured by the surveillance car. Additionally, we provide a comprehensive review of the ongoing research and advancements in this field. Existing surveillance systems often output low-resolution images, but through the implementation of Enlighten GAN, we can achieve high-resolution results. The innovative integration of GAN technology empowers the surveillance car system to respond quickly to known situations with improved image clarity, enabling effective monitoring and investigation. This paper contributes to the advancement of surveillance capabilities by providing valuable insights into the potential of GANs for enhancing image quality and super-resolution in surveillance applications.

Keywords: Generative Adversarial Network (GAN) · Surveillance Car · Super Resolution Generative Adversarial Networks (SRGAN)

1 Introduction

Recently, in the world of machine learning, generative adversarial networks (GANs) have become a very exciting innovation. GANs are called generative models because they generate new data instances, which in turn resemble the training data used for training purposes. For example, GANs are used to generate images that resemble human faces that don't have any real identity or don't belong to a real person. GANs come under unsupervised learning that can automatically discover and learn about the patterns and

P. Pareek et al. (Eds.): IC4S 2023, LNICST 536, pp. 433–445, 2024.
https://doi.org/10.1007/978-3-031-48888-7_37

regularities in the testing data involved so that it can also generate the output, which is new examples that match the original dataset. GANs have a very unique and smart way of training the training dataset of the generator model by framing the issue as a supervised learning problem with two sub-models: generator and discriminator. The generator is used to create or generate fake data that resembles the original one, and the discriminator model tries to classify or separate the fake and the original data. Both the models are trained simultaneously until the discriminator is unable to tell the difference between real and fake images; hence, we can understand that the generator is trained and is generating plausible images enough to fool the discriminator. Now, when trying to understand what super-resolution is, which is based on the idea that images with low resolution (noisy) or containing some disturbances are used to generate HR images, it creates original images with HR when it gives it images in LR. Now achieve super-resolution in the images that it obtains from the surveillance car. With the ESPCAM32 camera module, the surveillance car can take photographs and videos, and the output can be displayed on any screen via the hotspot connection to the ESP. With the help of a Bluetooth module and an Android device, the car can be controlled. This surveillance car's purpose is to capture videos and images from closed areas, where normally CCTV cameras don't look good or suffer from black zones. It is a mobile device, hence does not have any black zones, and can be very handy in the surveillance of any suspicious activity in the area. Now, with the help of GAN, it can construct HR images from the output of the surveillance car. CCTV output can also be processed using GAN to get highly resolved images.

2 Related Work

Ronneberger et al. [1] have developed a more advanced architecture called the "fully convolutional network" to improve the accuracy of image segmentation with very few training images. They added a continuous layer to the regular contract network and replaced the pooling operator with an upsampling operator to increase the resolution of the output. Goodfellow et al. [2] discussed an adversarial learning framework for training generative and discriminative models, where both models are multi-layer perceptrons. This approach, called adversarial nets, can be trained using back propagation and dropout algorithms, and sampling from the generated models can be done using only forward propagation. Yuntan et al. [3] applied adversarial learning to semi-supervised segmentation training and achieved performance comparable to fully supervised training with only half of the labelled data. Kulkarni et al. [4–6] discussed advanced techniques for non-linear filters, diffusion processes, wavelet denoising methods, thresholding techniques, and filters for fractional arithmetic. Zhang et al. [7] try to exploit the recently introduced conditional generative adversarial networks (CGAN) excellent generative modelling capabilities by applying an extra constraint that the de-rained image must be indistinguishable from its equivalent ground truth clean image. Xiang et al. [8] presented a deep neural network architecture for single-image rain removal called feature-supervised generative adversarial network (FS-GAN). Vaishnave et al. [9] identified gaps in the literature in order to spur the development of new data-driven algorithms and presented quantitative metrics for evaluating scene classification in satellite imagery. Zhang et al.

[10] suggested a GPS integration and semantic estimation method for decreasing per-pixel noise in region-like tasks that is simple to integrate into various backbone GAN designs. Devabalan [11] discussed the challenge of processing and analyzing remote sensing data quickly, and Isola [12] suggested conditional adversarial networks as a promising approach for many frame-to-frame translation tasks. Laxman [13] proposed a multi-scale gradient-based U-Net (MSG U-Net) for high-resolution frame-to-frame transformation. Park et al. [14] discuss the application of a loss function in the training of a fused network to reduce GAN-generated artefacts, restore fine details, and maintain colour components. The SRGAN method employs spectral normalization techniques to ensure Lipschitz continuity and mitigate the vanishing gradient issue. RaGAN, used in ESR-GAN, predicts a relative decision instead of a definitive one. Wassertein GAN, or WGAN, utilises IPM metrics modules for divergence minimization and provides a nonactivation value as output, which can be taken as a high Lipschitz metric. Gradient clipping can be used to modify weight parameters in one batch, making it an effective technique for GANs as well.

3 Methodology

GANs (Generative Adversarial Networks) consist of two deep networks: one is a discriminator network and the other is a generator network. Some parameters and Variables are as.

D = Discriminator.
G = Generator
θ_d = ParametersofDiscriminators
θ_g = ParametersofGenerator
$P_z(z)$ = InputNoiseDistribution
$P_{data}(x)$ = Originaldatadistribution
$P_g(x)$ = Generateddistribution

The discriminator is given both real and phony images, and it attempts to distinguish between the two. Returns a chance of the image being "true" in the range of 0 and 1. The generator tries to deceive the discriminator into believing the bogus images it produces are real images. Educating a GAN.

The binary entropy loss can be given as –

$$\log(\hat{y}, y) = [y.log\hat{y} + (1 - y).\log(1 - \hat{y})] \tag{1}$$

where y = Original Data
$\hat{y}$ = Reconstructed Data
Discriminator Loss:

$$L^{(D)} = \max[\log(D(x)) + \log(1 - D(G(z)))] \tag{2}$$

Generator Loss:

$$L = min_G max_D[\log(D(x)) + \log(1 - D(G(z)))] \tag{3}$$

3.1 Training a GAN

Part 1: While the Generator is not in use, the discriminator is trained. The network is only propagated forward; no back propagation is carried out. The discriminator is tested to determine if it can accurately identify them as real after being trained on real data for n epochs. Also, the Discriminator is trained on the fictitious generated data from the Generator at this phase to determine if it can correctly identify them as fictitious. Discriminator training is shown in Fig. 1.

Part 2: While the Discriminator isn't in use, the Generator is being taught. After the Discriminator has been trained using the Generator's fabricated data, we may utilize the predictions to train the Generator and improve the Discriminator's prior state. Generator training is shown in Fig. 2.

Training Discriminator

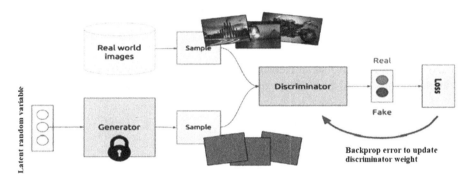

Fig. 1. Training Discriminator

Training Generator

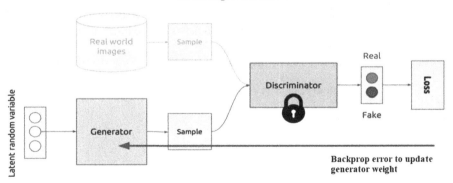

Fig. 2. Training Generator

3.2   SRGAN (Super Resolution Generative Adversarial Networks)

As part of an ongoing effort, one of the generative adversarial networks applications: SRGAN has been put into practice. By upscaling an image from a specified low-resolution to a comparable high-resolution image with greater visual quality, a process known as "super-resolution" is used. High-resolution photographs offer better reconstructed details of the settings and individual objects, or enhanced details as they may appear at any contemporary crime scene. An example of SRGAN is shown below in Fig. 3.

Fig. 3. Example of SRGAN

3.3   SRGAN Architecture

The generator architecture, which is used to produce excellent super-resolution images, is essentially a fully convolutional SRResNet model. To ensure that the overall architecture adapts appropriately to the quality of the photos, the discriminator model, which also serves as an image classifier, has been added. The images produced as a result are significantly better. Generator architecture is shown in Fig. 4. Given that the neural network can select the best suitable value on its own in this situation, parametric ReLU is preferred.

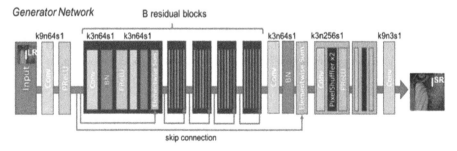

Fig. 4. SRGAN Generator Architecture

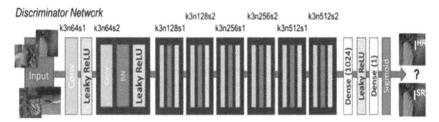

Fig. 5. SRGAN Discriminator Architecture

The discriminator architecture in the figure above functions to distinguish between genuine images and super-resolution images. The adversarial min-max problem is addressed by the discriminator model that is built. Discriminator architecture is shown above in Fig. 5.

A large number of residual blocks are used in the following layer of the feed-forward fully coevolutionary super-resolution reconstruction model (SRResNet) model. Following a batch normalization layer, a Parametric ReLU activation function, a convolutional layer with batch normalization, another convolutional layer with batch normalization, and a final elementwise sum technique, each residual block has a convolutional layer with 33 kernels and 64 feature maps. The feed-forward output and the skip connection output are both used by the elementwise sum method to produce the final output. After 4x upsampling the convolutional layer in this model architecture, super-resolution images are created using the pixel shuffler. The categorization action is carried out via the sigmoid activation function.

3.4   SRGAN Model Training

To build the SRGAN model and train it as needed for this project, we will use the VS Code platform.

Bringing in the necessary libraries Implementing all the crucial libraries needed to complete the ensuing tasks is the initial step in starting our SRGAN project. We have set up each and every necessary library.

Perceptual Loss:
Loss function for perception the effectiveness of our generator network depends on how our perceptual loss function, 1 SR, is defined and design a loss function that evaluates a solution with regard to perceptually meaningful properties, even if 1 SR is frequently represented based on the MSE. The weighted sum of a content loss (1 SR X) and an adversarial loss component is how it define the perceptual loss as follows in Eq. 4:

$$l^{SR} = l_X^{SR} + 10^{-3} l_{Gen}^{SR} \tag{4}$$

where $l_X^{SR} = ContentLoss$, $10^{-3} l_{Gen}^{SR} = AdversarialLoss$

3.5   Preparing the Dataset

The datasets are in the form of individual zip files. These files contain training and validation files for both low-resolution and high-resolution images. Once the download is complete, it extracts them accordingly. For low-resolution images, the size is 96 pixels by 96 pixels, and for high-resolution images, it is 396 pixels by 396 pixels. For training purposes, the same pixel size is required. The low-resolution image data set must also be equal to the high-resolution image data set. This means if it takes 500 LR images, it must use 500 h images; the number should be the same. Training the SRGAN Model After successfully setting up the libraries and collecting the datasets, it will construct the SRGAN architecture and start to train the model [15–17].

Sample outputs of the proposed trained Model are shown in Fig. 6 and 7.

Fig. 6. Model training Result 1

Fig. 7. Model training Result 2

4 Dataset

It uses two types of datasets: one that was obtained from the surveillance car and created, and another from Image Pair (Microsoft).

1. **Manual Dataset:** It is obtained as an output from around 1000 images from our surveillance car. These images were resolved and obtained as HR and LR images for the training of the model so that the model becomes familiar with the output of our car. It also obtained testing data from the car that was used for the input of the GAN to generate its highly resolved reconstruction. The images were captured in all formats to ensure proper training for the model.
2. **Image Pair Dataset:** Realistic Super Resolution refers to generating HR images from some or all the other low-quality images of the same scenery. The problem here is that low-resolution images completely erase the HR visual frequency information in the image.

Many deep and machine learning methods have been put into use or proposed to demonstrate the development of a model that is used to recover lost characteristics of a particular image in the generated one. Hence, using this strategy, deep and machine learning methods have been able to solve super resolution problems.

Data is very important to machine learning, especially when the DL algorithms are data-driven, where data drives the whole algorithm. Now, to solve the problem, gather and form the information and acquire and generate data, which may be just as important to the solution as solving the issue. This research aims to present a new way of collecting real data using various novel data acquisition techniques. Super resolution, noise deprecation, and quality improvement techniques can use this as an input. Click the same image with the same scenery with a low-resolution camera and a high-resolution camera. It took around 11000 images that are low and high-resolution as training data. And it can use testing data to convert those images into SR images.

5 Result Analysis

The performance of the model can be evaluated by using following parameters:

1. **SNR:** The square root of the quantity of photons in the image's brightest region is used to calculate the signal-to noise ratio (SNR) of a digital microscopic image. The ideal range is from 20 to 40.
2. **PSNR:** This ratio is often used to compare the original and compressed images. The ideal range is 30–50 dB.
3. **Jaccard's index:** A statistic used to assess the similarities and diversity of samples is the Jaccard index. It is commonly referred to as the Jaccard similarity coefficient.

Image compression 1 and image compression 2 are shown in Figs. 8 and 9, respectively. The result of the analysis is mentioned in Table 1.

Fig. 8. Image comparison 1

Fig. 9. Image comparison 2

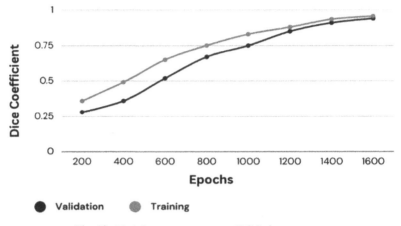

Fig. 10. Training accuracy verses Validation accuracy

Table 1. Result Analysis

	Image Pair 1	Image Pair 2	Image Pair 3	Image Pair 4	Image Pair 5
Jaccard's Index	0.9668	0.8929	0.9759	0.9742	0.9474
PSNR	44.20	43.08	44.57	42.06	44.75
SNR	25.20	19.76	25.39	24.65	26.84

6 Advantages

The advantages of the proposed system can be summarized as follows:

1 With the help of super resolution, it can identify the culprits and submit the results as evidence.
2 A surveillance car being mobile means it does not need to move, and the car can also go places where humans would not dare.
3 It can quickly identify trees, roads, bikers, people, and parked automobiles, and even compute the distance between various objects using GAN and machine.
4 GANs generate data similar to the original data.
5 GANs go into the details of the data and can be easily interpreted into different versions. This is useful when working with machine learning.
6 It can improve model generalization.

The hardware is shown in the Fig. 11.

Fig. 11. Hardware model

7 Comparison

The advanced surveillance car represents a substantial leap forward from older surveillance methods. Integrating cutting-edge components like the ESP CAM 32, a Bluetooth module, and an Android application for control, our system showcases significant advancements. Particularly noteworthy is the incorporation of SRGAN (Super-Resolution Generative Adversarial Network) technology, which enhances image quality to unparalleled levels. This integration introduces several transformative features and improvements to the surveillance process.

1. **Real-time Remote Control:** Operators can precisely navigate the car using the Android app, responding swiftly to emerging situations.
2. **Cost-effectiveness:** The system's integration of readily available components reduces hardware costs without compromising image clarity.
3. **Adaptable and Versatile:** With a modular design and open-source components, the surveillance car can be upgraded with emerging technologies.
4. **Real-time Monitoring and Analysis:** The Android app provides instant access to live video feeds and data insights, optimizing responsiveness and efficiency.

8 Conclusion

In this paper, a cutting-edge surveillance system that combines the power of Enlighten GAN with a mobile robot car is described to address blind spots in existing CCTV and surveillance setups. The system is designed to venture into areas where traditional surveillance systems might be limited, ensuring comprehensive coverage and enhanced security. At the core of the solution lies a highly versatile robot car equipped with state-of-the-art technology that can be remotely operated for secure and efficient surveillance. The robot car's mobility allows it to access remote or hard-to-reach locations, enabling a broader scope of monitoring and investigation. To facilitate seamless control and communication with the robot car, it incorporated a robust Bluetooth module. This feature ensures reliable and real-time data exchange between the mobile surveillance unit and the controlling device, enhancing responsiveness and usability. The Android app complements the system, providing a user-friendly interface for operators to control and manage the robot car remotely. Through the app, users can initiate surveillance sessions, receive live video feeds, and access real-time data insights, bolstering the system's overall surveillance capabilities. In addition to the mobility and control aspects, the system addresses the challenge of low-resolution imagery commonly encountered in surveillance scenarios. It leverages the power of Enlighten GAN to achieve super-resolution images captured by the surveillance system or robot car. This sophisticated image enhancement technique enables us to obtain clearer, high-resolution images, thereby improving the accuracy of person and number plate recognition. By integrating the robot car, Bluetooth module, Android app, and super-resolution image processing, the surveillance system stands as a comprehensive and effective solution for tackling blind spots in surveillance coverage. The combination of advanced components ensures seamless operation, enhanced data quality, and increased surveillance efficiency, empowering law enforcement and security personnel with a powerful tool to maintain public safety and security effectively.

References

1. Ronneberger, O., Fischer, P., Brox, T.: U-Net: convolutional networks for biomedical image segmentation. Int. Conf. Med. Image Comput. Comput-Assis. Interv. Springer Int. Publishing **18**, 234–241 (2015)
2. Goodfellow, I., et al.: Generative adversarial nets. Adv. Neural. Inf. Process. Syst. **27**, 2672–2680 (2014)
3. Tan, Y., Wu, W., Tan, L., Peng, H., Qin, J.: Semi-supervised medical image segmentation based on generative adversarial network. J. of New Media, 4, 3, 155, (2022)
4. Kulkarni, P., Madathil, D.: A review on echocardiographic image speckle reduction filters. Biomed. Res. **29**(12), 2582–2589 (2018)
5. Kulkarni, P., Madathil, D.: A review of echocardiographic image segmentation techniques for left ventricular study. ARPN J. Eng. Appl. Sci. **13**(10), 3536–3541 (2018)
6. Kulkarni, P., Madathil, D.: Fully automatic segmentation of LV from echocardiography images and calculation of ejection fraction using deep learning. Int. J. Biomed. Eng. Technol. **40**(3), 241–261 (2022)
7. Zhang, H., Sindagi, V., Patel, V.M.: Image de-raining using a conditional generative adversarial network. IEEE Trans. Circuits Syst. Video Technol. **30**, 3943–3956 (2019)
8. Xiang, P., Wang, L., Wu, F., Cheng, J., Zhou, M.: Single-image de-raining with feature-supervised generative adversarial network. IEEE Signal Process. Lett. **26**, 650–654 (2019)
9. Vaishnnave, M.P., Devi, K.S., Srinivasan, P.: A study on deep learning models for satellite imagery. Int. J. Appl. Eng. Res. **14**, 881–887 (2019)
10. Zhang, Y., Yin, Y., Zimmermann, R., Wang, G., Varadarajan, J., Ng, S.K.: An enhanced GAN model for automatic satellite-to-map image conversion. IEEE Access **8**, 176704–176716 (2020)
11. Devabalan, P.: Satellite image processing on a grid based computing environment. Int. J. Comput. Sci. mobile Comput., 3, 1039–1044, (2014)
12. Isola, P., Zhu, J. Y., Zhou, T., Efros, A. A.: Image-to-image translation with conditional adversarial networks. In: IEEE Conference on Computer Vision and Pattern Recognition, pp 1125–1134, (2017)

13. Kumarapu Laxman, Laxman, K., Dubey, S.R., Kalyan, B., Kojjarapu, S.R.V.: Efficient high-resolution image-to-image translation using multi-scale gradient U-Net. In: International Conference on Computer Vision and Image Processing, Springer International Publishing, (2021)
14. Park, J., Han, D.K., Ko, H.: Fusion of heterogeneous adversarial networks for single image dehazing. IEEE Trans. Image Process. **29**, 4721–4732 (2020)
15. Kulkarni, P., Madathil, D.: Echocardiography image segmentation using semi-automatic numerical optimisation method based on wavelet decomposition thresholding. Int. J. Imaging Syst. Technol. **31**(4), 2295–2304 (2021)
16. Leclerc, S., et al.: Deep learning for segmentation using an open large-scale dataset in 2D echocardiography. IEEE Trans. Med. Imaging **38**(9), 2198–2210 (2019)
17. Kushwaha, A.K., Khatavkar, S.M., Biradar, D.M., Chougule, P.A.: Depth estimation and navigation route planning for mobile robots based on stereo camera. Lect. Notes Inst. Comput. Sci. Soc. Inform. Telecommun. Eng. **472**, 180–191 (2023)

Brain Tumor Classification Through MR Imaging: A Comparative Analysis

G . Prasanna Kumar[1] , K. Kiran[1] , Kanakaraju Penmetsa[2] ,
K . Indira Priyadarsini[3] , Prudhvi Raj Budumuru[1(✉)] , and Yallapu Srinivas[1]

[1] Department of ECE, Vishnu Institute of Technology, Bhimavaram, India
prajece2005@gmail.com
[2] Department of ECE, SRKR Engineering College, Bhimavaram, India
[3] Department of ECE, DNR College of Engineering and Technology, Bhimavaram, India

Abstract. Tumor in brain is one of the serious diseases throughout the world and it leads to death around 300 thousand people in 2020. Hence, Brain tumor diagnosis is a sensible and important task in clinical and medical field. Identification of illness, area, depth and severity of the disease are major challenges encountered before technological improvement in the clinical field. These major challenges are fulfilled few decades ago by acquiring images of human body parts with collaborations of electronic and mechanical devices. The familiar medical images are Magnetic Resonance Imaging (MRI) Scan, Computed Tomography (CT) Scan and Positron Emission Tomography (PET) Scan. Manual observation of aforementioned scans may lead error in the treatment. Hence, various image processing algorithms and pre-trained methods have been employed on medical images to identify the accurate location, area, depth and severity of the disease, which effectively improvise the treatment. The evolution process has several stages such as: preprocessing; segmentation; future extraction; and classification. Therefore, this work presents a detailed report of CNN based brain tumor classification methods through MR imaging scans. Finally, the performance measures of brain tumor classification methods have been presented and compared.

Keywords: Brain tumor · Diagnosis; Illness · Medical images · Classification

1 Introduction

Global Cancer Statistics 2020 announces, twenty million new incident cancerous cases and nearly 10 million new deaths are happened throughout the world in 2020. According to this, 2.5% deaths were due to brain tumor disease in 10 million cancerous death cases [1]. Hence, diagnosis of tumor in brain has been significant in the medical field. Disease detection, classification and severity estimation are prominent in diagnosis steps of tumor [2–15]. In disease diagnosis process; first capture the skeleton image of required human body part; familiarly it's named as medical image. Most familiar medical image types are X-ray, MRI scan, CT scan and PET scan [1–20, 22]. Out of this MRI scan is suitable to capture the soft tissues of the body part. Manual identification and classification of

P. Pareek et al. (Eds.): IC4S 2023, LNICST 536, pp. 446–458, 2024.
https://doi.org/10.1007/978-3-031-48888-7_38

diseased/defected region may difficult, if raw medical image affected with noise and no variations in the intensity levels of diseased and non-diseased regions. It has been due to longer scan time by the sensors and age of the equipment also. Therefore, for accurate diagnosis of the particular disease, various automated algorithms and methods were designed [3–13]. Figure 1. Shows general flow of automated approach [2, 15]. First employ prepossessing scheme (noise removal) on raw medical image. Further, segmentation is employed on preprocessed image to locate/identify the Region of Interest (ROI). Image segmentation plays a vital role in the automation approach to differentiate diseased and non-diseased portions [21] for identifying the tumor/cancerous region and it helps to classify further. This information helps the diagnosis experts to treat patients effectively and accurately. The various segmentation methods are edge-based, threshold-based, region-based, cluster-based and watershed segmentation. Feature extractions are done from the extracted segmented portions of the image. Finally, classifier has to classify the disease based upon the extracted features. Now a days, CNN models are alternative for traditional algorithms/methods of image restoration, image dehazing [23, 24] along with classification.

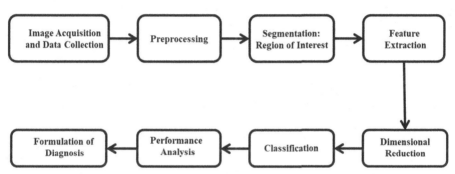

Fig 1. General flow of cancer detection system (Source: Chahal P K et al.[2] and P.R.Budumuru et al.[15]).

2 Related Work

S. Mohsen et al. (2023), proposed two intelligence models for the classification of tumor categories glioma and pituitary in brain. In these two models, VGG19 is first one and single-image super-resolution (SSIR) technique with ResNext101_32 × 8d is the second. Proposed SSIR technique based on GAN algorithm is employed on input MRI scan to produce high resolution images before classification. 344 layers in ResNext101_32 × 8d model includes, 104 layers each for batch normalization and convolution, 100 layers of ReLU, 33 bottleneck layers. Proposed VGG19 model in this work consists nineteen 2-D layers, in that three are fully connected layers and remaining 16 are convolution layers each of which is followed by max-pooling layers [3].

   H. A. Hafeez et al. (2023), come up with a low-grade and high-grade glioma type brain tumor classification model with CNN consisting of less number of layers, size and learnable parameters. This work has been designed in two ways; Feature extraction done by separately with resnet18, squeezenet, alexnet, and proposed CNN. Further, classification has been done with SVM classifier. Second approach is extraction and classification both are done by aforementioned CNN models used in first approach [4].

Assam, M et al. (2021), presented a four-step hybrid model with various stages pre-processing, feature extraction, feature reduction and classification for tumor classification in brain through MRI scans. Median filter, being one of the state-of-art and traditional method used to remove fixed valued impulse noise and unnecessary structures such as skull and scalp, it's a main block in pre-processing stage, further it converted to colored image. Feature extraction has been done in stage-2 with discrete wavelet transform (DWT). Feature reduction and optimal characteristics set is generated in the third stage with the help of color moments (CMs). Image classification is done by passing the reduced optimal characteristic set through various classifiers; FF-ANN, RSwithRF and RSwithBN [5].

Rehman, A et al. (2020), in this framework, three studies were conducted for classification of brain tumor using three popular CNN architectures (AlexNet, GoogleNet, and VGGNet). Each study investigated on MRI brain tumor dataset Figshare and explores it with transfer learning methods freeze and fine-tune. Chance of over-fitting is reduced by increasing the data set samples by employing augmentation on MRI slices [6].

Ali, M et al. (2020), proposed straight forward ensemble method for segmentation of tumor by processing the image through two individual networks 3D CNN and a U-Net. These two networks trained individually with BraTS-19 challenge dataset and estimate the segmentation maps which considerably differed from each other in sub-regions tumor segmentation. Final prediction of tumor segmentation is achieved by ensemble these two individual segmented maps [7].

Kumar, S. and Mankame, D.P., (2020), discussed the tumor classification based on the optimized deep learning mechanism in which fuzzy deformable fusion model used for the segmentation of the images. The statistical features are used to classify the tumor by deep convolutional neural networks (DCNN). Dolphin Echolocation Sine Cosine Algorithm (Dolphin-SCA) is also implemented in this work for the segmentation. BRATS and SimBRATS databases were used for the validation of this network [8].

Hasan, S.K. and Linte, C.A (2018), proposed a deep learning U-Net CNN model for characterization and segmentation of tumor in brain MR images. In this method, up-sampling with nearest neighbor algorithm is introduced instead of de-convolution component in the U-net model. Segmentation accuracy is improved by extracting low grade tumors with the help of data augmentation performed on dataset by employing elastic transformation. This frame work trained with BRATS 2017 MR dataset of 285 patients affected with glioma [9].

Seetha, J. and Raja, S.S., (2018), proposed an automatic classification of tumor in brain using CNN. Manual classification of tumor from the MRI data is challenging at particular times. So, low complexity CNN system proposed by training the last layer of the network and considered a pre-trained model brain data set for classification. Hence, validate with training accuracy and computational time; it examines with support vector machine (SVM) and deep neural networks (DNN) [10].

Selvapandian, A. and Manivannan, K., (2018), implemented tumor detection in brain with contourlet transform of non-sub sampled (NSCT) and neuro fuzzy inference adaptive system (ANFIS). Image enhancement is done with NSCT by combining low and high frequency sub-bands of MR image and extracted features from enhanced image are used to classify the normal and glioma type tumor images with ANFIS [11].

Bahadure, N.B et al. (2017), investigated Berkeley wavelet transformation (BWT) on brain MR images to extract the tumor region in segmentation process. The appropriate features are extracted from the segmented tissues by employing very well-known classifier support vector machine (SVM) [12].

Gopal, N.N. and Karnan, M (2010), proposed a two-stage intelligent system to recognize tumor tissues in brain MRI. In this system design optimization is employed along with Fuzzy c Means Clustering for segmentation. The adopted optimizing methods are Genetic Algorithm (GA) and Particle Swarm Optimization (PSO). In this, Preprocessing and Enhancement are the methods in the first phase; segmentation and classification are the methods in second phase [13].

3 Performance Measures

True Positive (T_p), True Negative (T_n), False Positive (F_p) and False Negative (F_n) are the words used in the evaluation parameters of tumor classification methods such as Accuracy, Sensitivity, Specificity [3–6, 8, 10–14]. Hence, these metrics are formulated from Eq. (1) to (3). Dice coefficient [7, 9] is another metric formulated as Eq. (4) to evaluate the similarity between predicted tumor pixels with ground truth pixels.

$$Accuracy(A) = \frac{Number\ of\ faithful\ detections}{Number\ of\ all\ assessments} = \frac{T_p + T_n}{T_p + T_n + F_p + F_n} \tag{1}$$

$$Sensitivity(Sen) = \frac{Number\ of\ true\ positive\ detections}{Number\ of\ all\ positive\ assessments} = \frac{T_p}{T_p + F_n} \tag{2}$$

$$Specificity(Spec) = \frac{Number\ of\ true\ negative\ detections}{Number\ of\ all\ negative\ assessments} = \frac{T_n}{T_n + F_p} \tag{3}$$

$$Dice\ Coefficient\ (DSC) = \frac{2 * |X \cap Y|}{|X| + |Y|} \tag{4}$$

where X : set of predicted pixels, Y : ground truth

4 Dataset Description and Evolution of Brain Tumor Classification Methods

Dataset summary and performance evaluation of each method discussed in previous section summarized in Table 2. Table 3 compare the detection accuracy of methods described in [3–6, 11, 12] and [13] evaluated on non- BRATS datasets and Table 4 compare the detection accuracy of methods described in [4, 8] and [11] evaluated on BRATS datasets.

Table 1. Summary of Brain Tumour Classification Methods using MRI Scans

Author(s), Year and Source	Methodology
[3] S. Mohsen et al. 2023 IEEE Access	Pituitary and glioma type tumor classification has been carried out by intelligence systems, single image super-resolution (SISR) technique with classification networks ResNext101_32 × 8d and VGG19 SISR method developed in two stages; generator is in first stage and discriminator in the second stage. Hence, it produces the super resolution image of size 256x256x3 from low and high-resolution images of same size. The entire process has been done on required input medical MRI scans of pituitary and glioma. Classification is done with CNN's such as ResNext101_32 × 8d and VGG19 after SISR method
[4] H. A. Hafeez et al. 2023 IEEE Access	It's a 12-layer CNN model to classify the grade I – II (low) and grade III – IV (high) glioma type brain tumor. In this, N4ITK [14] method was applied in the preprocessing stage to remove bias field distortion in MRI images. Further, feature extraction and classification done by proposed CNN
[5] Assam, M., et al. 2021 IEEE Access	Pre-processing: Median filter Features extraction: Discrete Wavelet Transform (DWT) Features Reduction: Color Moments (CMs) Classification: Individual classifier: FF-ANN (Feed Forward – ANN) Hybrid Classifiers: RSwithRF (Random Subspace with Random Forest) RSwithBN (Random Subspace with Bayesian Network)
[6] Rehman, A., et al. 2020 Circuits, Systems, and Signal Processing	Three studies have been conducting with different CNN's namely AlexNet, GoogleNet and VGGNet to classify the meningioma, glioma, and pituitary types
[7] Ali, M., et al. 2020 IEEE Access	It's a combination of 3D-CNN and U-Net, both are trained individually and finally ensemble the individual outputs of these two networks

(continued)

Table 1. (*continued*)

Author(s), Year and Source	Methodology
[8] Kumar, S. and Mankame, D.P., 2020 Biocybernetics and Biomedical Engineering	It's a optimized Deep-CNN model trained with Dolphin-SCA Preprocessing: Non-Local Means (NLM) filter is used to remove artifacts in the different modalities of medical MR images Segmentation: To Extract ROI from the preprocessed image, employed a fuzzy deformable model with Dolphin Echolocation based Sine Cosine Algorithm (Dolphin-SCA) Feature Extraction: Feature vector has built with statistical parameters such as mean, variance, and skewness being extracted from segmented regions. Power LBP model has been adapted to extract useful features to train the classifier Classification: Deep CNN has been trained with Dolphin-SCA
[9] Hasan, S.K. and Linte, C.A., 2018 2018 IEEE Western New York Image and Signal Processing Workshop, IEEE	An improved version of Conventional U-net model by introducing up-sampling with nearest neighbor algorithm instead of de-convolution component and employing elastic transformation named as Nearest-Neighbor Re-sampling Based Elastic-Transformed to increase the segmentation accuracy
[10] Seetha, J. and Raja, S.S., 2018 Biomedical & Pharmacology Journal	It's a low complexity CNN based brain tumor classification method. In this, only last layer of the network has been trained and classification steps are done by a pre-trained model brain dataset. These two changes were made in traditional CNN for reducing computation time and performance improvement
[11] Selvapandian, A. and Manivannan, K., 2018 Computer Methods and Programs in Biomedicine	It is a fusion based glioma brain tumor classification approach. Low and high frequency sub-bands of input MR image scan are the components of fused image. Decomposing of sub-bands from input image has been done in spatial domain with the help of Pyramid Filter Banks (PFB) and Directional Filter Banks (DFB). Further, extracted features from fused image trains the classification approach and ANIFS classifier identifies the non-glioma and glioma images
[12] Bahadure, N.B., et al. 2017 International Journal of Biomedical Imaging	Tumor segmentation being done by Berkeley Wavelet Transformation (BWT). Features that are intensity and texture have been extracted from segmented image by employing GLCM, SFTA and IBF along with area of the tumor and dice coefficient similarity index are the two more features. PCA is used to select optimized relevant features. Finally, SVM is used for the brain tumor classification

(*continued*)

Table 1. (*continued*)

Author(s), Year and Source	Methodology
[13] Gopal, N.N. and Karnan, M., 2010 2010 IEEE International Conference on Computational Intelligence and Computing Research	It's a two-stage intelligent tumor detection system Stage 1: Preprocessing and Enhancement, Stage 2: Segmentation and Classification. Methods involved in this, Fuzzy C-Means Clustering Algorithm (FCA) with optimization tools Particle Swarm Optimization (PSO) and Genetic Algorithm (GA). In this, the overall detection accuracy is sum of accuracy of tumor pixels (75%) and position accuracy (25%)

Table 2. Summary of Brain Tumour Classification Methods using MRI Scans

Reference Number	Dataset	Performance Evaluation
[3]	Kaggle Dataset: It consists 1800 brain MRI images, out of these 900 images of each glioma tumor and pituitary tumor. They were resized to 224 x 224 and increase this dataset by three times with the help of data augmentation methods rotation, width and height shift. Further, dataset has to divide with 75% for training VGG19 and 85% for training ResNext101_32 × 8d	Testing Accuracy (%) a. VGG19: 99.89 b. SISR + ResNext101_32 × 8d: 100
[4]	BRATS 2017, BRATS 2018 and BRATS 2019 Dataset BHVB has been developed by acquiring 159 high grade and 176 low grade glioma MRI scan images from B V Hospital, Bahawalpur, Pakistan	a) Deep learning Method + SVM Classifier Testing Accuracy (%) BRATS 2017: 97.87 BRATS 2018: 97.67 BRATS 2019: 89.77 BHVB: 98.89 b) CNN Model Testing Accuracy (%) BRATS 2017: 97.85 BRATS 2018: 97.15 BRATS 2019: 97.15 BHVB: 97.99
[5]	70 T2-weighted standard image dataset (Normal: 45 and Abnormal: 25)	Testing Accuracy (%) a) DWT + RSwithRF: 97.14 b) DWT + RSwithBN: 95.71 c)DWT + CMs + FF-ANN: 95.83
[6]	Figshare dataset: It contains 3064 various types of brain tumor MRI scan images in which, meningioma: 708, glioma: 1426 and pituitary: 930	Testing Accuracy (%) a) AlexNet: 97.39 b) GoogleNet: 98.04 c) VGG16: 98.69

(continued)

Table 2. (*continued*)

Reference Number	Dataset	Performance Evaluation
[7]	BraTS2019 challenge dataset: In this dataset total 335 patients' information has been used for training. Out of these, 259 high-grade and 76 low-grade type glioma cases information. Validation process has been done with 125 cases of unknown grade	Dice Scores Enhancing Tumor(ET): 0.750 Whole Tumor(WT): 0.906 Tumor Core: 0.846
[8]	BRATS Database: It consists 65 tumor images of T1, T1c, T2 and flair modalities. Out of which, 51 are high-grade glioma patients' information SimBRATS database: It's a simulated image dataset with 50 images of all four varieties as same as BRATS database. Out of which 25 high-grade and 25 low-grade glioma images	Testing Accuracy (%) BRATS database: 95.3 SimBRATS database: 96.3 Specificity (%) BRATS database: 0.953 SimBRATS database: 0.910 Sensitivity (%) BRATS database: 0.977 SimBRATS database: 0.992
[9]	BRATS 2017 MR dataset: 285 glioma patients' information is used for training and 146 patients' information for testing	Dice Similarity Coefficient (DSC): LGG: 0.8976 HGG: 0.8459 Intersection over Union (IoU): LGG: 0.8869 HGG: 0.826 LGG: Low-graded gliomas HGG: High-graded gliomas
[10]	Tumor images are accessed from Radiopaedia and BRATS 2015 dataset accessed for testing	Training Accuracy (%) CNN: 97.5
[11]	Low-grade and high-grade type brain tumor images of BRATS 2015 dataset sub-database has been used for training the classification stage alone. Brain MRI images from LeaderBoard and Challenge sub datasets also considered along with BRATS 2015 dataset have been for the evaluation of different performance metrics	Testing Accuracy (%) BRATS 2015 dataset: 99.30 LeaderBoard dataset: 95.9 Challenge dataset: 96.4 Specificity (%) BRATS 2015 dataset: 99.71 LeaderBoard dataset: 96.2 Challenge dataset: 95.1 Sensitivity (%) BRATS 2015 dataset: 70.25 LeaderBoard dataset: 92.3 Challenge dataset: 96.2

(*continued*)

Table 2. (*continued*)

Reference Number	Dataset	Performance Evaluation
[12]	DICOM Dataset: 22 infected brain tumor tissue images are considered Brain Web dataset: It has simulated three-dimensional MR imaging data of modalities T1-weighted, T2-weighted and proton density weighted. In this 13 out 44 are infected brain MR images Third dataset: It consists, 135 images of 15 patients collected from the expert radiologists	Testing Accuracy (%) SVM Classifier: 96.51 Specificity (%) SVM Classifier: 94.2 Sensitivity (%) SVM Classifier: 97.72
[13]	A Set of 120 MR images	Detection Accuracy (%) a. GA + FCM: 89.6 b. PSO + FCM: 98.87

Table 3. Testing Accuracy of Brain Tumor Classification Methods Tested on Brats Datasets

Reference	Alphabet [Method Name]	Accuracy (%)
[3]	a [VGG19]	99.89
[3]	b [SISR + ResNext101_32 × 8d]	100.00
[4]	a [Deep learning Method + SVM Classifier]	98.89
[4]	b [CNN Model]	97.99
[5]	a [DWT + RSwithRF]	97.14
[5]	b [DWT + RswithBN]	95.71
[5]	c [DWT + CMs + FF-ANN]	95.83
[6]	a [AlexNet]	97.39
[6]	b [GoogleNet]	98.04
[6]	c [VGG16]	98.69
[11]	a[Image Fusion + ANIFS]	96.40
[12]	a [SVM]	96.51
[13]	a [GA + FCM]	89.60
[13]	b [PSO + FCM]	98.87

Figure 2 and Fig. 3 show the bar graph representation of various methods corresponding reference number on X-axis and with their testing accuracy (%) on Y-axis. Most of the related works discussed in this paper were implemented with more than one methodology. Hence, represent the methods with alphabets a, b and c with method name in Table 3 and Table 4. Methods proposed by the models represented in Fig. 2, [3, 4] and [13] with two methods each, [5] and [6] with three methods each, [11] and [12] each one method. Performances of all these models were evaluated with different brain tumor

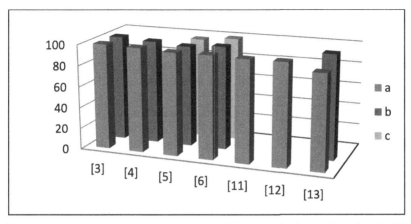

Fig. 2. Accuracy comparison of Brain Tumor Classification Methods tested on other than BRATS datasets

Table 4. Testing Accuracy of Brain Tumor Classification Methods Tested on Brats Datasets

Reference	Alphabet [Method Name]	Accuracy (%)
[4]	a [Deep learning Method + SVM Classifier]	97.87
[4]	b [CNN Model]	97.85
[8]	a [Proposed Dolphin-SCA based Deep CNN]	96.30
[11]	a [Image Fusion + ANIFS]	99.30

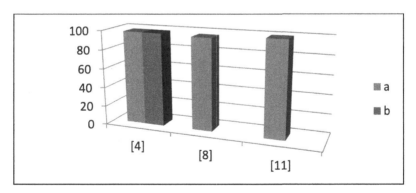

Fig. 3. Accuracy comparison of Brain Tumor Classification Methods tested on BRATS datasets

datasets. Figure 3 shown model [4] proposed with two methods and each one from [8] and [11]. These three models were tested with either of BRATS 2015, BRATS 2017, BRATS 2018, BRATS 2019 and SimBRATS database.

5 Conclusion

The availability of open-source image datasets of various diseases has enabled the automatic disease classification systems since past two decades. In this study the performance analysis of brain tumor classification methods through MRI scans has been analyzed. The intelligence systems proposed in [3–6, 8, 11, 12] and [13] are Deep-learning and CNN models. All these classify the various brain tumor types: meningioma; pituitary; low-graded and high-graded glioma. Tumor detection accuracy: In [3], SISR + ResNext101_32 × 8d achieves 100% on Kaggle Dataset. In [5], a hybrid classifier DWT + RSwithRF achieves 97.14% on standard dataset of 70 images. In [6], VGG16 achieves 98.69% on Fighshare dataset. SVM classifier [12] achieves 96.51% on dataset created with 135 images of 15 patients collected from expert radiologists. In [14], Particle Swarm Optimization with Fuzzy c Means Clustering (PSO + FCM) achieves 98.87% on a set of 120 images. Hence, all these methods performed well on various datasets in terms detection accuracy. In [4], Deep learning Method + SVM Classifier and CNN Model validated with BRATS 2017, BRATS 2018 and BRATS 2019 along with BHVB. It has been developed by acquiring images from Bahawal Victoria Hospital, Bahawalpur, Pakistan. Deep learning Method + SVM Classifier and CNN Model achieved 97.87%, 97.85% and 98.89%, 97.99% detection accuracies on BARTS 2017 and BHVB datasets in order. In [11], Image Fusion with ANIFS classifier validated with BRATS 2015, LeaderBoard and Challenge datasets. It achieved 99.30%, 95.9 and 96.4% of detection accuracies respectively. Proposed Dolphin-SCA based Deep CNN [8] achieves 96.3% on SimBRATS database. In separate datasets validation process, performance of super resolution image with ResNext101_32 × 8d model achieved highest detection rate of brain tumor type's pituitary and glioma. CNN model [4] classifies the Low grade and High-grade glioma brain tumor types efficiently with more than 97% detection rate on BRATS datasets 2017, 2018, 2019 and non-BRATS dataset. Image Fusion + ANIFS [11] outperformed on BRATS 2015 dataset for the classification of glioma type brain tumor.

References

1. Sung, H., et al.: Global cancer statistics 2020: GLOBOCAN estimates of incidence and mortality worldwide for 36 cancers in 185 countries. CA: a cancer journal for clinicians, 71(3), pp.209–249 (2021)
2. Chahal, P.K., Pandey, S., Goel, S.: A survey on brain tumor detection techniques for MR images. Multimedia Tools Appl. **79**, 21771–21814 (2020)
3. Mohsen, S., Ali, A.M., El-Rabaie, E.S.M., ElKaseer, A., Scholz, S.G., Hassan, A.M.A.: Brain tumor classification using hybrid single image super-resolution technique with ResNext101_32× 8d and VGG19 Pre-trained models. IEEE Access **11**, 55582–55595 (2023). https://doi.org/10.1109/ACCESS.2023.3281529
4. Hafeez, H.A., et al.: A CNN-model to classify low-grade and high-grade glioma from MRI images. IEEE Access **11**, 46283–46296 (2023). https://doi.org/10.1109/ACCESS.2023.327 3487
5. Assam, M., Kanwal, H., Farooq, U., Shah, S.K., Mehmood, A., Choi, G.S.: An efficient classification of MRI brain images. IEEE Access **9**, 33313–33322 (2021)

6. Rehman, A., Naz, S., Razzak, M.I., Akram, F., Imran, M.: A deep learning-based framework for automatic brain tumors classification using transfer learning. Circuits Syst. Sign. Proces. **39**(2), 757–775 (2020)
7. Ali, M., Gilani, S.O., Waris, A., Zafar, K., Jamil, M.: Brain tumour image segmentation using deep networks. IEEE Access **8**, 153589–153598 (2020)
8. Kumar, S., Mankame, D.P.: Optimization driven deep convolution neural network for brain tumor classification. Biocybernetics Biomed. Eng. **40**(3), 1190–1204 (2020)
9. Hasan, S.K. and Linte, C.A.: A modified U-Net convolutional network featuring a nearest-neighbor re-sampling-based elastic-transformation for brain tissue characterization and segmentation. In: 2018 IEEE Western New York Image and Signal Processing Workshop (WNYISPW), pp. 1–5. IEEE (2018)
10. Seetha, J., Raja, S.S.: Brain tumor classification using convolutional neural networks. Biomed. Pharmacol. J. **11**(3), 1457–1461 (2018)
11. Selvapandian, A., Manivannan, K.: Fusion based glioma brain tumor detection and segmentation using ANFIS classification. Comput. Methods Programs Biomed. **166**, 33–38 (2018)
12. Bahadure, N.B., Ray, A.K., Thethi, H.P.: Image analysis for MRI based brain tumor detection and feature extraction using biologically inspired BWT and SVM. Int. J. Biomed. Imaging **2017**, 1–12 (2017). https://doi.org/10.1155/2017/9749108
13. Gopal, N.N., Karnan, M.: Diagnose brain tumor through MRI using image processing clustering algorithms such as fuzzy C Means along with intelligent optimization techniques. In: 2010 IEEE International Conference on Computational Intelligence and Computing Research, pp. 1–4. IEEE (2010)
14. Kumar, S., Kumar, D.: Human brain tumor classification and segmentation using CNN. Multimedia Tools Appl. **82**(5), 7599–7620 (2023)
15. Budumuru, P.R., Varma, A.K.C., Satyanarayana, B.V.V., Srinivas, Y., Raju, B.E. and Kumar, G.P.: preprocessing analysis of medical image: a survey. In: 2022 4th International Conference on Advances in Computing, Communication Control and Networking, pp. 2369–2375. IEEE (2022)
16. Vishnuvarthanan, A., Rajasekaran, M.P., Govindaraj, V., Zhang, Y., Thiyagarajan, A.: An automated hybrid approach using clustering and nature inspired optimization technique for improved tumor and tissue segmentation in magnetic resonance brain images. Appl. Soft Comput. **57**, 399–426 (2017)
17. Pitchai, R., Supraja, P., Helen Victoria, A., Madhavi, M.: Brain tumor segmentation using deep learning and fuzzy K-means clustering for magnetic resonance images. Neural. Process. Lett. **53**(4), 2519–2532 (2020). https://doi.org/10.1007/s11063-020-10326-4
18. Krishna, T. G., Abdelhadi, M. A.: optimization segmentation and classification from MRI of brain tumor and its location calculation using machine learning and deep learning approach. Recent Adv. Math. Res. Comput. Sci. Vol (2022)
19. Sandhya, M., et al.: Deep Neural Networks with Multi-class SVM for Recognition of Cross-Spectral Iris Images. In: Thampi, S.M., Piramuthu, S., Li, K.-C., Berretti, S., Wozniak, M., Singh, D. (eds.) SoMMA 2020. CCIS, vol. 1366, pp. 29–41. Springer, Singapore (2021). https://doi.org/10.1007/978-981-16-0419-5_3
20. Subbarao, M.V., Ram, G.C., Kumar, D.G., Terlapu, S.K.: Brain tumor classification using ensemble classifiers. In: 2022 International Conference on Electronics and Renewable Systems (ICEARS), pp. 875–878. IEEE (2021)
21. Shaik, A.R., Chandra, K.R., Raju, B.E., Budumuru, P.R.: Glaucoma identification based on segmentation and fusion techniques. In: 2021 International Conference on Advances in Computing, Communication, and Control (ICAC3), pp. 1–4. IEEE (2021)

22. Satyanarayana, B.V.V., Prasanna Kumar, G., Varma, A.K.C., Dileep, M., Srinivas, Y., Budu-muru, P.R.: Alzheimer's Disease Detection Using Ensemble of Classifiers. In: Gupta, N., Prakash Pareek, M.J.C.S., Reis, (eds.) Cognitive Computing and Cyber Physical Systems: Third EAI International Conference, IC4S 2022, Virtual Event, November 26-27, 2022, Proceedings, pp. 55–65. Springer Nature Switzerland, Cham (2023). https://doi.org/10.1007/978-3-031-28975-0_5

23. Ravi Sankar, M., et al.: Performance Evaluation of Multiwavelet Transform for Single Image Dehazing. In: Gupta, N., Prakash Pareek, M.J.C.S., Reis, (eds.) Cognitive Computing and Cyber Physical Systems: Third EAI International Conference, IC4S 2022, Virtual Event, November 26-27, 2022, Proceedings, pp. 125–133. Springer Nature Switzerland, Cham (2023). https://doi.org/10.1007/978-3-031-28975-0_10

24. Sravanthi, I., et al.: Performance Evaluation of Fast DCP Algorithm for Single Image Dehazing. In: Gupta, N., Prakash Pareek, M.J.C.S., Reis, (eds.) Cognitive Computing and Cyber Physical Systems: Third EAI International Conference, IC4S 2022, Virtual Event, November 26-27, 2022, Proceedings, pp. 134–143. Springer Nature Switzerland, Cham (2023). https://doi.org/10.1007/978-3-031-28975-0_11

An Efficient Denoising of Medical Images Through Convolutional Neural Network

K. Soni Sharmila[1] ⓘ, S. P Manikanta[2] ⓘ, P. Santosh Kumar Patra[3] ⓘ,
K. Satyanarayana[4] ⓘ, and K. Ramesh Chandra[5]([✉]) ⓘ

[1] Department of CSE, Shri Vishnu Engineering College for Women, Bhimavaram,
Andhra Pradesh, India
[2] Department of ECE, St. Martin's Engineering College, Secunderabad, Telangana, India
[3] Department of CSE, St. Martin's Engineering College, Secunderabad, Telangana, India
[4] Department of AIML, Aditya College of Engineering, Surampalem, Andhra Pradesh, India
[5] Department of ECE, Vishnu Institute of Technology, Bhimavaram, India
rameshchandra001@gmail.com

Abstract. Denoising medical images is a critical step in enhancing image quality and improving diagnostic accuracy. In this work, an efficient denoising method has been proposed for medical images using convolutional denoising autoencoders. The proposed approach leverages the power of CNNs to learn complex patterns and features from a large dataset of clean and noisy medical images. To train the denoising network, a dataset has created consisting of pairs of clean medical images and their corresponding noisy versions. Various types and levels of noise are introduced to generate a diverse training set. The network architecture is carefully designed to effectively capture and extract relevant features from the noisy medical images. Multiple convolutional layers are used for feature extraction, followed by pooling, normalization, and non-linear activation layers. The final layers of the network focus on reconstructing the clean version of the input image. During the training phase, the network learns to map the noisy images to their corresponding clean versions. A suitable loss function, such as mean squared error or structural similarity index loss, is employed to guide the training process, and minimize the discrepancy between the network output and the ground truth clean image. The trained network is evaluated on a separate test dataset, and performance metrics such as peak signal-to-noise ratio and visual inspection are used to assess the denoising effectiveness. The experimental results demonstrate that the proposed CNN-based denoising method achieves superior performance compared to traditional denoising techniques. The network effectively reduces noise artifacts while preserving important image details and structures. The denoised medical images generated by the CNN can potentially lead to improved diagnosis and decision-making in medical applications.

Keywords: Medical image denoising · Convolutional denoising autoencoders · Deep learning · Image reconstruction · Quantitative evaluation · Visual inspection · Diagnostic accuracy

© ICST Institute for Computer Sciences, Social Informatics and Telecommunications Engineering 2024
Published by Springer Nature Switzerland AG 2024. All Rights Reserved
P. Pareek et al. (Eds.): IC4S 2023, LNICST 536, pp. 459–470, 2024.
https://doi.org/10.1007/978-3-031-48888-7_39

1 Introduction

Denoising medical images using Convolutional Neural Networks (CNNs) is an effective approach to enhance image quality and remove noise artifacts. CNNs have demonstrated remarkable performance in various image processing tasks, including medical image denoising. The denoising of medical images have certain phases include: data preparation; dataset augmentation; architecture; training; optimization; evaluation; and post pre-processing. The success of the denoising process relies on the quality and diversity of the training dataset, as well as the design of the network architecture and the hyperparameter settings. Iterative refinement, cross-validation, and assembling techniques can be employed to improve performance further [1]. Additionally, various advanced techniques have been developed to address specific challenges in medical image denoising, such as incorporating attention mechanisms, using generative adversarial networks (GANs), or leveraging self-supervised learning. These techniques can provide further improvements in denoising accuracy and efficiency. Furthermore, denoising medical images is a crucial task to improve the quality and diagnostic accuracy of medical imaging. Medical images often suffer from various types of noise, including Gaussian noise, Poisson noise, motion artifacts, and electronic noise. Several denoising techniques can be applied to medical images such as: filtering techniques; wavelet-Based techniques; non-local means (NLM) denoising; deep learning-based techniques; patch-based techniques [2].

These are some of the commonly used techniques for denoising medical images. The choice of method depends on the specific characteristics of the noise, the desired level of denoising, and the trade-off between noise reduction and preservation of important image features. It is important to evaluate the denoising results using appropriate metrics and, when necessary, involve medical experts to assess the impact on diagnosis and clinical decision-making. However, these techniques need large computational time and huge data set requirements in order to achieve better performance [3]. Hence, in tis work the denoising of medical images has been done based on Convolutional denoising autoencoders (CDAEs). These are a specific type of denoising autoencoder that leverage convolutional neural network (CNN) architectures for denoising tasks, particularly in image denoising applications. CDAEs follow the same basic principles as traditional autoencoders but incorporate convolutional layers in the encoder and decoder components. These convolutional layers allow the model to efficiently capture spatial features and patterns present in the input images. During the training process, a dataset of clean images is corrupted by introducing noise, such as Gaussian noise or random pixel perturbations, to generate the corresponding noisy images [4].

The CDAE is trained to encode the noisy images into a lower-dimensional representation and reconstruct the clean images using the decoder component. The loss function used during training typically measures the discrepancy between the reconstructed clean images and the original clean images. This encourages the CDAE to learn meaningful representations and effectively remove noise during the reconstruction process. The convolutional layers in CDAEs enable the models to capture local patterns and features in the input images, making them particularly suitable for denoising tasks where preserving spatial information is crucial. The hierarchical nature of CNNs allows for the extraction of complex and abstract image features, aiding in noise suppression while retaining important image details. Once trained, CDAEs can be applied to denoise new, unseen noisy images by passing them through the encoder to obtain the compressed representation and then using the decoder to reconstruct the denoised versions. Convolutional denoising autoencoders have demonstrated effectiveness in various domains, including medical image denoising [5, 6]. They have the potential to improve the quality of medical images, enhance diagnostic accuracy, and support medical professionals in making informed decisions. Finally, convolutional denoising autoencoders provide a powerful framework for effectively denoising images, leveraging the benefits of convolutional neural networks to handle spatially structured data, such as medical images, and removing noise while preserving important image features.

2 Literature Review

Medical image processing has recently attracted a lot of attention from researchers who are trying to diagnose and cure dangerous illnesses like cancer. Medical image analysis is thought to depend heavily on the process of denoising of images. Image denoising constitutes a fundamental undertaking within the realm of image processing. Its primary objective is to effectively eliminate undesired noise from an image, all the while safeguarding and retaining crucial details present in the visual content. This can effectively accomplish using a variety of Deep Learning approaches, such as autoencoders. A convolutional autoencoders-based method for medical picture denoising with short connections has been suggested in [7], which uses three datasets of medical imaging data for denoising. The outcomes showed that, across all three datasets, the suggested methodology outperformed the most advanced medical picture denoising techniques currently available. The authors of [8] has created a denoising convolutional autoencoder that first creates an encoded, lower-dimensional representation of the picture before recovering the original image from the lower-dimensional representation. By introducing random noise, the denoising autoencoders create corrupted copies of the input pictures and try to remove the noise from the noisy input to recreate an output that is very close to the original input. To compress and denoise grayscale medical pictures, a 3-layer autoencoder model has been suggested in [9].

A deep learning approach, to the denoising issue has been analyzed in [10]. The proposed algorithm eliminates unwelcome noise from a picture, autoencoders employ down- and up-sampling algorithms. The authors of [11], has investigated CNN architecture for sophisticated image denoising. A clustering technique for effective denoising of medical images based on CNN has been suggested by the authors in [12]. Convolutional neural networks (CNNs) are proposed [13] for the prediction of Parkinson's

Disease utilizing an autoencoder feature extraction approach. An autoencoder is used to extract features from input data and de-noise it. For categorization and forecasting, CNN is employed. The goal of image enhancement is to create a clear image from a noisy image, and ultrasound images are one application. In contrast to traditional picture enhancing techniques, deep learning was employed in the experimental investigation [14]. The convolutional denoising autoencoder network, one of the deep learning techniques, was used to attempt to eliminate various amounts of speckle noise that had been introduced to ultrasound pictures of the brachial plexus, also known as the big nerve community under the armpit. A significant number of noise sources significantly pollute extracellular recordings, making the denoising procedure a very difficult work that must be taken on for effective spike sorting. In order to do this, the authors of [15] suggested an end-to-end deep learning solution using a fully convolutional de-noising autoencoder, which learns to create a clear neural activity signal from a noisy multichannel input.

Using receiver operator characteristic (ROC) approaches, the authors of [16] suggested a task-based evaluation of reduced dosage reconstruction and denoising procedures. For the LDCT denoising challenge, the authors of [17] has suggested a brand-new 3D self-attention convolutional neural network. To train a domain-specific autoencoder as the perceptual loss function, the same authors also provide a self-supervised learning strategy. The authors of [18] has combined two approaches and use extensive experiments to show their efficacy on neural networks built using WGAN and CNN, respectively. A novel hybrid image compression-encryption scheme, has been introduced in [19] that leverages deep learning techniques, specifically stacked auto-encoders, in conjunction with the logistic map. The advanced encryption standard (AES) and data encryption standard (DES) are used to encrypt the DCT-based compression algorithm presented in [20]. In order to ensure safe data transfer, the biorthogonal discrete wavelet transform (DWT) approach is introduced in [21] for compression, followed by the advanced encryption standard (AES) and data encryption standard (DES) technologies. The unnecessary data is suggested to be removed using the fractal-based compression algorithm [22] because it does not depend on the image resolution. The grayscale images in [23] are compressed using the Huffman lossless compression technique. In addition, several algorithms have been proposed for encryption [24], encryption [25, 26], haze removal based on DCP [27], DWT [28], and feed forward ANN [29] and security [30, 31].

The literature review highlights the effectiveness of CDAEs in medical image denoising and their potential for further advancements. It emphasizes the importance of denoising for improving image quality and aiding in accurate diagnosis in medical imaging applications. Hence, this work focuses on medical image denoising based on CNN auto encoders for variety of noise sources. The performance measures have been estimated for different noise variance in terms of SSIM and PSNR.

3 Methodology

3.1 Autoencoders.

Autoencoders are a type of artificial neural network that is primarily used for unsupervised learning tasks such as data compression, feature extraction, and reconstruction. The basic autoencoder architecture consists of an encoder and a decoder, which work together to learn a compressed representation of the input data and reconstruct it as closely as possible. Given a collection of unlabeled training inputs $(x^1, x^2, x^3 \ldots ., x^n)$, which is represented using (Eqs. (1). It initially employs deterministic mapping using (Eqs. (2) via sending an input, where any nonlinear function, s, may be used [32].

$$z^{(i)} = x^{(i)}, i = 1 \; to \; n \qquad (1)$$

$$y = s(Wx + b) \qquad (2)$$

The encoder takes an input data sample and maps it to a lower-dimensional representation or code. The encoder network typically consists of multiple layers, such as fully connected layers or convolutional layers, that progressively reduce the dimensionality of the input. Each layer of the encoder learns to extract increasingly abstract features from the input data. The basic architecture of convolutional architecture has depicted in Fig. 1, where, layer L1 serves as the input layer. Layer L2 uses latent representation to encode layer L1, while layer L3 reconstructs layer L1.

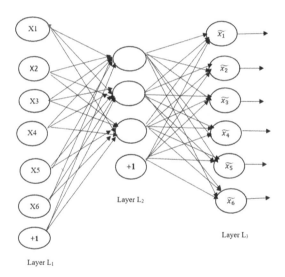

Fig. 1. Architecture of a basic autoencoder.

3.2 Denoising Autoencoders

Denoising autoencoders are a type of autoencoder specifically designed to remove noise or artifacts from input data. They are trained on corrupted data samples and learn to reconstruct the original, clean data by capturing the underlying structure and removing the noise components. Denoising autoencoders have been widely used in various domains, including image denoising, speech denoising, and signal denoising. Denoising autoencoders take corrupted or noisy input data as their input. The corruption process can involve adding random noise, introducing occlusions, or applying other forms of data corruption. The encoder network in a denoising autoencoder processes the corrupted input and maps it to a latent or compressed representation. The encoder learns to extract meaningful features from the noisy input, while filtering out the noise components. The decoder network reconstructs the clean data from the compressed representation obtained from the encoder. The decoder aims to generate output that closely resembles the original, noise-free input. Denoising autoencoders provide an effective approach for removing noise from corrupted data [32, 33]. They learn to reconstruct the original, clean data by capturing the underlying structure and filtering out the noise components. By training on corrupted data, denoising autoencoders can enhance the quality and utility of data in various domains. The basic architecture of denoising autoencoders are represented in Fig. 2.

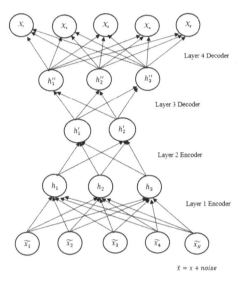

Fig. 2. Architecture of a denoising autoencoder.

3.3 Convolutional Autoencoders

Convolutional autoencoders are a type of autoencoder architecture that utilizes convolutional neural networks (CNNs) as both the encoder and decoder components. They are specifically designed for processing and reconstructing high-dimensional data, such as images, while preserving spatial information and capturing local features. Weights are distributed among all input regions in convolutional autoencoders, preserving local spatiality. The ith feature map's representation is provided in (Eqs. (3).

$$h^i = s(x * W^i + b^i) \tag{3}$$

where * stands for convolution (2D), s is an activation, and bias is broadcast to the whole map. Utilizing a single bias per latent map, reconstruction yields which is represented using (Eqs. (4). If W is a flip operation across both weight dimensions, H is a collection of latent feature maps, and c is bias per input channel.

$$y = s\left(\sum_{i \in H} h^i * W^i + c\right) \tag{4}$$

3.4 Data Set

A data set consisting 731 images with 512x420 resolution of X-ray images of human body parts. Some of the random images from dataset has depicted in Fig. 3.

Instead, adding noise to one image at a time, a flattened dataset has been corrupted, each row representing each image, so disrupting all photos at once. Datasets that have been corrupted were then utilized for modelling. Convolutional denoising autoencoder (CNN DAE), seen in Fig. 4, has been built using a rather straightforward design [31].

4 Simulation Results

4.1 Subjective Analysis

Subjective analysis in image processing refers to the evaluation and interpretation of images based on human perception and judgment. For the random images taken from dataset, the subjective analysis has been done which is depicted in Fig. 5 for gaussian noise with different noise proportions (*np*).

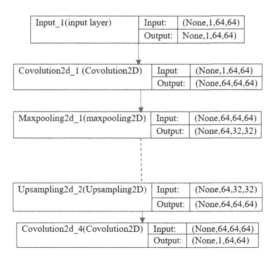

Fig. 4. Architecture of CNN DAE.

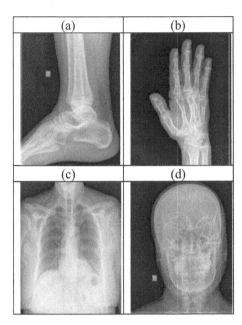

Fig. 3. Random samples of medical x-ray images from the dataset.

4.2 Objective Analysis

Objective analysis in image processing refers to the quantitative assessment and measurement of various characteristics, properties, and metrics of an image. The proposed methodology has undergone an objective analysis in terms of PSNR, MSE, under various noise (Gaussian and Poisson) patterns. SSIM has been analyzed using the Eqs. (5), where l, c, s are brightness, disparity and basic components which are calculated using Eqs. (6), (7), and (8) respectively. Where μ_x and μ_y stand for the mean of the original and coded pictures, σ_x and σ_y for their standard deviations, and σ_{xy} for their covariance.

$$SSIM(x, y) = \left[l(x, y)^\alpha\right]\left[c(x, y)^\beta\right]\left[s(x, y)^\gamma\right] \tag{5}$$

$$l(x, y) = \frac{2\mu_x\mu_y + C_1}{\mu_x^2 + \mu_{y+}^2 C_1} \tag{6}$$

$$c(x, y) = \frac{2\sigma_x\sigma_y + C_2}{\sigma_x^2 + \sigma_{y+}^2 C_2} \tag{7}$$

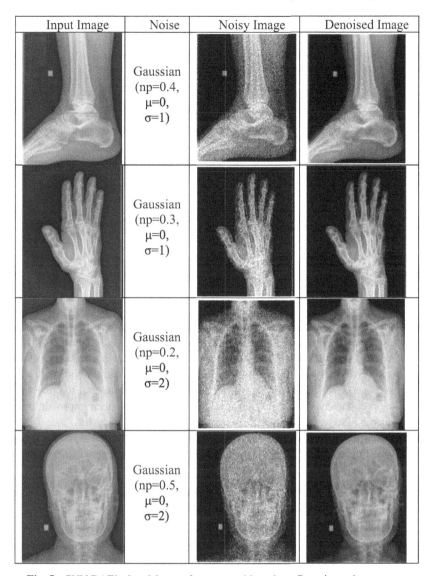

Input Image	Noise	Noisy Image	Denoised Image
	Gaussian (np=0.4, μ=0, σ=1)		
	Gaussian (np=0.3, μ=0, σ=1)		
	Gaussian (np=0.2, μ=0, σ=2)		
	Gaussian (np=0.5, μ=0, σ=2)		

Fig. 5. CNN DAE's denoising performance with various Gaussian noise patterns.

$$s(x, y) = \frac{2\sigma_{xy} + C_3}{\sigma_x \sigma_y + C_3} \tag{8}$$

The detailed analysis of performance measure for CNN DAE under various noise sources has been summarized in Table 1

Table 1. Comparison of SSIM values for CNN DAE and median filter under various noise parameters

Image	Noise Type	Noise Parameters	SSIM	
			Filtering	CNN DAE
Ankle X-ray	Gaussian	$np = 0.3, \mu = 0, \sigma = 1$	0.3156	0.8543
		$np = 0.5, \mu = 0, \sigma = 1$	0.2367	0.8478
Chest X-ray	Gaussian	$np = 0.2, \mu = 0, \sigma = 2$	0.1762	0.7643
		$np = 0.1, \mu = 0, \sigma = 2$	0.1934	0.7456
Skull X-ray	Poisson	$np = 0.2, \lambda = 1$	0.2467	0.4678
		$np = 0.4, \lambda = 2$	0.1219	0.5647
Hand X-ray	Poisson	$np = 0.1, \lambda = 1$	0.1565	0.4878
		$np = 0.5, \lambda = 5$	0.1725	0.70

5 Conclusion

In this work a denoising autoencoder has been erected through the convolutional layers for the purpose of serving efficient denoising of medical images. A dataset of having 731 medical X-ray images have been trained and introducing various noise types such as Gaussian and Poisson, by altering various noise parameters. The simulation results depict that, the proposed CNN DAE achieves high accuracy when compared to conventional denoising approaches based on filtering. The image quality has been improved by 60% after denoising based on CNN DAE, where as 19% of accuracy has been obtained using median filtering techniques. Overall, the performance of proposed CNN DAE has attained 50% improvement in image visual accuracy when compared to conventional filtering-based image denoising techniques.

References

1. Buades, A., Bartomeu, C., Jean-Michel, M.: A review of image denoising algorithms, with a new one. Multiscale Model. Simul. **4**(2), 490–530 (2005)
2. Burger, H.C., Schuler, C.J., Harmeling, S.: Image denoising: Can plain neural networks compete with BM3D? IEEE Conf. Computer Vis. Pattern Recogn. Providence RI USA **2012**, 2392–2399 (2012)
3. Cho, K.: Boltzmann Machines for Image Denoising. In: Mladenov, V., Koprinkova-Hristova, P., Palm, G., Villa, A.E.P., Appollini, B., Kasabov, N. (eds.) ICANN 2013. LNCS, vol. 8131, pp. 611–618. Springer, Heidelberg (2013). https://doi.org/10.1007/978-3-642-40728-4_76
4. S. Sharmila, K., Thanga Revathi, S., K. Sree, P.: convolution neural networks based lungs disease detection and severity classification.In: International Conference on Computer Communication and Informatics (ICCCI), Coimbatore, India, pp. 1–9 (2023)
5. Zhou Wang, A., Bovik, C., Sheikh, H.R., Simoncelli, E.P.: Image quality assessment: from error visibility to structural similarity. IEEE Trans. Image Process. **13**(4), 600–612 (2004)

6. Portilla, J., Strela, V., Wainwright, M.J., Simoncelli, E.P.: Image denoising using scale mixtures of Gaussians in the wavelet domain. Image Process. IEEE Trans. **12**(11), 1338–1351 (2003)

7. Gupta, M., Goel, A., Goel, K., Kansal, J.: medical image denoising using convolutional autoencoder with shortcut connections.In: 5th International Conference on Smart Systems and Inventive Technology (ICSSIT), Tirunelveli, India, pp. 1524–1528 (2023)

8. Thomas, J M., A. P. E.; Bio-medical image denoising using autoencoders. In: Second International Conference on Next Generation Intelligent Systems (ICNGIS), Kottayam, India, pp. 1–6 (2022)

9. Senapati, R K., Badri, R., Kota, A., Merugu, N., Sadhul, S.: Compression and denoising of medical images using autoencoders.In: International Conference on Recent Trends in Microelectronics, Automation, Computing and Communications Systems (ICMACC), Hyderabad, India, pp. 466–470 (2022)

10. Kulkarni, K., et al: Image denoising using autoencoders. : denoising noisy imgaes by removing noisy pixels/grains from natural images using deep learning and autoencoders techniques. In: IEEE 8th International Conference for Convergence in Technology (I2CT), Lonavla, India, pp. 1–6 (2023)

11. Li, B., Xu, K., Feng, D., Mi, H., Wang, H., Zhu, J.: Denoising convolutional autoencoder based B-mode ultrasound tongue image feature extraction.In: IEEE International Conference on Acoustics, Speech and Signal Processing (ICASSP), Brighton, UK, pp. 7130–7134 (2019)

12. Zhang, D., et al.: Unsupervised Cryo-EM images denoising and clustering based on deep convolutional autoencoder and K-Means++. IEEE Trans. Med. Imaging **42**(5), 1509–1521 (2023). https://doi.org/10.1109/TMI.2022.3231626

13. Hema, M.S., Maheshprabhu, R., Nageswara Guptha, M., Mary, P.A.G., Sharma, A.: Prediction of parkinson disease using autoencoder convolutional neural networks. In: International Interdisciplinary Humanitarian Conference for Sustainability (IIHC), Bengaluru, India, pp. 236–239 (2022)

14. Karaoğlu, O., Bilge, H.Ş., Uluer, İ.: Reducing speckle noise from ultrasound images using an autoencoder network.In: 28th Signal Processing and Communications Applications Conference (SIU), Gaziantep, Turkey, pp. 1–4 (2020)

15. Kechris, C., Delitzas, A., Matsoukas, V., Petrantonakis, P.C.: Removing noise from extracellular neural recordings using fully convolutional denoising autoencoders. In: 43rd Annual International Conference of the IEEE Engineering in Medicine & Biology Society (EMBC), Mexico, pp. 890–893 (2021)

16. Hendrik Pretorius, P., et al.: Assessment of defect detection in post-filtering and deep learning denoising strategies for reduced dose myocardial perfusion spect employing human and polar map observers. In: IEEE Nuclear Science Symposium and Medical Imaging Conference (NSS/MIC), Piscataway, NJ, USA, pp. 1–3 (2021)

17. Li, M., Hsu, W., Xie, X., Cong, J., Gao, W.: SACNN: self-attention convolutional neural network for low-dose CT denoising with self-supervised perceptual loss network. IEEE Trans. Med. Imaging **39**(7), 2289–2301 (2020)

18. Saranya, A., Kottilingam, K.: An efficient combined approach for denoising fibrous dysplasia images. In: International Conference on System, Computation, Automation and Networking (ICSCAN), Puducherry, India, pp. 1–6 (2021)

19. Gupta, N., Vijay, R.: Hybrid image compression-encryption scheme based on multilayer stacked autoencoder and logistic map. China Commun. **19**(1), 238–252 (2022)

20. Gupta, N., Vijay, R., Hemant Kumar, G.: Performance analysis of DCT based lossy compression method with symmetrical encryption algorithms. EAI Endorsed Trans. Energy Web **7**(28) 13(2020)

21. Gupta, N., Vijay, R., Hemant Kumar, G.: Performance evaluation of symmetrical encryption algorithms with wavelet based compression technique. EAI Endorsed Trans Scalable Inf. Syst. **7**(28) e-8 (2020)

22. Gupta, N., Vijay, R.: Effect on reconstruction of images by applying fractal based lossy compression followed by symmetrical encryption techniques. In: IEEE 11th International Conference on Computing, Communication and Networking Technologies (ICCCNT), Kharagpur, India, 2020, IEEE Xplore, pp. 1–7 (2020)

23. Gupta, N., Vijay, R.: Efficient Approach for Encryption of Lossless Compressed Grayscale Images. In: Sharma, H., Saraswat, M., Yadav, A., Kim, J.H., Bansal, J.C. (eds.) CIS 2020. AISC, vol. 1334, pp. 397–409. Springer, Singapore (2021). https://doi.org/10.1007/978-981-33-6981-8_32

24. Ramesh Chandra, K., Prudhvi Raj, B., Prasannakumar, G.: An efficient image encryption using chaos theory. In: International Conference on Intelligent Computing and Control Systems (ICCS), Madurai, India, pp. 1506–1510 (2019)

25. Raju, E.B., Sankar, R.M., Kumar, V.T., Chandra, R.K., Durga, B.V., Kumar, P.G.: modified encryption standard for reversible data hiding using AES and LSB steganography. In: International Conference on Computer Communication and Informatics (ICCCI), Coimbatore, India, pp. 1–5 (2023)

26. Chandra, K.R., Donga, M., Budumuru, P.R.: Reversible Data Hiding Using Secure Image Transformation Technique. In: Suma, V., Chen, J.-Z., Baig, Z., Wang, H. (eds.) Inventive Systems and Control. LNNS, vol. 204, pp. 657–668. Springer, Singapore (2021). https://doi.org/10.1007/978-981-16-1395-1_49

27. Ravi Sankar, M., et al.: Performance Evaluation of Multiwavelet Transform for Single Image Dehazing. In: Gupta, N., Pareek, P., Reis, M. (eds) cognitive computing and cyber physical systems. Lecture Notes of the Ins Comput. Sci. Soc. Inf. Telecommun. Eng., vol 472. Springer, Cham (2023)

28. Sravanthi, I., et al.: Performance evaluation of fast DCP algorithm for single image dehazing. In: Gupta, N., Pareek, P., Reis, M. (eds) cognitive computing and cyber physical systems, IC4S 2022,Lecture Notes of the Inst. for Comput. Sci. Soc. Inf. Telecommun. Eng. vol 472. Springer, Cham (2023)

29. Sharmila, K.S., Asha, A.V.S., Archana, P., Chandra, K.R.: Single Image Dehazing through feed forward artificial neural network. In: Gupta, N., Pareek, P., Reis, M. (eds) cognitive computing and cyber physical systems. IC4S 2022. Lecture Notes of the Inst. Comput. Sci. Soc. Inf. Telecommun. Eng. vol 472. Springer, Cham (2023)

30. Elisha Raju, B., Ramesh Chandra, K., Budumuru, P.R.: A Two-Level Security System Based on Multimodal Biometrics and Modified Fusion Technique. In: Karrupusamy, P., Balas, V.E., Shi, Y. (eds.) Sustainable Communication Networks and Application. LNDECT, vol. 93, pp. 29–39. Springer, Singapore (2022). https://doi.org/10.1007/978-981-16-6605-6_2

31. Vijjapu, A., Vinod, Y. S., Murty, S., V. S. N. Raju, B., E. Satyanarayana B. V. V., Kumar. G. P.: Steganalysis using convolutional neural networks-yedroudj net.In: International Conference on Computer Communication and Informatics (ICCCI), Coimbatore, India, pp. 1–7 (2023)

32. Gondara, L.: Medical image denoising using convolutional denoising autoencoders. In: IEEE 16th International Conference on Data Mining Workshops (ICDMW), Barcelona, Spain, pp. 241–246 (2016)

33. Jahangeer, G.S.B., Thambidurai, D.R.: Detecting breast cancer using novel mask R-CNN techniques. Expert. Syst. **39**(9), e12954 (2022)

Author Index

Printed in the United States
by Baker & Taylor Publisher Services